LAND APPLICATIONS OF RADAR REMOTE SENSING

Edited by **Francesco Holecz, Paolo Pasquali, Nada Milisavljević and Damien Closson**

Land Applications of Radar Remote Sensing

http://dx.doi.org/10.5772/55833

Edited by Francesco Holecz, Paolo Pasquali, Nada Milisavljević and Damien Closson

Contributors

Simone Atzori, Stefano Salvi, Bouaraba Azzedine, Paolo Pasquali, Paolo Riccardi, Alessio Cantone, Marco Defilippi, Fumitaka Ogushi, Stefano Gagliano, Masayuki Tamura, Damien Closson, Najib Abou Karaki, Mattia Crespi, Paola Capaldo, Andrea Nascetti, Martina Porfiri, Francesca Fratarcangeli, Francesca Pieralice, Loris Copa, Daniela Poli, Fabio Remondino, Nada Milisavljevic, Francesco Collivignarelli, Francesco Holecz, Andy Nelson, Tri Setiyono, Mirco Boschetti, Giacinto Manfron, Massimo Barbieri, Luca Gatti, Lorena Villano, Parvesh Chandna, Edmond Nezry, Josef Kellndorfer

CBS, Edition **2017**

Published by InTech

Janeza Trdine 9, 51000 Rijeka, Croatia

Publishing Process Manager Iva Lipovic

Technical Editor InTech DTP team

Cover Designer InTech Design Team

Additional hard copies can be obtained from orders@intechopen.com

Land Applications of Radar Remote Sensing, Edited by Francesco Holecz, Paolo Pasquali, Nada Milisavljević and Damien Closson
p. cm.
ISBN 978-953-51-1589-2

INTECH

open science | open minds

Contents

Preface VII

Section 1 A Review on Speckle 1

Chapter 1 **Adaptive Speckle Filtering in Radar Imagery 3**
Edmond Nezry

Section 2 Land-Cover Applications 57

Chapter 2 **Large Scale Mapping of Forests and Land Cover with Synthetic Aperture Radar Data 59**
Josef Kellndorfer, Oliver Cartus, Jesse Bishop, Wayne Walker and Francesco Holecz

Chapter 3 **Estimation of Cultivated Areas Using Multi-Temporal SAR Data 95**
Nada Milisavljević, Francesco Collivignarelli and Francesco Holecz

Chapter 4 **Combining Moderate-Resolution Time-Series RS Data from SAR and Optical Sources for Rice Crop Characterisation: Examples from Bangladesh 121**
Andrew Nelson, Mirco Boschetti, Giacinto Manfron, Franceco Holecz, Francesco Collivignarelli, Luca Gatti, Massimo Barbieri, Lorena Villano, Parvesh Chandna and Tri Setiyono

Chapter 5 **Change Detection and Classification Using High Resolution SAR Interferometry 149**
Azzedine Bouaraba, Nada Milisavljević, Marc Acheroy and Damien Closson

Contents

Section 3 Topographic Applications 165

Chapter 6 **High Resolution Radargrammetry – 3D Terrain Modeling 167**
Paola Capaldo, Francesca Fratarcangeli, Andrea Nascetti, Francesca Pieralice, Martina Porfiri and Mattia Crespi

Chapter 7 **Fusion of Interferometric SAR and Photogrammetric Elevation Data 191**
Loris Copa, Daniela Poli and Fabio Remondino

Section 4 Land Motion Applications 231

Chapter 8 **Mapping of Ground Deformations with Interferometric Stacking Techniques 233**
Paolo Pasquali, Alessio Cantone, Paolo Riccardi, Marco Defilippi, Fumitaka Ogushi, Stefano Gagliano and Masayuki Tamura

Chapter 9 **SAR Data Analysis in Solid Earth Geophysics: From Science to Risk Management 261**
S. Atzori and S. Salvi

Chapter 10 **Dikes Stability Monitoring Versus Sinkholes and Subsidence, Dead Sea Region, Jordan 281**
Damien Closson and Najib Abou Karaki

Preface

In 1978, for a period of three months, the Seasat Synthetic Aperture Radar (SAR) system provided the first spaceborne data. This limited amount of scenes, together with airborne data (primarily AIRSAR, E-SAR, and Convair-580 SAR) acquired during sporadic campaigns, constituted, in the eighties, the main source of data for the development of focusing, calibration, despeckling, and radargrammetry algorithms, but also for the understanding of information (tonal, textural, bio-, geophysical) contained in intensity and polarimetric phase data. Thirteen years later, with the launch of the ERS-1 SAR system, followed four years later by ERS-2 SAR, the attention shifted to the interferometric phase with the development of interferometric processors for the derivation of elevation and displacement maps. At the same time, first analyses performed on intensity and coherence time-series revealed the broad information – the interpretation of which represents a challenge - contained in multi-temporal data, opening new frontiers particularly for all those applications where the spatial-temporal component is dominant. Given the availability of large temporal data stacks (unfortunately acquired over limited regions) and new spaceborne SAR sensors and constellations (unfortunately acquiring in a non-systematic mode), the most recent and appreciable progresses are in the development of advanced methods in differential interferometry, including the exploitation of geophysical modeling. Concomitantly, for countrywide applications, the core of the data fusion research moves towards semi-automated classification algorithms and the inference of key biophysical parameters (particularly for forest and agricultural purposes) including the incorporation of microwave semi-empirical scattering and ecophysiological modeling. Finally, experiments based on polarimetric SAR interferometry provided in the last decades first attempts to gain 3-D structure information from semi-transparent volume scatterers in single-pass.

Looking at the past four decades regarding the evolution of algorithm developments and applications, it is worth noting a symptomatic divergence: while algorithm developments have reached a consensus and, to some extent, a maturity, large scale land applications are still struggling to take off. Omitting the political and institutional aspects (which beyond doubt play a crucial role), three are the identified reasons:

Remote sensing user – In the past two decades the divide between algorithm developments and remote sensing users has considerably amplified., Today, the available software solutions are still not perfect, however, compared to twenty years ago, they are more robust, sophisticated, user-friendly and the SAR data are of higher quality. A contradiction. Unfortunately, the misuse of prepackaged solutions, which are still for specialized users only, is not rare, and this is mainly due to the limited knowledge of SAR basics of the user. Academies, research institutes and space agencies can play a key role by introducing appropriate basic SAR courses, in order to educate adequately remote sensing users.

End-user requirements – The information requirements are more stringent compared to twenty years ago. It is not uncommon that the requested products are not feasible, due either to the lack of suitable data or to the fact that they are still at an experimental stage or not at all possible by using SAR systems. Moreover, despite the type of information, nowadays there is a clear trend that maps must cover the whole country at large scale. SAR systems are predestined, because, due to their characteristics, they can assure the mapping and monitoring over large coverage at a resolution of twenty metres and higher, which is essential to meet end-user requirements.

SAR data – So far, a key challenge is that systematic acquisitions are still non-existent, despite the availability of a few consolidated products. This hampers applications. Positive attempts to obtain application-oriented SAR data archive have been carried out by ESA (ERS-Tandem), NASA/JPL (SRTM), JAXA (ALOS PALSAR-1 and to be continued with ALOS-2), and DLR/EADS Astrium (TanDEM-X). The positive impact with respect to topographic and forest applications is unquestionable. Our hope is that space agencies in primis will follow this model for a wider spectrum of land applications and, that, in the medium term, common strategies will be pursued.

The aim of this book is to demonstrate the use of SAR data in three application domains, i.e. land cover (Part II), topography (Part III), and land motion (Part IV). These are preceded by Part I, where an extensive and complete review on speckle and adaptive filtering is provided, essential for the understanding of SAR images. We have deliberately omitted other fundamental arguments such as focusing and calibration (in geometric, radiometric, interferometric and polarimetric terms), because existing methods are consolidated and they have been already extensively covered in other publications. Part II is dedicated to land cover mapping. Here, the focus is set on the large scale mapping (requested from end-users), on the multi-temporal aspect (fundamental when using SAR data) and, finally on the use of complementary data sources (crucial for the provision of accurate and detailed information). In synthesis, it is shown that data synergy based on a multi-temporal approach is a pre-requisite for the provision of land cover/change maps with high level of detail (in terms of spatial resolution, information content, and temporal variations), particularly where the spatial-temporal component or the biophysical aspect is dominant. Part III is devoted to the generation of Digital Elevation Models based on radargrammetry and on a wise fusion (by considering sensor characteristic and acquisition geometry) of interferometric and photogrammetric elevation data. Even if the elevation accuracy derived from radargrammetry is not comparable to the interferometric and photogrammetric one, this technique became appealing after the launch of very high resolution SAR sensors overcoming some key limitations of interferometry and photogrammetry. Part IV provides a contribution to three applications related to land motion. Here, particular emphasis has been set on the combination of interferometric SAR data acquired at different frequencies using complementary techniques and their impact regarding deformation information and accuracy. Finally, the exploitation of geophysical modeling based on differential interferometry displacement maps evidences its usefulness for risk management purposes in the seismic domain.

Francesco Holecz
Sarmap SA,
Purasca, Switzerland

Paolo Pasquali
Sarmap SA,
Purasca, Switzerland

Nada Milisavljević
Department of Communication,
Information Systems & Sensors (CISS),
Royal Military Academy,
Brussels, Belgium

Damien Closson
Department of Communication,
Information Systems & Sensors (CISS),
Royal Military Academy,
Brussels, Belgium

Section 1

A Review on Speckle

Chapter 1

Adaptive Speckle Filtering in Radar Imagery

Edmond Nezry

Additional information is available at the end of the chapter

http://dx.doi.org/10.5772/58593

1. Introduction

Historically, foundations in the domain of radar speckle properties have been laid down from the 1940's to the 1980's. Decisive theoretical advances were made by teams led by Professor Fawwaz Ulaby at the University of the Michigan (USA), by Professor Christopher Oliver at the Defence Research Agency in Great Malvern (UK), and by Professor Keith Raney at the Canadian Centre of Remote Sensing (Canada) since the 1970's. Then, the domain of speckle filtering in SAR images matured in the period 1976-2000, mostly under the impulsion of Dr. Jong Sen Lee of the Naval Research Center, Washington D.C. (USA). Since 1986, the team led by Dr. Armand Lopès at the Centre d'Etude Spatiale des Rayonnements in Toulouse (France), has carried out and then inspired the development of the most efficient speckle filters existing today. Since 2000, with speckle filters having reached a satisfactory level of performance, no significant advances have been made. Nevertheless, in this period, the use of speckle filters in a wide range of applications using SAR imagery has become generalized.

A radar wave can be considered, with a good approximation, as plane, coherent and monochromatic. It is emitted by an antenna towards a target. The target backscatters partially the radar wave in the direction of a receiving antenna. In the vast majority of spaceborne Synthetic Aperture Radars (SAR), a single antenna assumes the two functions of emission and reception (monostatic radar).

The complete radar measurement is the combination of the horizontally (H) and vertically (V) linearly polarised radar waves, at emission and at reception after backscattering by the observed target. After signal calibration, this measurement, affected by noise, enables to restitute for each resolution cell a polarimetric backscattering matrix S:

$$S = [S_{pq}] = \begin{pmatrix} S_{HH} & S_{HV} \\ S_{VH} & S_{VV} \end{pmatrix} \tag{1}$$

whose coefficients $S_{pq}=|S_{pq}|.exp(j.\phi_{pq})$ are the complex backscattered field amplitudes for p-transmit and q-received wave polarisations. They relate the backscattered wave vector $\vec{Ed}$ to the incident wave vector $\vec{Ei}$: $\vec{Ed} = exp(j.k.r).S.\vec{Ei}$. This complete radar measurement is said "fully polarimetric", often abbreviated in "polarimetric".

Let consider the interaction of the radar wave with an extended random surface, *i.e.* a surface containing a sufficiently great number of scatterers per resolution cell, with no preponderant scatterer with respect to the others. For whatever configuration pq of polarisation, the signal S_{pq} received from such a surface by the antenna becomes, after quadratic detection in intensity I_{pq}:

$$I_{pq} = |S_{pq}|^2 = S_{pq}.S_{pq}^* \text{ (where}^*\text{denotes complex conjugation)} \quad (2)$$

This detected signal intensity I is proportional in average to the radar "backscattering coefficient" $\sigma°$. The backscattering coefficient $\sigma°=4.\pi.|S_{pq}|^2$ is the average radar cross-section per surface unit [1]. $\sigma°$, expressed in m^2/m^2, is a dimensionless physical quantity. It is a physical property of the sensed surface, which depends principally on its roughness, its dielectric properties, its geometry, and the arrangement of its individual scatterers.

Carrying the radiometric information with regard to the sensed target, $\sigma°$ is a function of the frequency of the radar wave, of its angle of incidence upon the target, and of the configuration of polarisation. In terms of physical meaning, the radar backscattering coefficient is analogous to the bidirectional reflectance in the domain of optical wavelengths: $\sigma° \# 4\ cos^2\theta$, where is the incidence angle of illumination on the sensed target.

Nevertheless, detected radar images look visually very noisy, exhibiting a very characteristic salt-and-pepper appearance with strong tonal variations from a pixel to the next. Indeed, since radar imaging systems are time-coherent, radar measurements over random rough surfaces are corrupted by "speckle" noise due to the random modulation of waves reflected by the elementary scatterers randomly located in the resolution cell. Then, coherent summation of the phases of elementary scatterers within the resolution cell results in a random phase of the complex pixel value.

This speckle "noise" makes both photo-interpretation and the estimation of $\sigma°$ extremely difficult. Actually, speckle is a physical phenomenon, which is inherent to all coherent imaging systems (radar, lidar, sonar, echography). In most remote sensing applications using radar/SAR imagery, speckle is generally considered a very strong noise that must be energically filtered to obtain an image on which classic and proven information extraction techniques could be further applied, in particular the techniques used for optical imagery acquired in the visible and near-infrared part of the electromagnetic spectrum.

Therefore, speckle filtering and radar reflectivity restoration are among the main fields of interest in radar images processing for remote sensing. Speckle filtering is a pre-processing aiming at the restoration of $\sigma°$ value in the first place. This pre-processing must account for both the particular properties of the speckle, and those of extended imaged targets (often called "clutter"). It must also account for the radar imaging system that has sensed the target and for the processor that has generated the images. For stationary targets of infinite size, speckle filtering is equivalent to a simple smoothing using a moving processing (averaging) window. An ideal filter must nevertheless avoid image degradation through excessive smoothing of the signal. To this end, it must respect structural image information (road and water networks, etc.), and the contours of radiometric entities. In addition, it must also respect the particular texture of the clutter, in forested or urban environments for example. Last, it must also identify the strong local radiometric variations due to the presence of strong scatterers (often artificial in nature) from those due to spatially rapid speckle fluctuations.

Therefore, an ideal speckle filter must satisfy to the following specifications:

1. Preserve accurately the local mean value of the radar reflectivity (*i.e.* the quantity actually measured by the radar, which is proportional to $\sigma°$) to enable, for example, the comparison of radar reflectivities in the framework of a multitemporal analysis of radar acquisition series.

2. Smooth as much as possible homogeneous image areas and therefore reduce the speckle to increase the Equivalent Number of Looks (ENL) of the radar image (*cf.* § 2.1.4). The minimum ENL depends on the desired radiometric accuracy. For example, a 1 dB accuracy with 90% confidence level (*i.e.* less than 25% variation of the radar intensity) requires an ENL value around 230, and a 2 dB accuracy with 90% confidence level (*i.e.* less than 60% variation of the radar intensity) needs an ENL value around 40.

3. Preserve texture as much as possible where it exists in the image (forests, non-homogeneous fields, etc.) to avoid confusions among radiometrically similar areas exhibiting different texture. Therefore, a speckle filter must be able to discriminate heterogeneity effects due to texture from those due to speckle.

4. Both preserve and denoise image structures (contours, lines) as well as the quasi-deterministic responses due to corner reflector effects within strongly textured areas such as urban environments. Indeed, the energy of artificial radar reflectors responses must be preserved to enable radiometric calibration, in particular when calibration targets are dispersed in the radar image.

5. Minimise, and whenever possible prevent loss in useful spatial resolution during the speckle filtering process.

2. Statistical properties of speckle and texture in radar images

In this section, the statistical properties of speckle in images produced by coherent imaging systems such as imaging radars, lidars or sonars, are exposed. Since a good speckle filter must restore the texture of the scene imaged by the radar, the statistical properties of texture in radar images are examined as well. This analysis intentionally restrains to the first order statistical properties, since only these are generally used by the estimation techniques involved in speckle reduction methods. Explicit use of second order statistical properties of both the speckle and the imaged scene in the filtering process is adressed in Section 4.

2.1. Speckle in SAR images

First, let consider a natural target with radar backscattering coefficient $\sigma°$ which remains constant from a resolution cell to the next. Such a target, said stationary, "homogeneous", or "textureless", may be found in a medium-resolution radar image within cultivated fields or forest cut areas for example. Spatial variations of the radar signal backscattered by such a target are therefore only due to speckle and the statistical properties of the radar signal are those of the speckle, which depends only on the radar system and the image production system.

2.1.1. First order statistics for "1-look" complex radar images

The radar imaging system is linear, spatially invariant, and can be characterised at each image pixel (x,r), where x is the azimuth and r is the radial distance with respect to the flight path of the sensor's carrier, by a complex impulse response $h(x,r)$, which defines the resolution cells. The 3dB widths of the impulse response, often used to quantify the spatial resolutions $\delta(x)$ in arimuth and $\delta(r)$ in radial distance.

If the number of individual scatterers within a resolution cell is large enough and none of them has absolute predominance in scattering the radar wave [2] [3], speckle can be modelled in output of the radar processor by a circular complex gaussian random process $u^c(x,r)$ at the corresponding image pixel of coordinates (x,r). The complex amplitude (complex radar signal, in a complex radar image) in output of the radar processor can be expressed as $A_c(x,r)=a(x,r)$ $+j.b(x,r)$, where $a(x,r)$ is proportional to the field backscattered by the surface element located in (x,r). In the literature, the terms $a(x,r)$ and $b(x,r)$ are generally called $i(x,r)$ ("in-phase" term) and $q(x,r)$ ("in-quadrature" or "out-of-phase" term), respectively.

These complex data in output of the radar processor are called a "1-look complex" image or equivalently, a "Single-Look-Complex" (SLC) image.

For a homogeneous area where $A(x,r)=A_0$ (constant), *i.e.* in the presence of speckle only, the complex amplitude has the form [4]:

$$A_c(x, r) = [A(x, r).u^c(x, r)] * h(x, r) = A_0.\, u^c(x, r) * h(x, r) = A_0.\, V(x, r) \quad (3)$$

where $A=A_0$ is the amplitude of the backscattered wave and * is the convolution operator. The random process $u^c(x,r)$ represents the circular white gaussian (by application of the central limit theorem over a large number of individual scatterers within the resolution cell) complex process responsible for the speckle. Then, $V(x,r)$ is a correlated complex gaussian process characterising the "gaussian fully developed speckle", which behaves as multiplicative "noise" with a very high spatial frequency dependent on $h(x,r)$.

Owing to the particular character of the speckle phenomenon in coherent imagery, which is the case of radar imagery, information extraction from a radar image results in a statistical estimation problem. It is therefore mandatory to have an as complete as possible statistical speckle model. The statistical model of the fully developed speckle has proven perfectly adapted to SAR images, at the spatial resolutions / wavelength frequencies combinations actually used in radar remote sensing. Note that this speckle model emphasises the importance of wave phase information under the condition that the phase has not been lost during signal detection.

Goodman hypotheses [2] [5] enable to calculate speckle statistics before (*i.e.* in a complex radar image) and after radar signal quadratic detection in amplitude A or in intensity I. Taking this hypotheses into account, one obtains:

$$E[i]=E[q]=0 \qquad \textit{and} \qquad Var[i]=Var[q]=\sigma_i^2=\sigma_q^2=\sigma^2 \tag{4}$$

where E[.] denotes the mean expectation and $Var[.]$ denotes the variance. Therefore, the mean backscattered intensity R is expressed as:

$$R=E[I]=E[A^2]=E[i^2+q^2]=2\sigma^2 \tag{5}$$

R=E[I] is proportional to the radar backscattering coefficient $\boldsymbol{\sigma}°$. R can be estimated by R≈<I>, where <.> denotes the averaging operator applied in a neighbourhood of the pixel under consideration. Then, from R, it is possible to retrieve the $\boldsymbol{\sigma}°$ value through the calibration parameters of the radar image. Problems related to radar image calibration are discussed in detail in [6] [7] [8]. R=E[I] can be estimated locally in a radar image through the spatial averaging <I> of the intensities of a number N of image pixels in a spatial neighbourhood of the image pixel of interest [9]. This estimate is the unbiased Maximum Likelihood estimate with minimal variance [6].

Besides, the in-phase and the out-of-phase components i and q of the complex radar signal are decorrelated, E[$i.q$]=0, and therefore independent of each other. It results of the above considerations that the probability density functions (pdf) of i and q are Gaussian distributions:

$$P(i)=1/\sqrt{2\pi.\sigma^2}.\,exp(i^2/2\sigma^2) \qquad \text{and} \qquad P(q)=1/\sqrt{2\pi.\sigma^2}.\,exp(q^2/2\sigma^2) \tag{6}$$

with a phase of the complex radar signal $\phi=Arctg(q/i)$, which is uniformly distributed in the interval $[0, 2\pi]$.

2.1.2. First order statistics for "1-look" detected radar images

Since i and q are independent variables, their joint pdf is P(i,q)=P(i).P(q). The pdf $P_u(I)$ of the speckle in intensity is obtained by a simple change of variable I=$a2$+$b2$ in Equation (6). Then, $P_u(I)$ results being an exponential pdf:

$$P_u(I) = (1/E[I])\,.\,exp(-I/E[I]) \qquad \text{for} \qquad I>0 \tag{7}$$

It is important to note that, since E[I] is proportional to the radar reflectivity R, $P_u(I)$ is the pdf of I conditional to the value of R, therefore $P_u(I)=P_u(I/R)$.

The two first first-order moments of this pdf can be expressed as a function of its standard-deviation as follows:

$$\text{mean } E[I] = 2\sigma^2 \quad \text{and} \quad \text{standard-deviation } \sigma_I = 2\sigma^2 \tag{8}$$

To characterise the strength of speckle in radar images, it is convenient to consider the normalised coefficient of variation of the intensity, C_I, *i.e.* the standard-deviation of the radar signal intensity I over its mean value:

$$E[C_I] = \sigma_I / E[I] = 1 \tag{9}$$

The value of the coefficient of variation of speckle only, sometimes called "contrast index", is a constant for every type of radar image product. Equation (9) means that, in a radar image, the dispersion (variance) of radiometry increases as the mean signal backscattered by the target increases. This justifies in part the qualification of "multiplicative noise" given to the speckle.

Clearly, with a signal-to-noise ratio of 1, the radiometric degradation due to speckle makes very difficult the discrimination of two homogeneous targets with very different radar backscattering coefficients. As an example, a theoretical computation [10] demonstrates that two textureless target classes of homogeneous radar reflectivities (*i.e.* exhibiting only speckle) radiometrically separated by 2.5 dB present a probability of confusion of 40% in a 1-look SAR image.

Therefore, a first radiometric enhancement is needed to achieve a reduction of the coefficient of variation of the speckle over homogeneous areas. It corresponds to an enhancement of the signal-to-noise ratio and to a preliminary speckle "noise" reduction.

2.1.3. *First order statistics in "multilook" images*

A first method of speckle reduction consists in averaging incoherently M independent samples of the intensity I (or "looks") obtained from the same target:

$$I = (1/M).\sum_{k=1}^{M} I_k \quad \textit{with each of the } Ik \text{ distributed according to Equation (7)} \tag{10}$$

The goal of this method is to reduce speckle enough to make radar image photo-interpretation possible. Indeed, experience has shown that M values of the order of 3 or 4 enable a photo-interpreter to use a SAR image [1]. Such values have been adopted for most spaceborne SARs (ERS, Almaz, JERS-1, Radarsat, Envisat, ALOS: 3-looks, Seasat, SIR-B: 4-looks), including recent ones (TerraSAR-X, Cosmo-Skymed, Sentinel-1, etc.).

This operation is realised by splitting the Doppler bandwidth of the backscattered signal into M sections. This is equivalent to divide the length of the synthetic antenna into M sub-antennas. Independent processing of each section results in an independent image called a "look". Nevertheless, this operation results in a degradation by a factor M of the spatial resolution in azimuth of each look [1].

Over the same target, the mean intensity resulting from this operation remains the same mean intensity of each of the individual looks. If the M individual looks were independent, the standard-deviation is divided by $\sqrt{M}$. Thus, the coefficient of variation of the speckle measured over a homogeneous area of the intensity multilook image becomes:

$$C_I \ (homogeneous\ area) \ = \ C_u \ = \ 1/\sqrt{M} \tag{11}$$

It is important to note that multilook radar image formation is at the expense of the spatial resolution in azimuth. In practice, the value of M does never exceed a few units (less than 16 for airborne SARs, and in general only 3 or 4 for spaceborne SARs). This remains insufficient to improve satisfactorily the signal-to-noise ratio of a radar image. Indeed, in our example of two target classes with homogeneous radar reflectivities radiometrically separated by 2.5 dB, the probability of confusion of 40% in a 1-look SAR image reduces only to 33% in a 4-look image.

If the individual looks are uncorrelated with each other, the pdf of the speckle, which is the sum of M independent exponential distributions, becomes a χ^2 distribution with $2M$ degrees of freedom:

$$\mathrm{P}_u(I) \ = \ \mathrm{P}_u(I/R) \ = \ (M/\mathrm{E}[I])^M \ / \ (M-1)! \ . \ exp(-M.I/\mathrm{E}[I]) \ . \ I^{(M-1)} \tag{12}$$

2.1.4. "Equivalent Number of Looks" of a SAR image

If the M sections of Doppler bandwidth used to produce the individual looks overlap, the averaged samples, and therefore the individual looks, are correlated. The coefficient of variation of the speckle C_u, measured over a perfectly homogeneous image area will therefore be always superior to what could have been expected from Equation (11). In this situation, speckle strength results as if the number of independent looks were equal to:

$$L \ = \ 1/C_u^2 \ < \ M \tag{13}$$

The L value, which is generally a non-integer value, is called the "Equivalent Number of Looks", or "ENL", of the multilook radar image.

Hence, the pdf of the speckle in intensity can be approximated, by extension of Equation (12), for whatever value of L, by a Gamma distribution with parameters E[I] and L:

$$\mathrm{P}_u(I) \ = \mathrm{P}_u(I/R) \ = (L \ /E[I])^L \ / \Gamma(L) \ . \ exp(-L \ .I/E[I]) \ . \ I^{(L-1)} \tag{14}$$

which is rigorously equivalent to Equation (12) when $L=M$ (L integer) for M independent looks.

All speckle models for multilook images (M correlated or uncorrelated averaged looks, *i.e.* L equivalent looks, with $M \geq L$) use this approximated distribution. Note that the pdf of the speckle as formulated in Equation (14) is slightly inexact if the M looks are correlated, as it is the case in general. Therefore, the pdf of the speckle for the average of M correlated looks is close to, but is not exactly, a Gamma pdf [11].

Let us consider that the M looks $(I_1, I_2, \ldots, I_M)$ are correlated to each other with correlation coefficients ρ_{ij}, and that the I_i are distributed according to the same exponential marginal pdf (1-look, *cf.* Equation (7)) with same parameter $R=\mathrm{E}[I]=\mathrm{E}[I_i]$, for whatever i. This last condition can be actually fulfilled by re-weighting the individual looks to make them contain an identical mean energy $\mathrm{E}[I]$.

For the intensity image resulting of the averaging of the M looks, $I=(\sum I_i)/M$, the set of looks correlations ρ_{ij} is taken into account to compute the ENL value L ($L<M$), which is related to the coefficient of variation of the speckle in intensity, C_u, by:

$$C_u{}^2 = 1\Big/L = \Big[\, 1 + (1\Big/M) \,.\, \sum_{i=1}^{M}\sum_{j=1}^{i-1} \sigma_{ij}^2 \Big] \Big/ \, M \tag{15}$$

The exact pdf of the average intensity I of M correlated looks, is extremely difficult to establish mathematically in a closed-form (for 2-looks, *cf.* [11]), and, when established, to manipulate. Nevertheless, Equation (14) is a very close and therefore satisfactory approximation when using the appropriate ENL value L calculated using Equation (15).

2.1.5. The logarithmic (homomorphic) transformation

Since the radar backscattering coefficient $\sigma°$ is generally expressed in dB in physics, and with the goal to make speckle an additive noise that would be independent on the level of the radar signal, some authors [12] [13] [14] chose to use a neperian logarithmic (homomorphic) transformation:

$$D = -\, Log\,(I) \tag{16}$$

Arsenault and April [15] [17] [19] demonstrated that, after this transformation, the pdf of the speckle for a multilook radar image, Equation (14), becomes a Fisher-Tippett distribution:

$$\mathrm{P}_u(D) = L^{\,L} \,/\, \Gamma(L)\,.\; exp\big[-L\,.(D-D_0)\big]\,.\,exp\big\{-L\;.\,exp\big[-(D-D_0)\big]\big\} \tag{17}$$

with $D_0 = -Log(E[I])$

The first order statistical moments of this distribution are as follows [15]:

1) Mean <D>:

$$<D> = D_0 - \Gamma'(L)/\Gamma(L) + Log(L) \tag{18}$$

Equation (18) shows that the logarithmic transformation causes a signal distortion, increasing with decreasing number L of independent speckle samples taken into account, and with decreasing R (therefore, $\boldsymbol{\sigma}°$) values. There is therefore a tendency to systematically underestimate the value of $\boldsymbol{\sigma}°$ [15] [16] [18].

2) Variance $\sigma_D{}^2$:

$$\sigma_D{}^2 = (\pi^2/6).[\gamma + \Gamma'(L)/\Gamma(L)]^2 + 2\sum_{k=1}^{L-2}[(1/(L-k)).\sum_{j=1}^{L-k-1}(1/j)] \tag{19}$$

Lopès [20] has shown that for N independent L-look radar data samples (NL independent speckle samples), the standard-deviation of the samples averaged after logarithmic transformation, and then retransformed into intensity, is always significantly larger than the standard-deviation σ_I of the same samples averaged in intensity. Thus, the logarithmic transformation degrades both the measurement accuracy of the backscattering coefficient $\boldsymbol{\sigma}°$ and its local range of fluctuations due to local scene texture.

2.2. Texture in SAR images

Texture concerns the spatial distribution of grey levels. It is important in the analysis of SAR images for a wide range of applications (*e.g.* [21] [22] [23] [24] etc.). In most of these applications, not only the radiometric, but also the textural information must be retrievable after adaptive speckle filtering of the SAR images.

2.2.1. *Texture of the imaged scene*

As seen above, within a homogeneous area image of a detected radar image, one can consider the observed speckle I as originating from a real non-gaussian random process $u(x,r)$, with unit-mean $<u>=1$, and proportional to the radar signal I (multiplicative noise). A simplification of Equation (3) enables to write, for every pixel (x,r) located within a homogeneous image area (*i.e.*$R(x, y)=\mathrm{E}[R]=R, \ \forall (x, y)$):

$$I(x, r) = u(x, r) . \mathrm{E}[I(x, r)] = u(x, r) . R \tag{20}$$

In most remote sensing applications, it is reasonable to consider the imaged scene as an arrangement of discrete objects superimposed to a background with mean reflectivity E[R]. The imaged scene is composed of classes of non-homogeneous ($R(x, y) \neq \mathrm{E}[R]$) objects characterised by the statistical properties and parameters of the variable R [25]:

If Tj is the random variable which represents the spatial fluctuations of the reflectance (let consider in the following its analog in radar images, the denoised/speckle-filtered radar reflectivity) $R(x,r)$ around its mean expectation $E[R(x,r)]$, in the terrain class j to which the pixel at coordinates (x,r) belongs, the following results are obtained:

1. $R(x,r)$ is proportional to the $\sigma°$ value of the resolution cell containing pixel (x,r);
2. $E[R(x,r)]$ is proportional to $\sigma°$ in a neighbourhood of pixel (x,r) belonging to class j;
3. $E[T_j]=1$ for every class j;

$$R(x, r) = E[R(x, r)] . T_j \tag{21}$$

The spatial structure of actual imaged scenes, called "texture", induces a mesurable spatial structure (after the properties of T_j) in the images of these scenes. This is the case, for both optical and radar images. Indeed, a simple visual interpretation of a radar image reveals radiometric spatial variations at a longer scale than those due to speckle only. These variations originate from the spatial fluctuations of the radar backscattering coefficient $\sigma°$ within a given terrain class, and are affected by the presence of speckle.

To characterise a given class j, one must be able to restore T_j, that means to separate the respective effects of speckle and texture in the spatial variations of the radar intensity signal. To this end, the multiplicative fully-developed speckle model exposed above will be used. It has been shown [26] [3] that, even when the scene is not stationary, speckle still remains multiplicative, eventually correlated, and independent on the imaged scene as long as Goodman's conditions [2] [5] are realised.

2.2.2. *First order statistics of texture in SAR images*

2.2.2.1. *Spatially uncorrelated speckle*

Considering the scene texture model of [25] in Equation (21), it is possible [27] [28] to model the radar image intensity at pixel of coordinates (x,r) by generalising the multiplicative speckle model for a wide-sense stationary scene (Equation (20)) as follows:

$$I(x, r) = R(x, r) . u(x, r) = \left(E[R(x, r)].T_j\right) . u(x, r) \tag{22}$$

This relationship is valid as long as the spatial variations of $R(x,r)$ happen at length scales longer than the size of the resolution cell of the radar imaging system.

The variance of the intensity I within a given target class is computed using:

$$\sigma_I{}^2 = E[I^2]-E^2[I] = E^2[R].\left(E[T^2.u^2]-E^2[T.u]\right) \tag{23}$$

Since the fluctuations of $\sigma°$, and therefore those of R, are independent on the speckle, one gets [27] [28]:

http://dx.doi.org/10.5772/58593

$$(\sigma_I / \mathrm{E}[I])^2 = (\sigma_R / \mathrm{E}[R])^2 \, .(\sigma_u / \mathrm{E}[u])^2 + (\sigma_R / \mathrm{E}[R])^2 + (\sigma_u / \mathrm{E}[u]) \tag{24}$$

and:

$$\mathrm{E}[I] = \mathrm{E}[R.u] = \mathrm{E}[R].\mathrm{E}[u] = \mathrm{E}[R] \tag{25}$$

Therefore, E[R] is locally estimated in the radar image by averaging pixel intensities in a neighbourhood of pixel (x,r):

$$\widehat{\mathrm{E}[R]} \quad \langle I \rangle \tag{26}$$

Introducing the coefficients of variation ($C_I = \sigma_I / \langle I \rangle$ for the radar intensity, $C_R = \sigma_R / \langle R \rangle$ for the imaged scene, $C_u = \sigma_u / \langle u \rangle$ for the speckle), Equations (24) and (25) lead to the important result [27] [28]:

$$C_R{}^2 = (C_I{}^2 - C_u{}^2) \; / \; (1 + C_u{}^2) \tag{27}$$

which characterises scene texture in terms of heterogeneity, with $C_u^2 = C_{uI}^2 = 1/L$ for a homogeneous/textureless area of a L-look intensity radar image. Indeed, Equation (27) shows that the more heterogeneous is the scene, the easier its texture can be restored.

For a better description of the scene, one must use the pdfs of its diverse classes (distribution of the random variable R or $\boldsymbol{\sigma}°$). Their knowledge may result, either from a priori knowledge, or from direct estimation in the image. If the general form of the pdf is known, its parameters can be estimated from the image data. In all cases, if one can establish the pdf of R, $\mathrm{P}_R(R)$, for a given terrain class in the scene, the unconditional pdf of the corresponding radar image intensity I is:

$$\mathrm{P}_I(I) = \int_0^{\infty} \mathrm{P}_u(I) \, . \, \mathrm{P}_R(R) \, . \, dR \tag{28}$$

where $\mathrm{P}_u(I)$ is the pdf of the speckle for multilook images (*cf.* Equation (14)). Equation (28) shows, as does also Equation (27), that the respective contributions of scene texture and speckle micro-texture can be separated in the radar image.

2.2.2.2. *Spatially correlated speckle*

To produce a SAR image, speckle samples at the output of the SAR processor are correlated through the SAR impulse response to obtain a sampling rate (pixel size) of about half the spatial resolution of the radar sensor, thus avoiding aliasing effects. Using the multiplicative noise model and Equations (24) and (25) [27] [28], the correlation coefficient $\rho_I(\Delta x, \Delta y)$ between two pixels of intensities I_1 and I_2 separated by (Δx,Δy) in the SAR image:

$$\rho_I(\Delta x, \Delta y) = (<I_1(x, y)-<I_1>> . <I_2(x, y)-<I_2>>) \ / \ (\sigma_{I1} . \sigma_{I2}) \tag{29}$$

can be related to the scene correlation coefficient $\rho_R(\Delta x, \Delta y)$ and to the speckle correlation coefficient $\rho_u(\Delta x, \Delta y)$ by [10]:

$$\begin{aligned} <C_I>^2 . \ \rho_I(\Delta x, \Delta y) &= <C_R>^2 .<C_u>^2 .\rho_R(\Delta x, \Delta y).\rho_u(\Delta x, \Delta y) \\ &+ <C_u>^2 .\rho_u(\Delta x, \Delta y) + <C_R>^2 .\rho_R(\Delta x, \Delta y) \end{aligned} \tag{30}$$

Therefore, if $\rho_u(\Delta x, \Delta y) > 0$ for $\Delta x>0$ and/or $\Delta y>0$, which is the case in correctly sampled SAR images, the local statistics estimated in the neighbourhood of a given pixel will have to be corrected of the effect of speckle spatial correlation properties (*cf.* Section 4).

2.2.3. *Local statistics computation and consequences of the speckle model*

Using the simplified multiplicative noise model for the speckle *u* (*cf.* Equation (22)), the first-order non-stationary statistics of the scene, $<R>$ and σ_R^2, can be deduced locally (*cf.* Equations (26) and (27)) from those of the radar image intensity, $<I>$ et σ_I^2.

In practice, C_I is estimated in the radar image, and its locally estimated value can be inferior to C_u. Since σ_R^2 and C_R^2 are positive quantities, finding cases where $C_I^2 < C_u^2$ must be attributed to the limited size of the neighbourhood of the pixel under processing over which $<I>$ and σ_I^2 are estimated [29]. Independently of the scene model $P_R(R)$, the multiplicative speckle model fixes therefore an inferior threshold $C_{I\ \min}=C_u$ to the possible values of the coefficient of variation C_I.

Below the $C_{I\ min}$ value, speckle filters based on the use of local statistics, it means all adaptive speckle filters, are no longer valid. In this case, the image area under processing must be considered homogeneous and textureless.

2.2.4. *The coefficient of variation as a heterogeneity indicator*

The coefficient of variation C_I is a heterogeneity measure particularly well appropriate to the case of radar images and for isotropic textures. Nevertheless, if the source of heterogeneity presents a particuliar orientation (contour, line...), additional detectors able to identify or retrieve this orientation are needed.

C_I increases with scene heterogeneity. Depending on the heterogeneity of the area under processing, it enables to discriminate among the following situations:

1. homogeneous areas with neither texture, nor structures.
2. areas with low or moderate heterogeneity where the presence of structures is improbable.
3. strongly heterogeneous areas where one must search for the possible presence of structures (contour, line, strong scatterers).

3. Adaptive speckle filters for single-channel detected radar images

This section is dedicated to a theoretical analysis of the most common and more efficient speckle filtering techniques developed for Synthetic Aperture Radar (SAR) images, and of their corresponding estimation techniques. These filters use variants of the statistical speckle model exposed in the preceding Section 1. They also use diverse statistical estimators to restore the radar reflectivity (in the sense of the CEOS's (Committee for Earth Observation Satellites) "radar brightness"). These statistical estimators are, either Minimum Mean Square Error estimators (*e.g.* Lee *et al.* filter, Kuan *et al.* filter, etc...), or autoregressive estimators (*e.g.* Frost *et al.* filter), or Bayesian estimators (*e.g.* Gamma-Gamma and DE-Gamma MAP filters). These estimators are discussed, and their behaviours are analysed.

3.1. Wiener Method: the Frost *et al.* filter (1982) [30]

3.1.1. Theoretical development

The SAR image model considered by [30] is as follows:

$$I(t) = [R(t) \,.\, u(t)] * h(t) \tag{31}$$

where $R(x,y)$ is a wide-sense stationary random process decribing the radar reflectivity of the observed scene at pixel of coordinates $t=(x,y)$ located in a homogeneous area of the radar image. $u(t)$ is the multiplicative noise due to speckle, modelled by a real white stationary random process with a Gamma pdf (*cf.* Equation (14)). $h(t)$ is the impulse response function of the system. This model is a simplification of the fully developed speckle model exposed above (*cf.* Equation (3); [31] [32].

To estimate the radar reflectivity $R(t)$ of the noise-free image, Frost *et al.* apply a linear filter whose impulse response $m(t)$ minimises the mean quadratic error (MMSE estimator):

$$\varepsilon = \mathrm{E}[\{ R(t) - I(t) * h(t) \}^2] \tag{32}$$

The MMSE least-squares solution is valid for homogeneous areas for which $R(t)$ can be represented by a stationary random process. To deduce $m(t)$, [30] assume that the transfer function $H(f)$ is constant over a certain bandwidth. This leads to an uncorrelated multiplicative speckle model:

$$I(t) = R(t) \,.\, u(t) \tag{33}$$

where $u(t)$ is a pseudo-white noise, independent of the signal, also used by other authors [33] [34]. The multiplicative speckle model is deduced from Equation (3) assuming that the bandwidth of the signal $R(t)$ is much narrower that that of the linear filter $h(t)$.

The impulse response of the filter is calculated adopting an autoregressive process model for $R(t)$ with an exponential autocorrelation function, a classic hypothesis for this family of models [35] [36]. Later studies showed that the choice of an exponential autocorrelation function is appropriate to SAR scenes [37].

These hypotheses enable to define an optimal MMSE filter (Wiener filter) with impulse response $m(t)$:

$$\hat{R} = \sum_{\forall t \in \text{ neighborhood of } N \text{ pixels}}^{t=(x,y)} m(t) \,.\, I(t) \tag{34}$$

with:

$$m(t) = K_1 \,.\, exp\left[-K \,.\, C_I{}^2 \,.\, d(t) \right] \tag{35}$$

where K is the filter parameter, and $d(t)$ is the distance between the pixel located in t and the pixel under processing. K_1 is a normalisation constant granting the preservation of the mean value of the radar reflectivity:

$$K_1 = (1/N). \sum_{\forall t \in \text{neighborhood of } N \text{ pixels}}^{t=(x,y)} m(t) \tag{36}$$

The final Frost *et al.* filter (Equation (36)) is not exactly a Wiener filter. It is designed using Yaglom's method [38]. In this method, contrarily to that of Wiener, one searches for the frequential characteristic, not for the impulse transfer function susceptible not to exist. This function is chosen empirically taking into account the hypotheses it must fit.

3.1.2. Behaviour of the Frost et al. filter

The behaviour of the filter depends on the values of the locally observed coefficient of variation, *i.e.* on the local heterogeneity:

- Extremely homogeneous areas, for which $C_I = 0$, then: $\hat{R} = \langle I \rangle$
- Very strong scatterers (extreme heterogeneity), where $C_I \to \infty$, then: $\hat{R} = I$
- Between these two situations, in textured natural areas, when C_I value increases, neighbour pixels located far away from the pixel under processing are given less weight and the neighbourhood on which the weighted mean value is actually estimated narrows.

Considering the requirements for an ideal speckle filter, the filter should:

- Restore the mean intensity value over homogeneous areas, where $C_I \leq C_u$. A consequence is that K might be close to 0, unless the value of C_u tends towards 0, which is obviously not possible for usual radar images.
- Preserve the observed intensity when the value of C_I is very high but obviously always finite. For high C_I values, the only pixel to weigh might be the pixel under processing, and K might therefore tend towards infinity since C_I has always a finite value.

Since it is impossible to fulfil simultaneously these two conditions (*cf.* [39] [40], one must make a difficult choice between, either an efficient speckle reduction, or the preservation of image structures and texture.

3.2. Locally adaptive linear MMSE estimators: Lee and Kuan filters

3.2.1. *Kuan et al. filter (1985) [41] [32]*

The radar image $I(t)$ is modelled as a function of the radar reflectivity image $R(t)$ to be restored and of a white uncorrelated noise $n(t)$ with 0 mean and dependent on $R(t)$. This is an additive noise model different of the multiplicative speckle model used in Equation (33):

$$I(t) = R(t) + n(t) \tag{37}$$

Kuan *et al.* [41] introduce a scene model $R(t)$ with non-stationary mean and non-stationary variance (NMNV model), where the non-stationary mean describes the general structure of the image, whereas the non-stationary variance characterises texture and the presence of local structures. With this model, the linear filter minimising the quadratic mean error (LMMSE) has the general form [32]:

$$\hat{R}_{LMMSE} = \mathrm{E}[R] + C_{RI} . C_I^{-1} . (I - \langle I \rangle) \tag{38}$$

where C_{RI} and C_I are the non-stationary spatial covariance matrices of the image. These matrices are diagonal if, as it is assumed in the model, $n(t)$ is a white noise and the image model is NMNV. Then, Equation (38) takes a scalar form [31]:

$$\hat{R} = \mathrm{E}[R] + (I - \langle I \rangle) . \sigma_R^2 \; / \; (\sigma_R^2 + \sigma_n^2) \tag{39}$$

The local statistics of the scene R are deduced from those of the intensity I. The Kuan *et al.* filter is obtained by replacing locally:

$$\mathrm{E}[R] \text{ by } \langle I \rangle \quad \text{and} \quad \sigma_R^2 \text{ by } [\sigma_R^2 - \langle I^2 \rangle . \sigma_n^2 \,] \,/\, [1 + \sigma_n^2] \tag{40}$$

Since the variance of the additive noise $\sigma_n{}^2$ is numerically equal to the coefficient of variation of the speckle C_u in the multiplicative model, Equation (40) is, in practice, exactly equivalent to Equations (26) and (27) established for the multiplicative speckle model.

It is noteworthy that, since the image model does not assume independence of the noise n and the signal R, the filter is theoretically able to take into account the effects of an eventual dependence between speckle and the radar reflectivity, when speckle is no longer fully developed as in the case of strong scatterers.

3.2.2. Lee filter (1980) [33] [42] [43] [44] [45]

Lee uses the unit-mean uncorrelated multiplicative speckle modele (*cf.* Equation (33)). A linear approximation is done by developing Equation (33) (I as a function of u for pixel located in t) in a first-order Taylor series with respect to $u(t)$:

$$I(t) = R(t).\,\mathrm{E}[u(t)] + \mathrm{E}[R(t)]\,.[u(t)-\mathrm{E}[u(t)]\,] \qquad \text{with} \qquad \mathrm{E}[u(t)] = 1 \tag{41}$$

This approximation enables to transform Equation (33) into a weighted sum of the signal and of a noise independant on the signal. The linear MMSE estimator, and Equations (26) and (27) that are consequences of the multiplicative speckle model, enable to establish the Lee filter [33], historically the first speckle filter designed to be adaptive to local radar image statistics:

$$\hat{R} = \langle I\rangle + (I - \langle I\rangle).\,(\sigma_I{}^2 - \sigma_u{}^2)/\sigma_I{}^2 \tag{42}$$

It is noteworthy that, assuming the independence of noise and signal in the model used by the Kuan *et al.* filter, this filter become identical to the Lee filter [41].

3.2.3. Behaviour of the Kuan et al. and Lee filters

The linear MMSE speckle filters of Lee and of Kuan *et al.*, based on the use of the local statistics of the observed intensity can be written under the same general form:

$$\hat{R}(t) = I(t).\,W(t) + \langle I(t)\rangle.\,[1-W(t)] \tag{43}$$

with:

$$W(t) = (1-[C_u{}^2/C_I{}^2(t)])/(1+C_u{}^2) \qquad \text{for the Kuan filter} \tag{44}$$

and

$$W(t) = 1 - [C_u{}^2 \,/\, C_I{}^2(t)] \qquad \text{for the Lee filter} \tag{45}$$

Both methods perform a linearly weighted average of the local mean of the intensity and of the observed pixel intensity. In both cases, weights depend on the ratio of the coefficients of variation of the observed intensity and of the noise [7] [40]. As for the Frost *et al.* filter, the locally observed heterogeneity governs the behaviour of these filters:

- Homogeneous areas where $C_I = C_u$, therefore $\hat{R} = \langle I \rangle$
- Very stong scatterers (extreme heterogeneity) where $C_I \to \infty$, therefore $\hat{R} = I$
- Between these two extreme situations, the weight of the observed value of the pixel intensity increases with the heterogeneity of its neighbourhood.

Nevertheless, note that the weight on the estimated mean intensity remains significant for high values of the coefficient of variation. Thus, responses of small impulse targets are cut-off by these filters and image structures suffer some amount of smoothing. To correct this drawback, Lopès *et al.* [40] have proposed a very efficient enhancement to the Lee and Kuan *et al.* filters.

Inversely, when $C_I < C_u$, some weighting on the observed intensity reappears, resulting in an amplification of the noise [13]. In this case, the local mean of its neighbour pixels must be assigned to the pixel under processing as filtered value.

The linear MMSE filters differ from the Wiener filters by the fact that the A Priori mean and variance, $\langle R \rangle$ and σ_R^2, are estimated using the local statistics of the original image I, and not implicitly estimated from an assumed autocorrelation model (*cf.* [30], Equation (35)). The goal of this approach is to obtain a better adaptivity to the local properties of the scene R, which verifies in pratice.

It is remarkable that the often used MMSE estimation ([33] [43] [44] [45] and [41] [32], among others) is nothing else than the mean of the conditional A Posteriori pdf P(R/I) [46]. Note that the complete evaluation of the terms enabling to obtain the A Posteriori pdf P(R/I) requires a description of P$I(I)$. This last pdf, which depends on the non-linear transformation leading from R to I is, in general a K-distribution. The solution adopted by Lee and Kuan *et al.* consists in a forced linearization of the problem assuming a Gaussian P$I(I)$ [33] [42] [44], which is unjustified for low (< 3) ENL values.

As an effect, this linearization restrains the validity of the linear MMSE estimator to situations where the noise level is not too high (multilook images with high values of L), or where the local heterogeneity C_I is not too high. These considerations justify one of the modifications proposed to the Lee and Kuan *et al.* filters par Lopès *et al.* [40]: an upper heterogeneity threshold $C_{I\max}$, above which the observed intensity of the pixel under processing is preserved.

3.3. Bayesian Maximum A Posteriori (MAP) speckle filters

3.3.1. Originality of the MAP approach in the case of SAR images

The Bayesian MAP approach [47] [48] consists in characterising the imaged scene and the speckle "noise" by their statistical description, using their associated pdfs.

In the Bayesian perspective, the theory of probabilities is extended to the logics of probabilistic inference. Probabilities are seen as a relationship between a formal hypothesis and a possible conclusion. This relationship corresponds to a certain degree of rational credibility and is limited only by the extreme relationships of certitude and impossibility [49]. The classic deductive logics considering only these extreme relationships (*à la* Sherlock Holmes: "When all impossible hypotheses have been excluded, what remains, even if unlikely, corresponds to reality") is nothing but a particular case of this general development [50].

As a general consequence, the theory of probabilities cannot base on the sole concepts of classic logics (frequentist as in the MMSE or in Wiener's approaches). In particular, the relationship of probability cannot be defined in terms of certitude, since certitude is viewed as a particular case of probability. The frequentist definition of probabilities based on relations of certitude related to the knowledge of a number of parameters (<*R*> for example, *cf.* § 3.3.5 below) is therefore no longer sufficient [50].

In this context, probabilities are used to describe stochastically an incomplete information on a global phenomenon (here, the radar reflectivity and the superimposed speckle), rather than to describe only the noise randomness that corrupts its comprehension. Probability relationships are viewed as being conditional to the context. This way, the pdf of the speckle $P_u(I)$ (*cf.* § 2.1.4, Equation (14)) is therefore formulated under the form $P(I/R)$. Taking into account a reasonable A Priori statistical model for the radar reflectivity $P(R)$, one must also take into account – which is new, with respect to the methods precedently exposed – the probability of R, given the information obtained through observation (the intensites I measured by the radar), formulated under the form of an A Posteriori pdf $P(R/I)$. This approach has the great advantage to enable the characterisation of speckled radar image formation while easing the description of non-linear effects.

The least error-cost inference mechanism leading from the observed intensity I to the A Posteriori deduced radar reflectivity R through Bayes's theorem [47] [48], allows a rigorous combination of the A Priori knowledge $P(R)$ and of the new knowledge provided by the observation I:

$$P(R/I) \quad = \quad P(I/R) \,.\, P(R) \;/\; P(I) \tag{46}$$

$P(R/I)$ depends on the pdf of R introduced as A Priori information about the scene to restore. Thus, the estimate is influenced by the A Priori statistical knowledge about R or, by default, the hypotheses made about R. The noisiest are the radar data, the less the A Priori information would contribute to estimate R [46].

In theory, the MAP method enables to avoid direct estimation of the mean of the conditional A Posteriori pdf, which is necessary to the MMSE estimation. This feature is of great interest in the resolution of non-linear problems where the evaluation of the conditional mean is, either difficult, or impossible [51].

The Maximum A Posteriori (MAP) estimation of R corresponds to the maximum value of P(R/I), *i.e.* to the mode of the A Posteriori pdf P(R/I). The mode and the mean of P(R/I) coincide if P(R/I) is symmetrical, in particular in the Gaussian case. Since the available knowledge is:

- P(I/R), the conditional pdf of speckle, which is a Gamma distribution (*cf.* Equation (14)),
- P(R), the A Priori unconditional pdf of the imaged scene, which is an assumed statistical distribution (not necessarily Gaussian),

there is no particular justification to prefer the MMSE estimation by minimisation of the mean quadratic error implicitly assuming a symmetrical P(R/I), rather than to use this more complete available knowledge through the Bayesian estimation technique, which is less subject to arbitrary assumptions [46].

It is of interest to notice that the "Maximum Likelihood" estimation is nothing more than a particular case of Maximum A Posteriori, in the situation where the conditional A Posteriori and A Priori pdf's are equal: P(R/I)=P(I/R), with P(R)=P(I). Note that the Maximum Likelihood estimation of the scene R is the radar image I itself, which is of interest in the case of strong scatterers whose radar response is deterministic.

Once the first-order statistical speckle model P(I/R) is either known, or reasonably approximated, (*cf.* Equation (14)), the form of the MAP estimate depends principally on the form of P(R).

The NMNV scene model [32] enables to solve locally, in an analytic manner, the estimation of R. The parameters of the pdf P(R) are locally estimated in a neighbourhood of the pixel under processing. This way, the complicated recursive forms of mathematical resolution as exposed by Hunt [52], Kuan *et al.* [53] and Geman & Geman [54] are avoided. This approach is particularly adapted to the high spatial resolution of airborne and modern spaceborne SARs. Indeed, it allows a good preservation of scene texture in an adaptive way by taking into account the local fluctuations of $\sigma°$ in the A Priori model, while being at the same time computationally efficient.

3.3.2. *Computation of the MAP estimate*

Since the logarithm function is a monotonically increasing function, Bayes's formula ([47] [48]; Equation (46)) can be rewritten as:

$$Log[\mathrm{P}(R/I)] = Log[\mathrm{P}(I/R)] - Log[\mathrm{P}(I)] + Log[\mathrm{P}(R)] \tag{47}$$

which gives the local MAP estimation of R when $Log[\mathrm{P}(R/I)]$ is maximum, *i.e.* when its first-order derivative with respect to R is locally equal to 0:

$$\partial/\partial R\ Log[\mathrm{P}(I/R)] + \partial/\partial R\ Log[\mathrm{P}(R)] = \partial/\partial R\ Log[\mathrm{P}(R/I)] + \partial/\partial R\ Log[\mathrm{P}(I)] \tag{48}$$

with:

$\partial / \partial R Log[P(I)]=0$ because P(I) does not depend on the local fluctuations of R.

P(R/I) reaches its maximum when:

$$\partial / \partial R \; Log[P(R/I)] = 0 \quad \text{for} \quad R = \hat{R}_{MAP}, \quad \text{where } \hat{R}_{MAP} \text{ is the MAP estimation of } R \tag{49}$$

Then, the general equation of the MAP speckle filter becomes, locally:

$$\partial / \partial R \; Log[P(I/R)] + \partial / \partial R \; Log[P(R)] = 0 \quad \text{for} \quad R = \hat{R}_{MAP} \tag{50}$$

The first term of Equation (50), the Maximum Likelihood term, accounts for the effects of the whole imaging system on the radar image and describes the detected radar intensity once the speckle statistical model is known. The second term, the Maximum A Priori term, represents the prior statistical knowledge with regard to the imaged scene.

In the Bayesian approach, probabilities are used to describe incomplete information rather than randomness. As Equation (50) shows, in the Bayesian inference process, induction is influenced by the prior expectations allowed by the prior knowledge of P(R) [46]. In addition, the non-linear system and scene effects are implicitly taken into account by the restoration process. Therefore, MAP speckle filtering can be considered as a controlled restoration of R, where A Priori knowledge controls the inference restoration process and allows an accurate estimation of the radar backscattering coefficient $\boldsymbol{\sigma}°$.

The pdf of the speckle in intensity for a L-looks radar image, P(I/R), is a Gamma distribution with parameters R and L (*cf.* Equation (14)). The Maximum Likelihood term for a L-looks radar image is then equal to:

$$\partial / \partial R \; Log[P(I/R)] \; = \; L \, . \left(I/R^2 - 1/R \right) \tag{51}$$

The Maximum A Priori term must be calculated according to the scene model chosen as A Priori knowledge, hypothesis, or belief.

3.3.3. Gaussian distributed imaged scene: The Gaussian-Gamma MAP filter (1987)

The hypothesis of Gaussian-distributed scene has been adopted as a natural hypothesis by a large number of authors who had worked, either on images from optical sensors, or on images from passive/active microwave sensors. These authors are comforted in this hypothesis by both the force of habit and by the mathematical ease in manipulating a Gaussian distribution.

Kuan *et al.* [32] have developed a MAP filter (Gaussian-Gamma MAP filter, with a Gaussian-distributed scene and Gamma-distributed speckle) for radar images under this hypothesis. Though, the hypothesis of a Gaussian-distributed scene, although widely spread in the literature, is inappropriate. Indeed, this hypothesis assumes implicitly the theoretical possibility of negative $\boldsymbol{\sigma}°$ values-which has no physical sense-in the extreme case of a large variance with a low mean value, therefore needing a regularisation of the filter behaviour in such a situation.

Therefore, one must preferably take into account a positive pdf as a realistic scene model. For reasons that are both experimental and theoretical, in the case of natural extended targets as it is most often the case dealt with in remote sensing, a Gamma distributed scene model is better appropriate.

3.3.4. Gamma distributed imaged scene: The Gamma-Gamma MAP filter (1990)

Natural textures observed, either by coherent radar imagery, or by incoherent optical imagery are due to a common contribution corresponding to the variability in spatial distribution of the objects within the scene. Even if the interaction mechanisms between the electromagnetic wave and the observed medium are very different in either case, the natural arrangement of the scene makes the second-order statistics very similar in either kind of imagery [55] [56]. At the scale of a large number of resolution cells, the pdfs of the cross-section variables corresponding to either mechanism belong to the same family of distributions, at least for the high radar frequencies in bands Ku, X, and C for which wave penetration into natural media is limited. This point is more arguable for radar bands L and P.

In a wide range of radar backscattering situations, the Gamma distribution is experimentally the one that best fits, not only the distribution of the radar backscattering coefficient [21] [57], but also the distribution of radiometries observed in incoherent optical images [55]. This scene model has been successfully used also for radar images of the sea [58] [59] [60].

The local pdf of a scene statistically described by a Gamma distribution, has the form:

$$\mathrm{P}(R) = (\alpha / \mathrm{E}[R])^{\alpha} / \Gamma(\alpha) \, . \, exp(-\alpha . R / \mathrm{E}[R]) \, . \, R^{\alpha-1} \tag{52}$$

with $\alpha = 1 / C_R^2$. The parameter is called the "heterogeneity coefficient".

Note that assuming a Gamma-distributed R, and by performing the integration in Equation (28), the pdf $\mathrm{P}(I)=\mathrm{P}(I/R)$ of the intensity is a K-distribution [57]. Introduced in 1976 [58] by British researchers of the RSRE (later DRA) to describe the non-Gaussian properties of waves backscattered by objects within a radar resolution cell, the K-distribution has been theoretically recognised as the pdf of the intensity I backscattered by a rough non-stationary surface [61] such as most natural scenes observed by a radar.

Nevertheless, in an illustration of the so-called "Cromwell's rule" [62], even hard A Priori conviction that the scene presents a Gamma-distributed texture must not be insensitive to counter-evidence. Therefore, the complete MAP filter is a set of three filters adapted to diverse situations locally encountered in a radar image: the application of this or that filter is decided depending on the degree of heterogeneity of the image part under processing, that is on the locally estimated value of the coefficient of variation C_I. This may eventually imply the determination of thresholds on C_I calculated as a function of a user-defined probability of false alarm with respect to the local presence of texture or of strong scatterers (*cf.* [63] [64] [10]). Note that these considerations apply to all MAP speckle filters, and can be extended to all other adaptive speckle filters.

3.3.4.1. Textured areas

Assuming that the pdf of the scene is a Gamma distribution, the Maximum A Priori term is locally equal to: $\partial/\partial R$ Log[P(R)]=(α-1)/R-α/<R>. Once E[R] is estimated locally by <I> in an ergodic stationary neighbourhood of the pixel under processing in the radar image, the equation of the Gamma-Gamma MAP filter ([65] [10] [63] [64]) is:

$$\alpha.R^2 + (1 + L - \alpha).<I>.R - L.I.<I> = 0 \quad (53)$$

This second-degree equation admits only one real positive solution R in the interval ranging between the mean intensity <I>=<R> and its observed value I. Therefore, the Gamma-Gamma MAP estimate of the radar reflectivity of the pixel under processing, is:

$$\hat{R} = \frac{<I>.(\alpha - L - 1) + \sqrt{<I>^2.(\alpha - L - 1)^2 + 4\alpha.L.I.<I>}}{2\alpha} \quad (54)$$

The integration of a heterogeneity/texture detector based on the coefficient de variation and of specific detectors (ratio-of-amplitudes – RoA; [66] [7]) for contours, linear structures and strong scatterers in the filtering process is described by the general algorithm presented in [63] [64] [10]. The integration of texture and structure detectors using second-order statistics (autocorrelation functions) of both the speckle and the radar reflectivity of the scene [67] is presented in Section 4. In all cases, mage areas identified as textured are filtered using Equation (54).

3.3.4.2. Homogeneous areas

In the particular case of a perfectly homogeneous (textureless) scene, with $C_I \le C_u$, the radar reflectivity R is a constant (R=E[R]), and can be statistically represented by a Dirac distribution:

$$P(R) = \delta(R) \quad (55)$$

This distribution is the limit of Gamma distributions when tends towards+∞. In such a case, the MAP estimate is equal to the local mean intensity:

$$\hat{R} = <I> \quad (56)$$

This case is taken into account when the local statistics calculated in the neighbourhood of the processed pixel show a nearly perfect homogeneity of the scene.

3.3.4.3. Strong scatterers and impulsional targets

The other extreme case regards strong scatterers, when speckle is no longer fully developed. The considerations that led to consider the Gamma pdf as an A Priori model for a textured natural scene are no longer valid: we no longer have any A Priori information about the scene. In this situation, the information content of every grey level of the image being A Priori the same, the pdf of the imaged scene can be represented by an uniform distribution:

$$\mathrm{P}(R) = 1/(R_{max} - R_{min}) \tag{57}$$

with undetermined extreme values R_{min} and R_{max}, which we may consider equal to respectively 0 and+∞. Thus, the Maximum A Priori term of the MAP equation becomes $\partial / \partial R[\mathrm{Log}(\mathrm{P}(R))]=0$, and the MAP estimate is:

$$\hat{R} = I \tag{58}$$

If the resolution cell contains only one isolated strong scatterer, the response (value I of the pixel) of this scatterer is deterministic and must therefore be preserved. This situation leads to the same conclusion and is therefore treated similarly.

3.3.5. Scene pdf estimated from the data: The Gamma/Distribution-Entropy MAP filter (1998)

SAR images of dense tropical forest, urban areas, or very strong and rapidly varying topography often show very strong or mixed textures. This is also the case of high-and very-high spatial resolution SAR images. In these situations, it may be hazardous to make an assumption about the probability density function of the radar reflectivity.

Indeed, the MAP technique does not account for any uncertainty in the value of the parameters of the A Priori pdf chosen as a Gamma distribution once it has been locally estimated on a given image area. Hence, in the presence of mixed (forests with underlying structures, for example) or rapidly varying (strongly textured area located on strong slopes, or very-high spatial resolution radar images, for example) texture, the MAP estimator will underestimate the variance of the predictive distribution. Indeed, this predictive distribution can hardly take into account the fact that it results of a compound of a mix of different distributions.

In this context, the A Priori knowledge with regard to the observed scene can hardly be an analytical first order statistical model, chosen on the base of prior scene knowledge. However, to retrieve local statistical scene knowledge directly from SAR image data, Datcu & Walessa [68] [69] proposed to introduce the local entropy of the radar reflectivity, $S(R)$, as a measure of local textural disorder. This concept originates from the theory of information. $S(R)$ is estimated on a neighbourhood (N_{pix} pixels) of the pixel under processing:

$$S(R) = -\sum_{k=1}^{Npix} [R_k \cdot log(R_k)] \tag{59}$$

Because the radar reflectivities R_k are non-negative and $exp[S(R)]/Z$ is normalised, and since $S(R)$ is a measure of the spread/dispersion of the radar reflectivities of the scene, characterising its heterogeneity, Equation (59) can be treated as a pdf whose entropy is $S(R)$ [70]:

$$P(R) = (1/Z) . exp[S(R)] = (1/Z). exp(-\sum_{k=1}^{Npix} [R_k . log(R_k)]) \tag{60}$$

For a single detected SAR image, the conditional pdf of the speckle can be, as long as speckle is fully developed, modelled as a Gamma distribution:

$$P(I/R) = (L/R)^L / \Gamma(L) . exp(-L . I/R) . I^{L-1} \tag{61}$$

Incorporating these scene and speckle models, the Gamma/Distribution-Entropy MAP (Gamma-DE MAP) speckle filter for single-channel detected SAR images is the solution of the following equation [71]:

$$L . I - L . R - R^2 . \sum_{k=1}^{Npix} [log(R_k) - 1/Ln(10)] = 0 \tag{62}$$

The radar reflectivites R_k in the neighbourhood of the pixel under processing are pre-estimated by a first speckle filtering pass.

Note that the local DE MAP estimation of R is contrained by the value of the entropy $S(R)$ retrieved from image data. Since this fixes upper/lower bounds to the local entropy $S(R)$, the DE MAP speckle filter combines both the Bayesian MAP and the Maximum/Minimum Entropy estimation techniques.

The DE-MAP filters adapt to a much larger range of textures than the other MAP filters ([10] [11] [72] [73]) developed under the assumption of K-distributed SAR intensity (*i.e.* Gamma-distributed scene). In particular, these filters are of particular interest in the case of very high-resolution SAR images and strongly textured scenes.

Compared to those of the other MAP filters, performances in terms of speckle reduction are identical. However, texture restoration and structures or point targets preservation, identical for moderate textures, are superior in strongly textured areas. These filters have proven a remarkable efficiency in operational remote sensing (*cf.* [74]).

3.3.6. *Behaviour of the MAP filters*

The local MAP estimate is the mode of the local A Posteriori pdf of the radar reflectivity R, under the hypothesis made for P(R). The MAP estimation is therefore probabilistic: it corresponds to the most probable value of radar reflectivity. This justifies to chose the local mean observed intensity as filtered value within areas identified as perfectly homogeneous (*cf.* § 2.4.4; [64] [10). Excellent speckle reduction and preservation of the mean value of the radar reflectivity are therefore granted over such areas.

In the presence of texture, the estimator takes into account the non-linear effects in getting from R to I via the imaging system. As for the Lee and Kuan *et al.* filters, the local A Priori mean and variance, <*R*> et σ_R^2, are estimated using the first-order non-stationary local statistics of the original image I (*cf.* Equations (26) and (27)), thus allowing adaptivity to the local isotropic properties of the scene R. At the scale of a terrain parcel, the MAP filter must theoretically restore the whole pdf of the radar reflectivity. This includes its mean E[R] and its variance (equivalent to parameter α), as well as the spatial relationships between pixels if one introduces speckle and scene second-order statistics in the filtering process (*cf.* [10] [75] [67]).

Nevertheless, in the presence of structures, the NMNV model may be used abusively on an inappropriate neighbourhood of the pixel under processing and therefore the local statistics <*R*> and σ_R are no longer those of an ergodic process. Thus, in order to use the filter in the conditions required by the NMNV model, *i.e.* in the statistical situation it has been designed for, one must introduce, as one might also do for the Lee and Kuan *et al.* filters, structure detectors. These detectors contribute to select a neighbourhood of the pixel under processing composed of pixels pertaining to the same thematic class. They may be, either geometrical improvements based on the use of ratio (RoA) detectors ([7] [66] [63] [64] [10]), or detectors based on the use of the second-order statistics of both the speckle and the scene ([67]; *cf.* Section 4). The desired result is a better preservation of image texture, structures, and responses of strong scatterers in the speckle filtered radar image.

4. Using speckle and scene spatial correlation to preserve structural information in SAR images

The assumptions usually made by adaptive speckle filters ([76] [30] [32] [10]) with regard to the first order statistics of the speckle are:

- the multiplicative speckle model;
- the pdf of the speckle which reflects only incompletely the spatial correlation of the speckle through the ENL.
- The assumptions regarding the imaged scene are:
- Local wide-sense stationarity and ergodicity;
- A formal Non-stationary Mean Non-stationary Variance (NMNV) model accounting for the spatial variation of the first order statistical properties [32];
- For the Maximum A Posteriori (MAP) filter, the general form of the pdf of the scene radar cross-section [10].

The main consequences of these assumptions are that:

- The formal NMNV model justifies local treatment (processing window) using the local statistics;

- Assuming spatially uncorrelated speckle leads to the design of scalar (single-point) filters;
- Local adaptivity to tonal properties (mean intensity <*I*> and scene radar reflectivity R) is achieved by controlling the filtering process through the local coefficient of variation C_R of the radar reflectivity R of the scene, through the local coefficient of variation C_I of the original SAR image intensity I [27] [28].

The major drawbacks of these assumptions are that:

- The coefficients of variation are statistically [10] sensitive to texture and speckle strength [40], but do not provide direct information on spatial correlation properties and texture directionality. Locally estimated, they can also be biased if speckle samples are not independent (*cf.* § 2.2.2), as it is often the case in SAR images. Indeed, speckle can have strong spatial correlation properties, a situation the filters are not designed to deal with;
- Compliance with the ergodicity and stationarity hypothesis requires a preliminary identification of the structural elements of the scene to fully exploit the NMNV model and to correctly estimate the local mean reflectivity around the pixel under processing;

In this Section, local second order properties, describing spatial relationships between pixels are introduced into single-point speckle adaptive filtering processes, in order to account for the effects of speckle and scene spatial correlations. To this end, texture measures originating from the local autocorrelation functions (ACF) are used to refine the evaluation of the non-stationary first order local statistics as well as to detect the structural elements of the scene [75].

4.1. Effects of the spatial correlation on the speckle

In practice, the usual single-point filters do preserve texture, only due to the spatial variation of the local first order statistics, using the NMNV formal model [32]. Fairly good preservation of textural properties and structural elements (edges, lines, strong scatterers) can be achieved by associating constant false alarm rate structure detectors such as the directional Ratio-Of-Amplitudes (RoA) detectors; ([7] [66] [10] [63] [64]) to the speckle filtering process. The combination of detection and speckle filtering allows to preserve scene structures, and to enhance scene texture estimation on a shape adaptive neighborhood of the pixel under consideration. When the conditions for which they have been developed (especially spatially uncorrelated or low-correlated speckle) are fulfilled, the best single-point adaptive speckle filters and their structure retaining associated processes based on RoA detectors retain enough scene texture to allow its use as an additional discriminator ([23] [24] [63] [64]).

However, the performances of the usual single-point filters, even when refined with associated RoA structure detectors, degrade when the actual spatial correlation of speckle samples becomes significant [77], *i.e.* generally when SAR images are far too much oversampled [31]. In fact, the uncorrelated speckle model approximation is seldom justified, according to sampling requirements of the radar signal and to the SAR system and processor which have produced the image. In multilook SAR images, spatial correlations are introduced by a series of weighting functions: complex coherent weighting related to data acquisition (antenna pattern, pre-sum filter and Doppler modulation in azimuth), coding of transmitted pulses in range, and looks formation (useful Doppler bandwidth selection, extraction window in azimuth). In practice, they are related to sampling and scale requirements in the design of SAR image products, up to the extreme case of signal oversampling [31].

As an example of the effects of speckle spatial correlation, the case of structure detection using the RoA detectors [10] implemented in the Gamma-Gamma MAP filter is illustrated in [67]. Detection is successfully performed when the correlation of speckle is low between pixels. However, the performance of RoA detectors substantially degrades when the spatial correlation of speckle samples becomes significant. It has been shown theoretically in [67] that spatial speckle correlation results in an increasing detection of non-existing structures, *i.e.* in the increase of the probability of false alarm (Pfa). The expected 1% Pfa for uncorrelated speckle raises to about 25% if the correlation of adjacent pixels is 0.7 in both range and azimuth (*cf.* ERS: about 0.55 in range and 0.65 in azimuth). In addition, the spatial speckle correlation also decreases the probability of detection. The combination of these two effects causes artefacts in the filtered image (visual "crumbled paper" effect), making photo-interpretation uneasy and reducing substantially the performance of further automatic image processing like segmentation [78] and classification [23] [24].

Therefore, second order statistical properties must also be considered, including both scene texture and resolution/sampling related properties, to achieve a more complete restoration of the radar reflectivity.

4.2. A possible solution: Spatial vector filtering

A possible solution is the implementation of vector (multiple-points) filters ([32] [79]) where the spatial covariance matrices of the speckle and of the scene are taken into account. The development of a filtering method using second order statistics, *i.e.* an estimation of the complete ACF of the scene through the speckled image ACF, results in a vectorial equation giving rise to a multiple-points filter, such as the LMMSE vector filter developed theoretically by Kuan *et al.* [32]. A practical implementation of multiple-point filters has been first suggested by Lopès *et al.* [10], and then detailed by Lopès & Séry [79]. Such filters have good spatial memory, but they result in a complicated implementation and very heavy computations.

4.3. Single-point speckle filtering using spatial second order statistics

To avoid the mathematical complexity and the heavy computational burden of multiple-point filters, an alternative solution is to introduce an appropriate description of both speckle correlation properties and spatial relations between resolution cells into a single-point filter.

Second order statistics have explicitly been used in the past, following the scheme proposed by Woods & Biemond [80] implemented later by Quelle & Boucher [81] in the adaptive Frost speckle (single-point) filter [30]. The filter, which belongs to the family of Wiener filters, is established using Yaglom's method [38], where a frequency characteristic is determined, instead of an impulse transfer function, which actually exists only within wide-sense stationary areas. This frequency characteristic is determined in the speckled image at a distance of one pixel only in range and azimuth, thus mainly limiting the description of correlation properties to those of the speckle.

Considering the NMNV model [32], a much better solution is to introduce locally estimated second order statistics to refine the computation of the local NMNV first order statistics. In this way, speckle correlation properties can be taken into account for filtering. In addition, the local mean radar reflectivity E[R] would be estimated, taking better account of the textural properties of the imaged scene (natural variability of the radar reflectivity, including spatial relations between resolution cells). On a given textured class of the scene, when scene correlation length is smaller than the processing window size, the mean radar reflectivity E[R] will then be modulated by the scene ACF, *i.e.* by a function of the correlation coefficients of the radar reflectivity in all possible directions. When the correlation length of the imaged scene is longer than the processing window size, the estimate for the non-stationary mean radar reflectivity tends towards to the classical Maximum Likelihood estimate E[R]=<I>.

4.4. Local ACF's and texture fields

Spatial relations between pixels are well described by the intensity ACF, defined by a set of correlation coefficients (Δz), where Δz=(Δa, Δr), the normalised ACF of an intensity SAR image, $\{\rho_I(\Delta z)\}$, is a composition of the underlying scene ACF, $\{\rho_R(\Delta z)\}$, convoluted with an overlap function depending on the point spread function (PSF).

The local estimates for the normalised intensity ACF, $\{\rho_I(\Delta z)\}$, are, either deduced from the spatial autocovariance, which is computed on the domain of interest D (N pairs of pixels separated from each other by Δz) as follows:

$$\hat{Cov}_I(\Delta z) = (1/N).\sum_D [I(z+\Delta z)-<I>].[I(z)-<I>] \tag{63}$$

or, equivalently, directly computed within the domain of interest D as follows:

$$\hat{\rho}_I(\Delta z) = \frac{\sum_D ([I(z+\Delta z)-<I>].[I(z)-<I>])}{\sum_D [I(z)-<I>]^2} \tag{64}$$

These estimates have the attractive property that their mean square error is generally smaller than that of other estimators; considerations on the estimation accuracy can be found in Rignot & Kwok [82].

The contribution of scene texture must be separated from that of the speckle for all displacements Δz: the scene ACF is deduced from the intensity image ACF, using the following transformation [28]:

$$\hat{\rho}_R(\Delta z) = \frac{[1 + \hat{\rho}_I(\Delta z).C_I^2]}{[1 + |G_c(\Delta z)|^2 / L]} \tag{65}$$

where L is the equivalent number of independent looks, C_I is the local coefficient of variation of the image intensity, and $G_c(\Delta z)$ is the normalised ACF of the individual 1-look complex amplitudes (*i.e.* the ACF of the SAR imaging system).

$G_c(\Delta z)$ depends only on the SAR complex PSF [83]. It is realistic to adopt an exponential form for the ACF of the speckle for the separated looks detected in intensity:

$$|G_c(\Delta z)| = exp[-(\Delta a.dx / rx + \Delta r.dy / ry)] \tag{66}$$

where dx, dy are the pixel dimensions, and rx, ry are the spatial resolutions in azimuth and range directions. However, it is preferable to use directly the true ACF of the speckle in the actual multilook SAR image, if it is available, or if its estimation is possible.

The computation of a local non-stationary estimate of E(R) is performed by a simple convolution in the domain of interest around the pixel under consideration of the normalised ACF of the textured scene $\{\rho_R(\Delta z)\}$ with the intensity:

$$\hat{E}(R) = \frac{\sum_D [\hat{\rho}_R(\Delta z).I(z + \Delta z)]}{\sum_D [\hat{\rho}_R(\Delta z)]} \tag{67}$$

for all $\rho_R(\Delta z)$ within D, and as long as the following condition applies:

$$[\hat{\rho}_R(z) / \hat{\rho}_R(z = (0, 0))] \quad > \quad 1/e \tag{68}$$

This means that, along a direction z, the pixels whose correlation $\hat{\rho}_R(z + \Delta z)$ to the pixel under process is less than 1/e are ignored for the computation of the first order NMNV local statistics through Equation (67). This way, the algorithm achieves powerful speckle reduction in homogeneous areas, preservation of the textural properties (heterogeneity and directionality) wherever they exist, and correct structure detection.

The implementation of this operator, which estimates the local non-stationary E[R] is inspired from the notion of "texture fields" introduced by Faugeras & Pratt [84]. It is important to notice the following properties of the operator:

1. in homogeneous areas, Equation (67) acts as a smoothing operator: $E[R] \rightarrow <I>$, which is the Maximum Likelihood estimate of E[R] for a perfectly homogeneous neighbourhood. (The local energy scaling factor in the denominator of Equation (67) ensures the preservation of the absolute level of the radar reflectivity).

2. for a strong scatterer, it acts as a Laplacian: $E[R] \rightarrow I(z)$.

3. if a perfect step edge between two homogeneous areas is present in the neighbourhood (correlations=1,0,-1), it acts as a gradient and automatically stops weighting further pixels. This indicates structure detection capability;
4. in all other cases, *i.e.* for a textured scene, the E[R] estimate is a weighted average of the neighbourhood, modulated by the scene ACF in all possible directions.

This concept of "texture fields", based on the analysis of second order statistical properties, generalises image processing concepts such as filtering and detection, which are usually considered as distinct. It provides the possibility to perform simultaneously structure detection and image pre-filtering. Structural elements such as edges are detected (value of the estimate of the correlation coefficient becoming less than 1/e) when the distance from the pixel under processing (central pixel of the processing window) to the edge is reached, going in the direction of the edge. Therefore, the position of the edge is known as soon as the edge enters the "field of view" of the processing window.

The local mean radar reflectivity estimation takes now into account the textural properties of the imaged scene (spatial variability of R between resolution cells, directionality) as well as the spatial correlation properties of the speckle. Introduction of the non-stationary estimates of E[R] and C_R into the scalar equation of a single-point speckle filter improves the restoration of R fluctuations in the filtered radar image [75]. As speckle and scene correlations are explicitly accounted for, this method can be considered as being complete.

The whole process acts as an adaptive focusing process in addition to the filtering process, since emphasis is put on the restoration and the enhancement of the small-spatial-scale fluctuations of the radar reflectivity. Full profit is taken of the useful resolution offered by the compound sensor and processing systems. Thus, speckle related features and thin scene elements (scene short-scale texture and thin structures) are automatically differentiated. The latter are restored with enhanced spatial accuracy, according to the local statistical context.

There is no geometrisation of the scene structural elements, as when using templates-based detectors used in structure retaining speckle filters ([76] [63] [64] [10]). The neighbourhood on which the NMNV local statistics are estimated is delimited by the estimates of $\rho_R(\Delta z)$ for all possible Δz orientations. The only limitation to the potential extension of the domain D is the spatial extent of the available ACF of the speckle.

5. MAP speckle filters for series of detected SAR images

Since the launch of ERS-1 satellite in 1991 temporal series of calibrated SAR images have been made available. This has stimulated the development of multichannel ("vector") filters especially dedicated to filter the speckle in a series of images, taking into account the correlation of the SAR signal between image acquisitions, thus opening the way to furher developments of change detection techniques specific to SAR series of images.

Interest in multichannel adaptive speckle filtering arises from their ability to combine both multi-image diversity through the exploitation of correlation between speckle and scene between images, and spatial diversity through the locally adaptive averaging of pixel values in the spatial domain. The objectives of this combination are the simultaneously achievement of both, a better restoration of existing texture, and a stronger speckle smoothing in textureless areas.

Although numerous work with regard to multi-channel speckle filtering in SAR images had already being done in a remote past, the introduction of A Priori knowledge or A Priori guess which implies the use of Bayesian methods in the processing of multi-SAR's images, multi-date SAR images or SAR and optical images began only in 1992. Bayesian MAP vector speckle filters developed for multi-channel SAR images incorporate statistical descriptions of the scene and of the speckle in multi-channel SAR images. These models are able to account for the scene and system effects, which result in a certain amount of correlation between the different channels.

To account for the effects due to the spatial correlation of both the speckle and the scene in SAR images, estimators originating from the local ACF's (*cf.* Section 4) are incorporated to these filters to refine the evaluation of the non-stationary first order local statistics on an ergodic wide-sense neighbourhood of the pixel under processing. The goal is to improve the restoration of scene textural and structural properties and the preservation of the useful spatial resolution in the filtered SAR image.

5.1. Multichannel vector MAP speckle filtering

In the case of multi-channel detected SAR images, let define the vector quantities of interest: I is the speckled intensity vector available in the actual SAR data; R is the radar reflectivity vector, which is the quantity to restore. The MAP filtering method bases on Bayes's theorem [47] [48] in its matricial form:

$$\mathrm{P}(R/I) = \mathrm{P}(I/R)\,.\,\mathrm{P}(R)\,/\,\mathrm{P}(I) \tag{69}$$

where $\mathrm{P}(I/R)$ is the joint conditional pdf of the speckle.

For each individual detected SAR image i, $\mathrm{P}(I_i/R_i)$ is known to be well approximated by a Gamma distribution for multilook intensity images (*cf.* § 2.1.4; Equation (14)).

$\mathrm{P}(R)$ is the joint pdf of the radar reflectivity, introduced as statistical A Priori information in the restoration process. To describe the first order statistical properties of natural scenes, it has already been shown that, in most situations, for each individual image i, a Gamma distribution (*cf.* § 3.3.4) is an appropriate choice for $\mathrm{P}(R_i)$. Nevertheless, when in doubt about the reliability of such a choice, $\mathrm{P}(R_i)$ can be estimated directly from the data in the corresponding radar image i.

For multi-channel detected SAR images, MAP filtering is a vector filtering method. For every channel i, the posterior probability is maximum if the following condition is verified:

$$\partial[Ln(\mathrm{P}(I/R))]/\partial R_i + \partial[Ln(\mathrm{P}(R))]/\partial R_i = 0 \quad \text{for} \quad R_i = \hat{R}_{i\,MAP} \tag{70}$$

The MAP speckle filtering process acts as a data fusion process, since the information content of the whole image series is exploited to restore the radar reflectivity in each individual image. Among other advantages, this allows a better detection and restoration of thin scene details.

5.2. Multichannel scene models

To describe the first order statistical properties of a natural scene viewed at the spatial resolution and at the wavelength frequency of a radar remote sensing sensor, it has been shown that a Gamma pdf would be a suitable representation (*cf.* § 3.3.4).

However, to describe the first order statistical properties of a natural scene as viewed by diverse radar sensors (different physical scene features), or at different dates (scene evolution over time), there is no analytic multivariate Gamma pdf available under closed-form. Therefore, for the sake of mathematical tractability, a multivariate Gaussian pdf is used as analytic "ersatz" of a multi-channel scene statistical model:

$$\mathrm{P}(R) = [(2\pi)^N \mid Cov_R \mid]^{-1/2}.\, exp[-{}^t(R - \langle R \rangle).Cov_R{}^{-1}.(R - \langle R \rangle)] \tag{71}$$

Cov_R is the local covariance matrix of the imaged scene in all image channels. For each pixel location, Cov_R is estimated from the local covariance matrix of the intensities observed in all image channels Cov_I, using the multiplicative speckle model (*cf.* § 2.2.2; [27] [28]).

This statistical scene model is convenient to preserve the mathematical tractability of the problem. In addition, the Gaussian model, commonly used to describe the statistical properties of natural scenes in processing of optical imagery, is appropriate as a first-order statistical description of moderate, not too heterogeneous, textures.

However, when scene texture becomes very heterogeneous, this multivariate Gaussian scene model is not any longer reliable, and in the absence of specific prior knowledge about the distribution of scene texture, its model has to be directly estimated from the SAR image.

5.3. Speckle uncorrelated between SAR images: The Gamma-Gaussian MAP filter

Let consider the case of a series of N images originating from different SAR systems (different frequencies, incidence angles, or spatial resolution for example, but same spatial sampling). In such a case, it is justified to consider that speckle is independent between the N image channels.

Under this assumption, since each one of the SARs senses slightly different physical properties of the same scene, the joint conditional pdf of the fully developed speckle, $\mathrm{P}(I/R)$, can reasonably be modelled as a set of N independent Gamma distributions $\mathrm{P}(I_i / R_i)$:

$$\mathrm{P}(I_i / R_i) = (L_i / R_i)^{L_i} / \Gamma(L_i).\, exp(-L_i I_i / R_i) .I_i^{(L_i - 1)} \tag{72}$$

where the L_i parameters are the ENLs of the individual SAR images.

The first MAP filtering algorithm for multi-channel detected SAR images results in a set of N scalar equations, with N independent speckle models and a coupled scene model (Equation (71)). The coupled scene model is justified by the fact that the physical properties of the scene do contribute, but not in an identical way, to the formation of the images provided by the different sensors. The set of equations describing the Gamma-Gaussian MAP filter for multi-channel detected SAR images is as follows ([73] [85]):

$$\begin{aligned} & L_i(I_i / R_i^2 - 1 / R_i) - {}^t(1_i).Cov_R^{-1}.(R - \langle R \rangle) \\ & - {}^t(R - \langle R \rangle).Cov_R^{-1}.(1_i) - 1/2\, \mathrm{Tr}[Cov_R^{-1}.\partial Cov_R / \partial R_i] \\ & + {}^t(R - \langle R \rangle).Cov_R^{-1}.\partial Cov_R / \partial R_i.Cov_R^{-1}.(R - \langle R \rangle) = 0 \end{aligned} \tag{73}$$

where (1_i) is a vector where all components, but the ith, are equal to zero. The set of Equations (73) is solved numerically.

The introduction of coupling between the scene statistical representations leads to a data fusion process taking profit of the correlation between texture as it is observed in all the images in the series. Indeed, replacing speckle noise model by optical noise (or film grain noise, or any other appropriate noise model) in some of the N image channels, this filter adapts also easily to the case of multi-channel optical and SAR images.

5.4. Speckle correlated between SAR images: The Gaussian-Gaussian MAP filter

To filter series of images originating, either from the same SAR system operating in repeat-pass mode, or from different SAR systems with relatively close properties, a second speckle filtering algorithm has been developed. The different SARs may be close in terms of frequency, angle of incidence, spatial resolution and sampling, geometry of images, with difference in polarisation configuration only, or small differences in incidence angle, for example. In such cases, speckle correlation between individual SAR images must be taken into account to deal optimally with system effects on the SAR images series.

Taking into consideration speckle correlation between image channels, the joint conditional pdf of the fully developed speckle $P(I/R)$ should theoretically be a multivariate Gamma pdf. Nevertheless, since there is no analytic multivariate Gamma pdf available in closed-form, another reasonable choice for $P(I/R)$ should be done for the sake of mathematical tractability.

To solve this problem, Lee's assumption [33] is adopted: for multilook SAR images (more than 3 looks), the joint conditional pdf of the speckle can be reasonably approximated by a Gaussian distribution. Therefore, for convenience, $P(I/R)$ is approximated by a Gaussian multivariate distribution in the case of multilook SAR images:

$$P(I/R) = [(2\pi)^N \mid Cov_S \mid]^{-1/2}.\, exp[-{}^t(I-R).Cov_S^{-1}.(I-R)] \tag{74}$$

where Cov_S is the covariance matrix of the speckle between images of the series. In practice, Cov_S is estimated within the most homogeneous area one can identify in the series of detected SAR images.

It is noteworthy that Cov_S takes into account possible different L_i values of the respective ENLs of the SAR images in the series as well as energy unbalance between images. This enables to compose the series combining images from different SARs operating at the same wavelength, or acquired by the same SAR at different times or in different configurations of polarization.

At this point, the second MAP filtering algorithm for multi-channel detected multilook SAR images results in a set of N scalar equations. The set of equations describing the Gaussian-Gaussian MAP filter for multi-channel detected multilook SAR images is ([73] [85]):

$$\begin{aligned} &{}^t(1_i).Cov_S^{-1}.(I-R) + {}^t(I-R).Cov_S^{-1}.(1_i) - 1/2\,\mathrm{Tr}[Cov_R^{-1}.\partial Cov_R/\partial R_i] \\ &+{}^t(R-<R>).Cov_R^{-1}.\partial Cov_R/\partial R_i.Cov_R^{-1}.(R-<R>) \\ &-{}^t(1_i).Cov_R^{-1}.(R-<R>) - {}^t(R-<R>).Cov_R^{-1}.(1_i) = 0 \end{aligned} \quad (75)$$

where (1_i) is a vector where all components, but the ith, are equal to zero. Again, this set of equations is solved numerically.

Including a coupled speckle model (Equation (71)) and a coupled scene model (Equation (74)), this speckle filter takes profit of both speckle and scene texture correlations, thus restoring the radar reflectivity through a complete data fusion process.

5.5. Prior knowledge gathered from SAR image series: DE-MAP filters

In the presence of very strong or mixed textures, eventually combined with the presence of strong topographic relief, it may be hazardous to make an A Priori assumption about the pdf of the radar reflectivity. This situation is often the case in SAR images of dense tropical forest or of urban environments located in slanted terrain with rapidly varying slopes and counter-slopes.

This is also often the case in high-and very-high spatial resolution SAR images, where strong texture is omnipresent. In addition, at such very fine scale, the textural properties of the scene vary strongly within the timeframe elapsed between successive image acquisitions by the same SAR, and exhibit strong differences when imaged by different SARs or using different SAR configurations in terms of wavelength, polarisation, or angle of incidence, for example.

5.5.1. Speckle correlated between SAR images: The Gaussian-DE MAP filter

As mentioned precedently (*cf.* § 5.4), for a series of N multilook (L_i>3 for each image i) detected SAR images, the joint conditional pdf of the speckle, P(I/R), can be modelled as a multivariate Gaussian distribution [33] when speckle is correlated between image channels with covariance matrix Cov_S:

$$\mathrm{P}(I/R) = [(2\pi)^N . \mid Cov_S \mid]^{-1/2}.\, exp[-{}^t(I-R).Cov_S^{-1}.(I-R)] \quad (76)$$

The entropy of the scene texture [68] [69] [70] is introduced (on a sample of N_{pix} pixels of an ergodic neighbourhood of the pixel under processing) as follows:

$$S(R_i) = -\sum_{k}^{Npix} [R_{ik}.\, log(R_{ik})] \qquad \text{for the } i^{\text{th}}\text{channel} \tag{77}$$

Because the radar reflectivities R_k are non-negative and $exp(S(R_i))/Z$ is normalized, and since $S(R_i)$ is a measure of the spread/dispersion of radar reflectivities in image channel i, Equation (77) can be treated as a pdf whose entropy is $S(R_i)$ [70]:

$$P(R_i) = (1/Z)\,.\,exp(S(R_i)) = (1/Z)\,.\,exp(-\sum_{k}^{Npix} [R_{ik}.log(R_{ik})]\,) \tag{78}$$

Under this assumption, the Gaussian/Distribution-Entropy MAP (Gaussian-DE MAP) filter for multi-channel multilook detected SAR images (N image channels) results in a set of N coupled scalar equations of the form [71]:

$${}^t(1_i).Cov_S^{-1}.(I-R) + {}^t(I-R).Cov_S^{-1}.(1_i) - R_i^2.\sum_{k}^{Npix} [log(R_{ik}) - 1/Ln(10)] = 0 \tag{79}$$

where (1_i) is a vector where all components, but the ith, are equal to zero.

To estimate $P(R_i)$, the radar reflectivities R_{ik} in a neighbourhood of the pixel under processing in the ith SAR image are pre-estimated in a first speckle filtering pass; indeed, the iterative estimation process does already converge after a first application of Equation (79).

Note that this filter does not need to introduce implicitly scene texture correlation. Indeed, the restoration of the radar reflectivity in each image channel bases only on the A Priori knowledge retrieved from the speckled image channel. Nevertheless, the filter takes profit of speckle correlation between image channels. An application of this filter is illustrated in Figure 1.

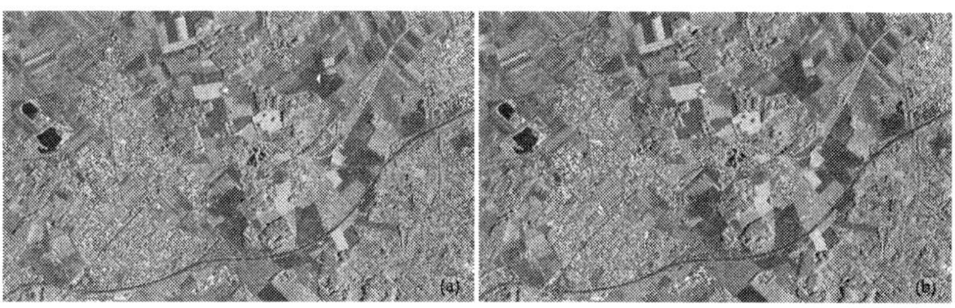

(a) color composition of the unfiltered HH (red) and VV (green and blue) amplitudes. (b) color composition of the speckle filtered HH and VV amplitudes obtained using the Gaussian-DE MAP filter for series of multilook detected SAR images (*cf.* § 5.5.1).

Figure 1. TerraSAR-X 4-looks SAR images (HH and VV polarisations) acquired on August 10, 2007, over Oberpfaffenhofen, Germany (Credits: Astrium and InfoTerra; filtered images: Privateers NV, using ®SARScape).

5.5.2. Speckle correlated between SAR images: The Gamma-DE MAP filter

Note that if speckle is not correlated between the N image channels, the Gamma/Distribution-Entropy MAP (Gamma-DE MAP) filter for multi-channel detected SAR images (N channels) results in the resolution of a set of N independent (uncoupled) scalar equations similar to Equation (62) (*cf.* § 3.3.5).

5.6. Vector MAP filtering and control systems

A number of advantages and improvements these filters offer is listed in this section. Most of these advantages arise from the use of the covariance matrices of the speckle and of the imaged scene (*i.e.* from the use of coupled models) in a Bayesian reconstruction of the radar reflectivity.

5.6.1. Imaging system's effects and preservation of high-resolution

In a series of SAR (but not only...) images, the resolution cells never overlap perfectly between the different individual images. The covariance matrix of the speckle Cov_S contains the information about these overlaps, and the filtering process accounts for these overlaps in order to preserve, and even improve the spatial resolution in the filtered SAR images, thus allowing us to restore the thinnest scene structures.

In addition, the simultaneous attempt to detect images structures and targets in all radar image channels improves the probability to detect such structures. As shown in Figure 1, such an improvement is already very substantial using only two SAR images.

5.6.2. Potential for change detection

In series of radar images acquired over time, the covariance matrix of the scene Cov_R contains the information about the temporal evolution of the imaged area and enables the detection of changes in scene physical properties (scene temporal evolution).

In a series of SAR images acquired by (not too...) different SAR's, the covariance matrix of the scene enables the detection of more aspects of the scene, as it is viewed by more SAR sensors.

5.6.3. MAP speckle filters and control systems

Let consider the mathematical form of the set of Equations (73), (75) and (79). Indeed, their formulation is that of a control system, since both equations can be rewritten under the form of Riccati's [86] continuous time algebraic matricial equations:

$$- A . X - X .^{t}A - Q + X .^{t}C . P^{-1}.C . X = 0 \quad (80)$$

Equation (10) represents the optimal state controlled reconstruction at constant gain of linear invariant processes (R and the textures of the channels) perturbed by white noises (speckle, pixel spatial mismatch between channels), from observed evolving state variables (I and Cov_I) [87]. The scene A Priori model acts as a command, and the covariance matrices act as controls.

It is also noticeable that the MAP elaboration of information generates automatically the feedback process, which enables to control the filtering process.

In addition, it is highly remarkable that Riccati's theorem [86] stipulates the existence of a unique positive definite solution for Equation (10). Therefore, this property holds also for the MAP Equations (73), (75) and (79).

The extension of this noise filtering technique to the case of SAR and optical images set is straightforward, as mentioned in § 5.3. It is also noticeable that, extending this method to the case of complex SAR images, this technique presents a very interesting potential for super-resolution. Such methodologies can be found in the literature (*cf.* [46], for example).

A major interest of control systems is that they offer wide possibilities for the choice and design of additional commands (statistical and physical models) for further data exploitation. In this view, speckle filtering should be regarded as a first step of integrated application oriented control systems.

6. MAP speckle filters for complex SAR data

SLC images are complex valued images. Hence, every image pixel is assigned a complex value, its complex amplitude $A_c = i + \mathrm{j}.q = \sqrt{I}.exp(\mathrm{j}.\phi)$, which corresponds to an intensity $I=i^2+q^2$ and a phase $\phi=Arctg(q/i)$.

Whereas the intensity I carries information related to the radar reflectivity R and the back-scattering coefficient $\sigma°$ of the imaged scene, the phase of a complex pixel of a single SLC image is totally random and does not carry any information about the scene when speckle is fully developed. (It is noteworthy that only, phase differences between SLCs acquired in interfero-metric conditions, or phase differences between configurations of polarisation in SAR polari-metric data, carry information about the imaged scene).

Therefore, the quantity of interest to restore in SLC images through speckle filtering is the radar reflectivity R. Thus, whereas the input of the speckle filters consists in complex radar data, their output consists in radar reflectivity images, *i.e.* speckle filtered intensity images.

6.1. Single-look complex (SLC) SAR image: The CGs-DE MAP filter

In SLC radar data the spatial correlation of the speckle can be dealt with by taking into account the spatial covariance matrix of the speckle in the filtering process. Indeed, the joint pdf of a local sample X (N_{pix}-dimention vector) of 1-look spatially correlated speckle is given by Equation (7) (*cf.* § 2.1.2), can be written as [88] [89] [90] [91]:

$$\mathrm{P}_u(X) = \mathrm{P}(X/R) = [1/(\pi^{Npix}.|C_S|)].exp[-{}^tX^*.C_S^{-1}.X] \qquad (81)$$

where C_S is the spatial covariance matrix of the complex speckle, estimated within the most homogeneous part of the SLC radar image.

In high and very-high spatial resolution SLC SAR images, which often exhibit strong textural properties, it becomes difficult to consider any theoretical statistical model as a reliable A Priori knowledge about scene texture. In such a situation, it seems reasonable to retrieve statistical scene knowledge directly from the SAR image data [68] [69]Introducing the local entropy of the radar reflectivity as A Priori scene knowledge [70] (*cf.* § 3.3.5), the Complex-Gaussian/ Distribution-Entropy MAP (CGs-DE MAP) filter for an SLC radar image is the solution of the following equation [71]:

$$(1/Npix)\,.\,{}^{t}X^{*}.C_{S^{-1}}.X - R - R^{2}.\sum_{k=1}^{Npix}[log(R_{k}) - 1/Ln(10)] = 0 \tag{82}$$

where N_{pix} is the effective number of neighbour pixels taken into account in the computation of local statistics. The radar reflectivites R_k are pre-estimated in this neighbourhood of the pixel under processing by a first pass of speckle filtering in intensity.

An example of application of this filter is shown in Figure 2.

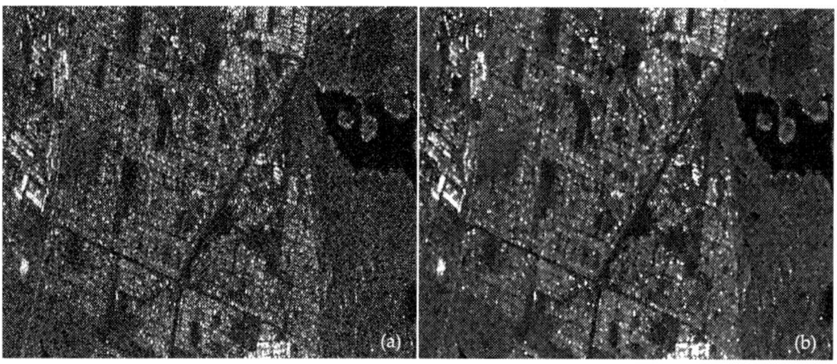

(a) unfiltered 1-look SAR image amplitude. (b) filtered SAR image amplitude obtained using the Complex-Gaussian-DE MAP filter for SLC SAR image (*cf.* § 6.1).

Figure 2. RADARSAT-2 1-look SAR image (HH polarisation) acquired on April 30, 2009, over Capetown, South Africa (Credits: Canadian Space Agency and MDA; filtered images: Privateers NV, using ®SARScape).

6.2. Separate complex looks, or series of SLC SAR images

Interest in adaptive speckle filtering of SLC image series arises from their ability to combine both temporal or spectral diversity, and spatial diversity. The objective is to obtain a multilook radar reflectivity image, where additional speckle reduction is obtained in the spatial domain by averaging pixel values through locally adaptive filtering.

The final objective is to obtain very strong speckle smoothing in textureless areas and to reach an ENL high enough (*i.e.* ENL>>30, and preferably ENL>150) to enable information extraction using classic techniques, or combined use with optical imagery, while better restoring existing or characteristic scene texture.

6.2.1. Compount statistical speckle model for series of SLC SAR images

A SAR sensor can produce series of SLC SAR images in three different situations:

1. L separate complex looks are usually extracted from the useful Doppler band of the radar signal, usually to produce a L-looks multilook image, but the individual looks may be conserved at this stage of SAR image production. Let remind that this operation is done at the cost of a loss in azimuth spatial resolution (*cf.* § 2.1.3). These looks generally overlap, in the sense that they have in common a part of the Doppler spectrum of the original full-resolution 1-look signal, in order to preserve as completely as possible the total amount of backscattered radar signal. This overlap introduces a certain amount of speckle correlation between looks extracted from adjacent portions of the Doppler spectrum.

2. In the case of a non-interferometric 1-look complex SAR data set, the individual SAR SLC images acquired independently of each other show low or inexistent speckle correlation between images.

3. However, in the case of interferometric 1-look complex SAR data, speckle correlation is close to 1 (corresponding to perfect correlation) or very high, since speckle correlation between SAR data acquisitions is exactly the fundamental requirement for interferometry.

As a consequence of these three possible configurations of a series of SLC images, a speckle filter dealing with the whole series must be able to take into account the complex correlation of speckle between the different images. The local complex covariance matrix Cov_S describes completely both, the radar intensities in each SLC image, and the correlation between images.

Considering the whole set of SLC images, the measurement vector for each pixel is $X=\{y_n\}$, where $y_n=i_n+j.q_n$. When speckle is fully developed, the (i_n, q_n) are statistically independent random processes. However, the y_n are correlated complex Gaussian random processes with a joint conditional pdf given by [88]:

$$P(X / Cov_S) = exp({}^tX^*.Cov_S{}^{-1}.X) / (\pi^L . | Cov_S |) \tag{83}$$

where $Cov_S=E[{}^tX. X^*]$ is the complex covariance matrix of the speckle between SLC images, $|Cov_S|$ is the determinant of Cov_S, and $(Cov_S{}^{-1}/|Cov_S|)$ is the inverse complex correlation matrix of the speckle. Theoretically, it is possible to compute the correlation matrix of the speckle from the sensor's and SAR processor's parameters. But, since these parameters are in general not easily available to SAR images users, Cov_S is estimated in practice over the most homogeneous area identified in the images.

6.2.2. Gamma distributed scene model: The Complex-Gaussian–Gamma MAP filter

In the case of separate complex looks corresponding to the same SAR data acquisition, or of series of SLC images acquired over time by the same SAR over a time-invariant scene, it is quite reasonable to assume that the textural properties of the scene are similar in all the SLC's.

Therefore, scene texture may be locally modelled for all SLCs in the series, as well as in the detected multilook image formed by incoherent pixel-per-pixel averaging, by the same Gamma pdf with parameters E[R] and α (*cf.* § 3.3.4, Equation (52)). Locally, E[R] is estimated, either by <I> or using Equation (67), and α is estimated by $1/C_R^2$ (*cf.* § 2.2.2, Equations (26) and (27)), in an ergodic wide-sense stationary neighbourhood of the pixel under processing in the (less noisy) multilook radar image.

In this situation, the Complex-Gaussian/Gamma MAP (CGs-Gamma MAP) filter for separate complex looks (L 1-look SAR images) has the following expression [11]

$$\hat{R} = \frac{<I>.(\alpha - L - 1) + \sqrt{<I>^2.(\alpha - L - 1)^2 + 4\alpha.<I>.({}^tX^*.(Cov_S^{-1}/|Cov_S|).X)}}{2\alpha} \tag{84}$$

Note that when the SLC images or the separate looks are independent, *i.e.* when the non-diagonal terms of Cov_S are equal to zero, then ${}^tX^*.(Cov_S^{-1}/|Cov_S|).X=L.<I>$, and the filter is equivalent to the Gamma-Gamma MAP filter for single detected SAR image (*cf.* § 3.3.4). Therefore, independent SLC images can be incoherently summed to produce an L-looks intensity image that is filtered using the Gamma-Gamma MAP filter (Equation (54)) with the same α and R local parameters, for the same final result.

Finally, it is noteworthy that the ENL L' of the multilook radar image can be computed from the complex correlation coefficients ρ_{mn} between images m and n, computed from the covariance matrix of the speckle, *i.e.* Cov_S estimated over the most homogeneous area identified in the images [11]:

$$L' = L.(1 + (2/L).\sum_{n=1}^{L-1}\sum_{m=n+1}^{L}|\rho_{mn}|^2)^{-1} \tag{85}$$

This relationship may prove useful if SLC images are incoherently summed to produce a L'-looks intensity image.

6.2.3. *Prior knowledge gathered from the data: The Complex-Gaussian-DE MAP filter*

High and very-high spatial resolution SLC SAR images often exhibit strong textural properties. As exposed above, in such a situation, statistical scene knowledge is estimated from the radar data [68] [69]Introducing the local entropy of the radar reflectivity as A Priori scene knowledge [70] (*cf.* § 3.3.5), the Complex-Gaussian/Distribution-Entropy MAP (CGs-DE MAP) filter for an SLC image series (L images) is the solution of the following equation [71]:

$${}^tX^*.Cov_S^{-1}.X - L.R_i - R_i^2.\sum_{k=1}^{Npix}[log(R_{ik}) - 1/Ln(10)] = 0 \tag{86}$$

where Cov_S is the covariance matrix of the speckle between the different SLC images [11] and N_{pix} is the effective number of neighbour pixels taken into account in the computation of local statistics. The radar reflectivites R_k are pre-estimated in this neighbourhood of the pixel under processing by a first pass of multichannel speckle filtering of the L images in intensity (using Equation (84) for example).

An application of this filter to a series of 18 (3x6) SLC SAR images is shown in Figure 3.

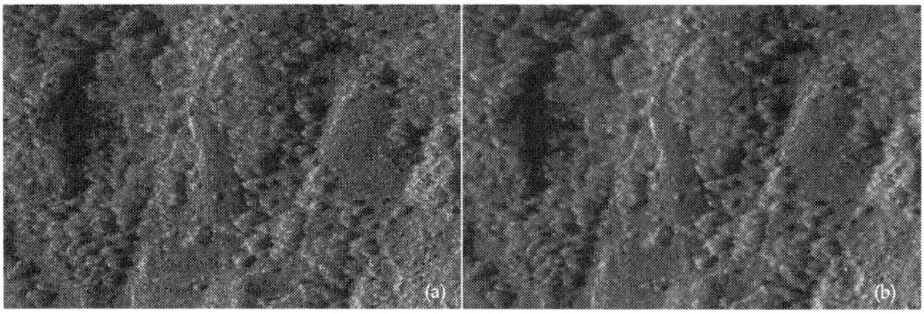

(a) color composition of three unfiltered 6-looks SAR images, each one obtained by pixel-per-pixel averaging of 6 individual 1-look images. (b) color composition of three speckle filtered images, each one obtained using the Complex-Gaussian-DE MAP filter for series (6 images) of SLC SAR images (*cf.* § 6.2.3).

Figure 3. Series of 18 (3x6) COSMO-SkyMed 1-look spotlight SAR images (HH and VV polarisations) acquired between February 2 and September 30, 2009, over Perito Moreno in Patagonia, Argentina (Credits: original images: Italian Space Agency; filtered images: Privateers NV, using ®SARScape).

7. MAP speckle filters for polarimetric radar data

7.1. Polarimetric radar data representation, and polarimetric speckle model

A polarimetric radar system produces, for each pixel location, a scattering measurement matrix, which is the scattering matrix S of the corresponding surface (*cf.* Section 1, Equation (1)) corrupted by speckle noise.

The first polarimetric speckle filter ever developed [92] resulted in an optimal summation of the intensities I_{HH}, I_{HV}, I_{VV} of the polarisation channels. The resulting image is called the "improved span" image:

$$I_{Span} = I_{HH} + (1 + |\rho_{HHVV}|^2) . I_{HV} . E(I_{HH}) / E(I_{HV}) + I_{VV} . E(I_{HH}) / E(I_{VV}) \tag{87}$$

where ρ_{HHVV} is the complex correlation coefficient between S_{HH} and S_{VV}.

However, Equation (87) clearly shows that the ENL achieved in the span image will barely reach 3, which remains largely insufficient to meet the noise reduction requirements of remote sensing applications. Besides, the physical interpretation of I_{Span}, which is a mixture of HH, VV, HV radar reflectivities is questionable. In addition, scene information contained in the polarimetric diversity of the radar measurement as well as in the complex correlations between polarimetric channels is lost in the process.

Lee *et al.* [93] developed a polarimetric optimally combining the polarisation channels to restore the radiometric information in the HH, VV, and HV channels. This polarimetric speckle filter has been shown to preserve polarimetric signatures [94].

Nevertheless, the ultimate aim of polarimetric speckle filtering is to restore the full polarimetric scattering matrix S, or at least the radar reflectivities in the HH, HV, VV configurations of polarisation, while achieving a very high ENL value (>100) in the filtered data. To achieve these objectives, fully polarimetric adaptive speckle filters take profit of both the entire polarimetric diversity (*i.e.* the full scattering matrix) and the spatial diversity (*i.e.* local statistics around the location under processing) [95].

7.1.1. Polarimetric covariance matrix

Under the assumption of reciprocity, $S_{HV}=S_{VH}$, the speckle-free (denoised) scattering matrix S can be transformed into its covariance matrix C_S, as follows:

$$C_S = \begin{pmatrix} |S_{HH}|^2 & S_{HH}.S_{HV}^* & S_{HH}.S_{VV}^* \\ S_{HH}^*.S_{HV} & |S_{HV}|^2 & S_{HV}.S_{VV}^* \\ S_{HH}^*.S_{VV} & S_{HV}^*.S_{VV} & |S_{VV}|^2 \end{pmatrix} \quad (88)$$

This representation puts emphasis on the radar reflectivities $R_{HH}=|S_{HH}|^2$, $R_{HV}=|S_{HV}|^2$ and $R_{VV}=|S_{VV}|^2$, as well as the covariances between HH, HV and VV configurations of polarisation.

If the polarimetric measurement is made by a monostatic radar system, the covariance matrix Σ_S of the actual polarimetric radar measurement is the speckle-corrupted version of C_S.

It is noteworthy that it is possible to produce multilook polarimetric data (L-looks) from the original 1-look data. For L-looks polarimetric data obtained by either spatial averaging (generally in azimuth) or spectral multilooking (*cf.* § 1.1.4), $\Sigma_{S\ multilook}$ is defined by:

$$\Sigma_{S\ multilook} = [\ (1/L).\sum_{1}^{L} (S_{pq}.S_{pq}{}^{t*})\ ] \quad \text{for all combinations } p,q \text{ of H,V} \quad (89)$$

7.1.2. Polarimetric vector, degrees of coherence, phase differences

For convenience, the polarimetric covariance matrix is often expressed, without changing its overall information content, under the form of a real valued vector, called the "polarimetric feature vector", or "polarimetric vector" X:

$$\begin{aligned} {}^tX = (\ & |S_{HH}|^2,\ |S_{HV}|^2,\ |S_{VV}|^2,\ Re(S_{HH}.S_{VV}^*),\ Im(S_{HH}.S_{VV}^*), \\ & Re(S_{HH}.S_{HV}^*),\ Im(S_{HH}.S_{HV}^*),\ Re(S_{VV}.S_{VV}^*),\ Im(S_{VV}.S_{VV}^*)\ ) \end{aligned} \quad (90)$$

Two quantities of great interest in polarimetric SAR applications can be obtained from the polarimetric vector:

1) The Phase Differences $\Delta\varphi_{KL\text{-}MN}$ between radar returns in configurations of polarisation KL and MN (where K,L,M,N ∈ {H,V} are preferably considered in most applications), which are computed as follows:

$$\Delta\varphi_{KL\text{-}MN} = Arg(S_{KL}.S_{MN}{}^{*}) = Arctan[Im(S_{KL}) / Re(S_{KL})] - Arctan[Im(S_{MN}) / Re(S_{MN})] \tag{91}$$

2) The (complex) Degrees of Coherences $\gamma_{KL\text{-}MN}$, which measure wave coherence between radar returns in two configurations of polarisation KL and MN, and express textural properties of an extended wide-sense stationary scene, are defined as follows:

$$\gamma_{KL\text{-}MN} = \frac{E[Re(S_{KL}.S_{MN}{}^{*})] + j.\,E[Im(S_{KL}.S_{MN}{}^{*})]}{\sqrt{(E[\,|S_{KL}|^2\,].E[\,|S_{MN}|^2\,]}} \tag{92}$$

However, in the actual polarimetric radar measurement, the observations $\Delta\varphi_{KL\text{-}MN}$ of the phases differences, and the observations $\rho_{KL\text{-}MN}$ of the complex degrees of coherence are the speckle-corrupted versions of $Arg(S_{KL}.S_{MN}{}^{*})$ and $\gamma_{KL\text{-}MN}$, respectivly. Therefore, these quantities, which represent important contributions of radar polarimetry must also be speckle filtered, justifying the development of fully-polarimetric speckle filters that do not limit themselves to the restoration of radar reflectivities.

7.1.3. Polarimetric speckle model

In the case of polarimetric radar data, Σ_S is the actually observed polarimetric covariance matrix, and C_S is the unspeckled polarimetric covariance matrix, *i.e.* the quantity to restore through speckle filtering. In the reciprocal case, and for low look correlation, the conditional pdf of Σ_S is a complex Wishart distribution of the form [72]:

$$P(\Sigma_S / C_S) = \frac{(det\ \Sigma_S)^{L-3}.L^{\ L-3}}{\pi^3.\Gamma(L).\Gamma(L-1).\Gamma(L-2).(det\ C_S)^L}\ .\ exp[-Tr(L\ .C_S{}^{-1}.\Sigma_S)] \tag{93}$$

where L is the ENL of the "improved span" image defined by Equation (87) [92].

7.2. Restoration of the whole polarimetric vector / covariance matrix

7.2.1. Gamma distributed scene model: The Wishart-Gamma MAP filter

Using physical backscattering models, assuming (as a rough approximation) that texture is identical in all polarizations, we get the following approximation [72]:

$$C_S = \mu \, . \, E[C_S] \tag{94}$$

where μ is the scalar texture parameter equal to the normalized number of scatterers within the resolution cell, and $E(C_S)$ is the mean covariance matrix [92].

$$P(\mu) = \alpha^{\alpha} / \Gamma(\alpha) \, . \, exp(-\alpha.\mu) \, . \, \mu^{\alpha-1} \qquad \text{and} \qquad E[\mu] = 1 \tag{95}$$

The characterisation of scene heterogeneity by only one textural random variable μ bases on the assumption that textural properties are identical in all configurations of polarisation. However, this has been shown to be inexact, both experimentally [96] and theoretically [97]. Therefore, μ should be regarded as an average textural characterisation of the scene observed in all possible configurations of polarisation. Besides the advantage of estimating α=1/ C_R^2 (using Equation (27)) in a less noisy multilook image is an additional justification to do it in the "improved span" image defined by Equation (87) [92].

Introducing the first-order statistical models for fully-polarimetric speckle (Equation (93)) and the polarimetric texture parameter μ in the MAP equation (Equation (50), *cf.* § 3.3.2), the fully-polarimetric Wishart-Gamma MAP filter restores the value of μ as:

$$\hat{\mu} = \frac{(\alpha - L - 1) \; + \; \sqrt{(\alpha - L - 1)^2 + 4.\alpha.L \; .Tr(E[C_S]^{-1}.\Sigma_S)}}{2\alpha} \tag{96}$$

where Tr(.) denotes the trace of a matrix.

Finally, the restored (speckle filtered) version of the covariance matrix C_S is obtained using the maximum likelihood estimator described in [72]:

$$\hat{C}_S = \hat{\mu} \, .E[C_S] \tag{97}$$

7.2.2. Prior knowledge gathered from polarimetric SAR data: The Wishart-DE MAP filter

In high and very-high spatial resolution polarimetric SAR data, strong and/or mixed textural properties justify to estimate statistical scene knowledge from the data themselves, rather than assuming a A Priori theoretical model. Assuming that textural properties are identical in all configurations of polarisation, the entropy constraint on scene texture ([68] [69] [70]) becomes:

$$P(\mu) = (1 / \mu). \, exp(-\sum_k [\mu_k.log(\mu_k)] \qquad \text{and} \qquad E[\mu] = 1 \tag{98}$$

In this case, the complex Wishart/Distribution-Entropy MAP (CW-DE MAP) filter for polarimetric multilook SAR data is expressed as [71]:

http://dx.doi.org/10.5772/58593

$$L\ .\mathrm{Tr}(\mathrm{E}[\mathrm{C}_S]^{-1}.\Sigma_S) - L\ .\mu - \mu^2.\sum_k [log(\mu_k) - 1/Ln(10)] = 0 \tag{99}$$

$E[C_S]$ is obtained using the maximum likelihood estimator (Equation (97)) described in Lopès *et al.* [72].

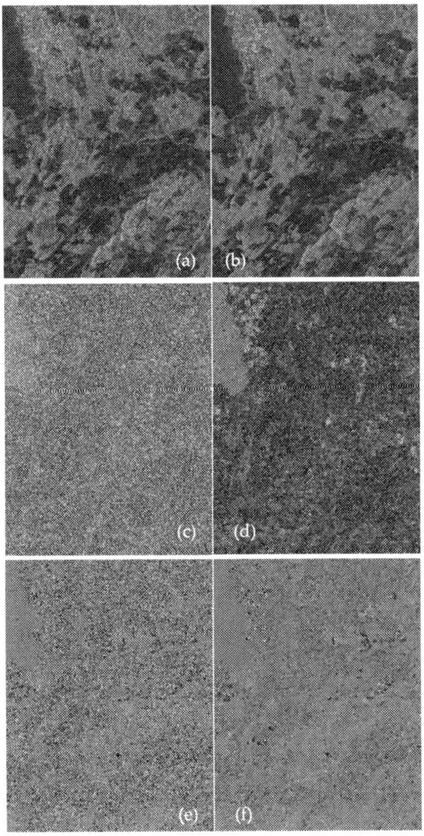

(a) color composition of the unfiltered HH (red), HV (green), and VV (blue) amplitudes. (b) color composition of the speckle filtered HH, HV, and VV amplitudes obatained using the fully polarimetric Wishart-DE MAP filter (*cf.* § 7.2.2). (c) color composition of the unfiltered HH-VV (red), HH-HV (green) and HV-VV (blue) degrees of coherence. (d) color composition of the HH-VV, HH-HV, and HV-VV degrees of coherence obtained from the Wishart-DE MAP speckle filtered polarimetric vector (*cf.*cf. § 7.2.2). (e) Unfiltered HH-VV phase difference. (f) HH-VV phase difference obtained from the Wishart-DE MAP speckle filtered polarimetric vector (*cf.*cf. § 7.2.2).

Figure 4. ALOS-PALSAR 6-looks polarimetric SAR imagery acquired on June 30, 2006 over Bavaria, Germany (Credits: JAXA and MITI; filtered images: Privateers NV, using ®SARScape).

The texture parameters μ_k in the neighbourhood of the pixel under processing are pre-estimated over an ergodic neighbourhood of the pixel under processing by a first speckle filtering pass using the Wishart-Gamma MAP filter (*cf.* § 7.2.1).

Although these fully polarimetric MAP filters assume identical texture properties in the HH, HV, and VV channels, which has been shown both experimentally [96] and theoretically [97] inexact, they have nevertheless been shown to preserve polarimetric signatures [98]. Fully polarimetric speckle filtering is illustrated in Figure 4.

Acronyms

ACF: spatial AutoCorrelation Function (of a signal)

DE: Distribution-Entropy (statistical distribution)

ENL: Equivalent Number of Looks (of a radar image)

LMMSE: Linear Minimum Mean Square Error (estimation)

MAP: Maximum A Posteriori (statistical estimation)

MMSE: Minimum Mean Square Error (estimation)

NMNV: Non-stationary Mean Non-stationary Variance (model)

pdf: Probability Density Function (random variable distribution)

Pfa: Probability of False Alarm (detection)

PSF: Point Spread Function (of an imaging system)

RoA: Ratio Of Amplitudes (structure detector)

SAR: Synthetic Aperture Radar (imaging sensor)

SLC: Single-Look-Complex (radar image)

Author details

Edmond Nezry[1*]

1 Privateers NV, Philipsburg, Netherlands Antilles

2 ParBleu Technologies Inc, St. Jean-sur-Richelieu, Québec, Canada

References

[1] F.T. Ulaby, A.K. Fung and R.K. Moore, 1982: "*Microwave Remote Sensing*", Vol.1 & 2, Artech House Publishers, 1982.

[2] J.C. Dainty, 1975: "Statistical properties of laser speckle patterns", in *Laser Speckle and Related Phenomena*, Springer Verlag, Berlin (Germany), 1975.

[3] M. Tur, K.C. Chin and J.W. Goodman, 1982: "When is speckle noise multiplicative?", *Applied Optics*, Vol.21, n°7, pp.1157-1159, April 1982.

[4] R.K. Raney, 1983: "Transfert function for partially coherent SAR system", *IEEE Trans. on AES*, Vol.AES-19, n°5, pp.740-750, September 1983.

[5] J.W. Goodman, 1976: "Some fundamental properties of speckle", *J. Opt. Soc. Am.*, Vol. 66, n°11, pp.1145-1150, November 1976.

[6] A. Lopès and R. Touzi, 1986: "Extraction of the backscatter coefficient of agricultural fields from an airborne SAR image", *Proc. of IGARSS'86*, Zürich (Switzerland), ESA SP-254, Vol.3, pp.621-627, 8-11 September 1986.

[7] R. Touzi, 1988: "*Analyse d'images radar en télédétection: améliorations radiométriques, filtrage du speckle et détection des contours*", Ph.D dissertation, Université Paul Sabatier (Toulouse III, France), n°258, 4 March 1988.

[8] A.L. Gray, P.W. Vachon, C.E. Livingstone and T.I. Lukowski, 1990: "Synthetic aperture radar calibration using reference reflectors", *IEEE Trans. on GRS*, Vol.GE-28, n°3, pp.374-383, May 1990.

[9] V.S. Frost and K.S. Shanmugan, 1983: "The information content of synthetic aperture radar images of terrain", *IEEE Trans. on AES*, Vol.AES-19, n°5, pp.768-774, September 1983.

[10] A. Lopès, E. Nezry, R. Touzi and H. Laur, 1993: "Structure detection and statistical adaptive speckle filtering in SAR images", *Int. J. Rem. Sens.*, Vol.14, n°9, pp.1735-1758, June 1993.

[11] A. Lopès, E. Nezry, S. Goze, R. Touzi and G. Aarbakke-Solaas, 1992: "Adaptive prossessing of multilook complex SAR images", *Proc. of IGARSS'92*, Clear Lake (USA), Vol.2, pp.890-892, 26-29 May 1992.

[12] H.H. Arsenault and M. Levesque, 1984: "Combined homomorphic and local-statistics processing for restoration of images degraded by signal dependant noise", *Applied Optics*, Vol.23, n°6, pp.845-850, March 1984.

[13] H.H. Arsenault, 1987: "Information extraction from images degraded by speckle", *Proc. of IGARSS'87*, Ann Arbor (USA), Vol.2, pp.1317-1320, 18-21 May 1987.

[14] Ping Fan Yan and C.H. Chen, 1986: "An algorithm for multiplicative noise in wide range", *Traitement du signal*, Vol.3, n°2, pp.91-96, 1986.

[15] H.H. Arsenault and G. April, 1976: "Properties of speckle integrated with a finite aperture and logarithmically transformed", *J. Opt. Soc. Am.*, Vol.66, n°11, pp. 1160-1163, November 1976.

[16] E. Nezry, 1988: "*Evaluation comparative des méthodes de filtrage du speckle sur les images radar (SAR)*", Rapport de DEA, CESR - Université Paul Sabatier (Toulouse, France), June 1988.

[17] H.H. Arsenault and G. April, 1985: "Information content of images degraded by speckle noise", *Proc. of SPIE*, Vol.556, *International conference on speckle*, pp.190-195, 1985.

[18] D.H. Hoekman, 1991: "Speckle ensemble statistics of logarithmically scaled data", *IEEE Trans. on GRS*, Vol.GE-29, n°1, pp.180-182, January 1991.

[19] H.H. Arsenault and G. April, 1986: "Information content of images degraded by speckle noise", *Optical Engineering*, Vol.25, n°5, pp.662-666, May 1986.

[20] A. Lopès, 1983: "*Etude expérimentale et théorique de l'atténuation et de la rétrodiffusion des micro-ondes par un couvert de blé. Application à la télédétection*", Ph.D. dissertation, Université Paul Sabatier (Toulouse III, France), n°852, 25 October 1983.

[21] H. Laur, 1989: "*Analyse d'images radar en télédétection: discriminateurs radiométriques et texturaux*", Ph.D. dissertation, Université Paul Sabatier (Toulouse III, France), n°403, 23 March 1989.

[22] G. Kattenborn, E. Nezry and G. DeGrandi, 1993: "High resolution detection and monitoring of changes using ERS-1 time series", *Proc. of the 2nd ERS-1 Symposium*, 11-14 Oct. 1993, Hamburg (Germany), ESA SP-361, Vol.1, pp.635-642.

[23] E. Nezry, E. Mougin, A. Lopès, J.-P. Gastellu-Etchegorry and Y. Laumonier, 1993: "Tropical vegetation mapping with combined visible and SAR spaceborne data", *Int. J. Rem. Sens.*, Vol.14, n°11, pp.2165-2184, 20 July 1993.

[24] E. Nezry, A. Lopès, D. Ducros-Gambart, C. Nezry and J.S. Lee, 1996: "Supervised classification of K-distributed SAR images of natural targets and probability of error estimation", *IEEE Trans. on GRS*, Vol.34, n°5, pp.1233-1242, September 1996.

[25] D.L.B. Jupp, A.H. Strahler and C.E. Woodcock, 1988: "Autocorrelation and regularization in digital images : I. Basic theory", *IEEE Trans. on GRS*, Vol.GE-26, n°4, pp. 463-473, July 1988.

[26] G. April and H.H. Arsenault, 1984: "Non-stationary image-plane speckle statistics", *J. Opt. Soc. Am. A*, Vol.1, n°7, pp.738-741, July 1984.

[27] F.T. Ulaby, A.K. Fung and R.K. Moore, 1986: "*Microwave Remote Sensing*", Vol.3, Artech House Publishers, 1986.

[28] F.T. Ulaby, F. Kouyate, B. Brisco and T.H. Lee Williams, 1986: "Textural information in SAR images", *IEEE Trans. on GRS*, Vol. GE-24, n°2, pp-235-245, March 1986.

[29] H. Laur, T. Le Toan and A. Lopès, 1987: "Textural segmentation of SAR images using first order statistical parameters", *Proc. of IGARSS'87*, Ann Arbor (USA), Vol.2, pp. 1463-1468, 18-21 May 1987.

[30] V.S. Frost, J.A. Stiles, K.S. Shanmugan and J.C. Holtzman, 1982: "A model for radar images and its application to adaptive digital filtering of multiplicative noise", *IEEE Trans. on PAMI*, Vol.PAMI-4, n°2, March 1982.

[31] R.K. Raney and G.J. Wessels, 1988: "Spatial considerations in SAR speckle simulation", *IEEE Trans. on GRS*, Vol.GE-26, n°5, pp.666-672.

[32] D.T. Kuan, A.A. Sawchuk and T.C. Strand, 1987: "Adaptive restoration of images with speckle", *IEEE Trans. on ASSP*, Vol.ASSP-35, n°3, pp.373-383, March 1987.

[33] J.S. Lee, 1980: "Digital image enhancement and noise filtering by use of local statistics", *IEEE Trans. on PAMI*, Vol.PAMI-2, pp.165-168, March 1980.

[34] J.S. Lim and H. Nawab, 1981: "Techniques for speckle noise removal", *Optical Engineering*, Vol.20, n°3, pp.472-480, May 1981.

[35] I. Ghikhman and A. Skorokhod, 1977-1980: "*Introduction à la théorie des processus aléatoires*", Chapt. 5, pp.290-305, Editions MIR, Moscow (USSR), in Russian 1977, in French 1980.

[36] D.S. Zrnic, 1977: "Mean power estimation with a recursive filter", *IEEE Trans. on AES*, Vol.AES-13, n°3, pp.281-289, May 1977.

[37] G. Horgan, 1994: "Choosing weight functions for filtering SAR", *Int. J. Rem. Sens.*, Vol. 15, n°5, pp.1053-1164.

[38] A. Yaglom, 1955: "Introduction to the theory of random stationary functions", in Russian, *Ouspekhi Matematicheskoie Naouk*, Vol. 7, n°5, pp.3-168, 1955.

[39] G.C. Deane, S. Quegan, et P.N. Churchill, A. Hendry, A. Evans and J.W. Trewett, 1986: "Land-use features detection in SAR images", ESA SP-257, 1986.

[40] A. Lopès, R. Touzi and E. Nezry, 1990: "Adaptive speckle filters and scene heterogeneity", *IEEE Trans. on GRS*, Vol.GE-28, n°6, pp.992-1000, November 1990.

[41] D.T. Kuan, A.A. Sawchuk, T.C. Strand and P. Chavel, 1985: "Adaptive noise smoothing filter for images with signal-dependant noise", *IEEE Trans. on PAMI*, Vol.PAMI-7, n°2, pp.165-177, March 1985.

[42] J.S. Lee, 1981: "Speckle analysis and smoothing of synthetic aperture radar images", *Computer Graphics and Image Processing*, n°17, pp.24-32, 1981.

[43] J.S. Lee, 1983: "A simple speckle smoothing algorithm for synthetic aperture radar images", *IEEE Trans. on SMC*, Vol.SMC-13, n°1, pp.85-89, January 1983.

[44] J.S. Lee, 1986: "Speckle suppression and analysis for synthetic aperture radar images", *Optical Engineering*, Vol.25, n°5, pp.636-643, May 1986.

[45] J.S. Lee, 1987: "Statistical modelling and suppression of speckle in synthetic aperture radar images", *Proc. of IGARSS'87*, Ann Arbor (USA), Vol.2, pp.1331-1339, 18-21 May 1987.

[46] S.P. Luttrell, 1991: "The theory of bayesian super-resolution of coherent images: a review", *Int. J. Rem. Sens.*, Vol.12, n°2, pp.303-314, 1991.

[47] T. Bayes and R. Price, 1763: "An essay towards solving a problem in the doctrine of chance. By the late Rev. Mr. Bayes, communicated by Mr. Price, in a letter to John Canton, A.M.F.R.S.", *Philosophical Trans. of the Royal Society of London*, Vol.53, pp.370–418.

[48] P.-S. Laplace, 1774: "Mémoire sur la probabilité des causes par les événements", *Mémoires de l'Académie royale des Sciences de Paris (Savants étrangers)*, Vol.4, pp.621-656.

[49] J.M. Keynes, 1929: "*A treatise on probability*", Editions Macmillan, London (UK), 1929.

[50] R.P. Cox, 1946: "Probability, frequency and reasonable expectation", *American Journal of Physics*, Vol.14, n°1, pp.1-13, January-February 1946.

[51] H.L. Van Trees, 1968: "*Detection, Estimation, and Modulation Theory*", Editions J. Wiley & Sons, New York, 1968.

[52] B.R. Hunt, 1977: "Bayesian methods in non-linear digital image restoration", *IEEE Trans. on Computers*, Vol.C-26, n°3, pp.219-229, March 1977.

[53] D.T. Kuan, A.A. Sawchuk, T.C. Strand and P. Chavel, 1982: "MAP Speckle reduction filter for complex amplitude speckle images", *The Proceedings of the Pattern Recognition and Image Processing Conference*, pp.58-63, IEEE, 1982.

[54] S. Geman and D. Geman, 1984: "Stochastic relaxation, Gibbs distribution, and the Bayesian restoration of images", *IEEE Trans. on PAMI*, Vol.PAMI-6, n°6, pp.721-741, November 1984.

[55] J.R. Garside and C.J. Oliver, 1988: "A comparison of clutter texture properties in optical and SAR images", *Proc. of IGARSS'88*, Edinburgh (Scotland), ESA SP-284, Vol.2, pp.1249-1255, 13-16 September 1988.

[56] G. Edwards, R. Landry and K.P.B. Thompson, 1988: "Texture analysis of forest regeneration sites in high-resolution SAR imagery", *Proc. of IGARSS'88*, Edinburgh (Scotland), ESA SP-284, Vol.3, pp.1355-1360, 13-16 September 1988.

[57] A. Lopès, H. Laur and E. Nezry, 1990: "Statistical distribution and texture in multilook and complex SAR images", *Proc. of IGARSS'90*, Washington D.C. (USA), Vol.3, pp.2427-2430, 20-24 May 1990.

[58] E. Jakeman and P.N. Pusey, 1976: "A model for non-Rayleigh sea echo", *IEEE Trans. on AP*, Vol.AP-24, n°6, pp.806-814, November 1976.

[59] C.J. Oliver, 1984: "A model for non-Rayleigh scattering statistics", *Optica Acta*, Vol.31, n°6, pp.701-722, 1984.

[60] C.J. Oliver, 1988: "Representation of radar sea clutter", *IEE Proceedings*, Vol.135, Pt. F, n°6, pp.497-500, December 1988.

[61] E. Jakeman and R.J.A. Tough, 1987: "Generalized K distribution: a statistical model for weak scattering", *J. Opt. Soc. Am. A*, Vol.4, n°9, pp.1764-1772, September 1987.

[62] D. Lindley, 1991: "*Making Decisions*", 2nd edition, J. Wiley & Sons Inc. Ed., p.104, 1991.

[63] E. Nezry, A. Lopès and R. Touzi, 1991: "Detection of structural and textural features for SAR images filtering", *Proc. of IGARSS'91*, Espoo (Finland), Vol.4, pp.2169-2172, 3-6 June 1991.

[64] E. Nezry, 1992: "*Restauration de la réflectivité radar pour l'utilisation conjointe des images radar et optiques en télédétection*", Ph.D. dissertation, Université Paul Sabatier (Toulouse III, France), n°1236, 237 p., 3 July 1992.

[65] A. Lopès, E. Nezry, R. Touzi and H. Laur, 1990: "Maximum a posteriori speckle filtering and first order texture models in SAR images", *Proc. of IGARSS'90*, Washington D.C. (USA), Vol.3, pp.2409-2412, 20-24 May 1990.

[66] R. Touzi, A. Lopès and P. Bousquet, 1988: "A statistical and geometrical edge detector for SAR images", *IEEE Trans. on GRS*, Vol.GRS-26, n°6, pp.764-773, November 1988.

[67] E. Nezry, G. De Grandi and M. Leysen, 1995: "Speckle and scene spatial statistical estimators for SAR image filtering and texture analysis: Some application to agriculture, forestry and point targets detection", *Proc. of SPIE*, Vol.2584, pp.110-120, 25-29 September 1995.

[68] M. Datcu and M. Walessa, 1997: "Maximum entropy methods for despeckleling and resampling of synthetic aperture radar images of rough terrain", *Proc. of SPIE*, Vol. 3217, pp.76-83, September 1997.

[69] M. Walessa and M. Datcu, 1997: "Bayesian approach to SAR image reconstruction", *Proc. of IGARSS'97*, Vol.2, pp.767-770, 03-08 August 1997.

[70] A.K. Jain, 1989: "*Fundamentals of digital image processing*", Chapt. 8, pp.316-317, Prentice Hall Ed., Englewood Cliffs (NJ), 1989.

[71] E. Nezry and F. Yakam-Simen, 1999: "Five new distribution-entropy MAP speckle filters for polarimetric SAR data, and for single or multi-channel detected and complex SAR Images", *Proc. of SPIE*, Vol.3869, pp.100-105, 20-24 September 1999.

[72] A. Lopès, S. Goze and E. Nezry, 1992: "Polarimetric speckle filters for SAR data", *Proc. of IGARSS'92*, Houston (TX), Vol.1, pp.80-82, 26-29 May 1992.

[73] E. Nezry, F. Zagolski, A. Lopès and F. Yakam-Simen, 1996: "Bayesian filtering of multi-channel SAR images for detection of thin structures and data fusion", *Proc. of SPIE*, Vol.2958, pp.130-139, September 1996.

[74] F. Yakam-Simen, E. Nezry and J. Ewing, 1998: "The legendary lost city 'Ciudad Blanca' found under tropical forest in Honduras, using ERS-2 and JERS-1 SAR imagery", *Proc. of SPIE*, Vol.3496, pp.21-28, September 1998.

[75] E. Nezry, H.G. Kohl and H. De Groof, 1994: "Restoration and enhancement of textural properties in SAR images using second order statistics", *Proc. of SPIE*, Vol.2316, pp. 115-124, Rome (Italy), 26-30 September 1994.

[76] J.S. Lee, 1981: "Refined filtering of image noise using local statistics", *Computer Graphics and Image Processing*, n°15, pp.380-389, 1981.

[77] R.G. Caves, P.J. Harley and S. Quegan, 1992: "Edge structures in ERS-1 and airborne SAR data", *Proc. of IGARSS'92*, Clear Lake (TX), Vol.2, pp.1117-1119, 26-29 May 1992.

[78] R.P.H.M. Schoenmakers, 1995: "*Segmentation of Remotely Sensed Imagery*", Ph.D. dissertation, Katholieke Universiteit Nijmegen (NL), EUR 16087 EN, Office for Official Publications of the E.C., Luxembourg, 168 p., 13 Sept. 1995.

[79] A. Lopès and F. Séry, 1994: "The LMMSE polarimetric Wishart and the LMMSE textural vector speckle filters for SAR images". *Proc. of IGARSS'94*, Pasadena (USA), Vol. 4, pp.2143-2145, 8-12 August 1994.

[80] J.W. Woods and J. Biemond, 1982: "Comments on 'A model for radar images and its application to adaptive digital filtering of multiplicative noise'", *IEEE Trans. on PAMI*, Vol.PAMI-4, n°2, pp.168-169.

[81] H.-Ch. Quelle and J.-M. Boucher, 1990: "Combined use of parametric spectrum estimation and Frost-algorithm for radar speckle filtering", *Proc. of IGARSS'90*, Washington D.C. (USA), Vol.1, pp.295-298, 20-24 May 1990.

[82] E. Rignot and R. Kwok, 1993: "Characterization of spatial statistics of distributed targets in SAR data", *Int. J. Rem. Sens.*, Vol.14, n°2, pp.345-363.

[83] C.J. Oliver, 1985: "Correlated K-distributed clutter models", *Optica Acta*, Vol.32, n°12, pp.1515-1547.

[84] O.D. Faugeras and W.K. Pratt, 1980: "Decorrelation methods of texture feature extraction", *IEEE Trans. on PAMI*, Vol.PAMI-2, n°4, pp.323-332.

[85] E. Nezry, F. Yakam-Simen, F. Zagolski and I. Supit, 1997: "Control systems principles applied to speckle filtering and geophysical information extraction in multi-channel SAR images", *Proc. of SPIE*, Vol.3217, pp.48-57, September 1997.

[86] J. Riccati, 1724: "Animadversiones in aequationes differentiales secundi gradus", *Actorum Eruditorum, quae Lipsiae publicantur, Supplementa,* Vol.8, pp.66-73.

[87] P. Lancaster and L. Rodman, 1995: "*Algebraic Riccati equations*", Oxford University Press, 1995.

[88] J.W. Goodman, 1985: "*Statistical optics*", J. Wiley & Sons Inc. Ed., New York (USA), 1985.

[89] R. Fjørtoft, A. Lopès, J. Bruniquel and P. Marthon, 1998: "Optimal edge detection in SLC images with correlated speckle", *Proc. of IGARSS'98,* Seattle (USA), Vol., pp., 6-10 July 1998.

[90] R. Fjørtoft, J.-C. Cabada, A. Lopès and P. Marthon, 1998: "Optimal edge detection and segmentation of SLC SAR images with spatially correlated speckle", *Proc.of SPIE,* Vol. 3497, Barcelona (Spain), 21-25 September 1998.

[91] R. Fjørtoft, A. Lopès, J. Bruniquel and P. Marthon, 1999: "Optimal edge detection and edge localization in complex SAR images with correlated speckle", *IEEE Trans. on GRS,* Vol.37, pp.2272-2281, September 1999.

[92] L.M. Novak and M.C. Burl, 1990: "Optimal speckle reduction in polarimetric SAR imagery", *IEEE Trans. on AES,* Vol.26, pp.293-305, March 1990.

[93] J.S. Lee, M.R. Grunes and S.A. Mango, 1991: "Speckle reduction in multipolarization multifrequency SAR imagery", *IEEE Trans. on GRS,* Vol.GE-29, n°4, pp.535-544, July 1991.

[94] J.S. Lee, G. De Grandi, M.R. Grunes and E. Nezry, 1996: "Polarimetric signature preservation in SAR speckle filtering", *Proc. of IGARSS'96,* Lincoln (USA), Vol.3, pp. 1574-1576, 27-31 May 1996.

[95] R. Touzi and A. Lopès, 1994: "The principle of speckle filtering in polarimetric SAR imagery", *IEEE Trans. on GRS,* Vol.32, nr.5, pp.1110-1114, 1994.

[96] G. De Grandi, E. Nezry and G. Kattenborn, 1993: "Experimental characterization of spatial statistics in polarimetric multifrequency airborne SAR images", *Proc. of IGARSS'93,* Tokyo, (Japan), Vol.2, pp.806-812, 18-21 August 1993.

[97] G. De Grandi, J.S. Lee, D. Schuler and E. Nezry, 2003: "Texture and speckle statistics in polarimetric SAR synthesized images", *IEEE Trans. on GRS,* Vol.41, n°9, pp. 2070-2088, September 2003.

[98] E. Nezry and F. Yakam-Simen, 1999: "On the preservation of polarimetric signatures and polarimetric texture signatures by fully polarimetric MAP speckle filters", *Proc. of IGARSS'99,* Vol.3, 3 p., Hamburg (Germany), 28 June-2 July 1999.

Section 2

Land-Cover Applications

Chapter 2

Large Scale Mapping of Forests and Land Cover with Synthetic Aperture Radar Data

Josef Kellndorfer, Oliver Cartus, Jesse Bishop, Wayne Walker and Francesco Holecz

Additional information is available at the end of the chapter

http://dx.doi.org/10.5772/58220

1. Introduction

1.1. A short review on SAR for forest mapping

Forests are a key natural resource providing a range of ecosystem services like carbon sequestration, natural habitats for biodiverse fauna and flora, and providing food and fiber for human consumption. To obtain sound information for management, protection, and restoration of forests, some core information needs are: 1) mapping of forest extent, 2) identification of areas of disturbance, 3) estimation of above ground biomass or growing stock volume, and 4) estimation of stand canopy height. While the first two categories are of thematic character, hence directly detectable from remote sensing data, the latter two variables need inference from models driven with remote sensing data.

Severe storms and fire are examples of major disturbance events and remote sensing has been used operationally to identify them with low spatial resolution (> 500 m) optical imagery from sensors such as NOAA AVHRR, MODIS, ERS ATSR-1/2 and SPOT-Vegetation. While identification and rapid monitoring of disturbance events is invaluable, higher resolution sensors are needed to map areal extent of the events for resource management purposes. Typical optical remote sensors used to date for responding to such needs are carried on the Landsat-5/-8, SPOT, RapidEye, IKONOS, QuickBird, GeoEye, and WorldView satellites. With a spatial resolution ranging from 30 m to better than 0.5 m, accurate information on forest area and disturbances can be retrieved. Nonetheless, optical remote sensing is limited in areas which have significant cloud cover for long periods of the year (e.g. tropical), and in those regions where sun light is an additional limiting factor (e.g. boreal). Spaceborne Synthetic Aperture Radar (SAR) data with their cloud-penetrating and day-night measurements capability provide a key data resource, complementing optical instruments for forest monitoring.

For the purposes of estimating forest biomass and growing stock volumes at stand scale, SAR systems have been shown as a valuable mapping technique due to their sensitivity to the vertical structure of forests. First results were achieved at the beginning of the '90s using the three frequency polarimetric airborne AIRSAR NASA/JPL SAR system. A well-established approach for generating forest biomass maps is to relate the observed backscattering coefficient to ground measurements. Using parametric exponential regressions functions or non-parametric ensemble regression tree models, forest biomass is then estimated from the SAR intensity data. Improvements in the estimation can be achieved by combining different polarizations and/or by rationing several frequencies. In general, the radar backscatter was shown to be positively correlated with some biophysical parameters such as the aboveground biomass (AGB), tree height, tree age, diameter at breast height, and basal area. Comparison of radar data acquired at C-, L-, and P-band frequencies showed that correlation of the radar backscatter with the AGB increases with increasing radar wavelength. At these frequencies, HH- and HV-polarization provide a greater sensitivity to AGB than VV-polarization [1]. Based on hundreds of studies in different ecological regions, it has been recognized that backscatter-only approaches reach a saturation level, i.e. an increase of the radar backscatter do not correspond to an increase of the AGB. The typical saturation level observed is around 300 Mg/ha, for P-Band at 100 Mg/ha for L-Band with observations of HH and HV polarizations. To overcome the saturation problem five approaches have been pursued in the past decades:

1. Ferrazzoli et al. [3] proposes to make use of bistatic radar at L-band in a specular configuration. In order to demonstrate its feasibility, a simulation analysis has been carried out by using a microwave model of vegetated terrain. The results demonstrated that woody volume up to 900 tons/ha could be inferred, hence enabling to completely solve the saturation problem. However, this approach still remains at theoretical level, since up to date no bistatic L-band SAR systems are available.

2. By using low frequency the attenuation is significantly reduced, and the large scale structures (of the order of the wavelength) dominate the backscatter. The response from non-forested areas is therefore drastically reduced, normally much below the system noise floor. The response from forested areas, on the other hand, is dominated by the large trunk and branch structures together with coherence ground reflection interactions. Since these are generally where most of the AGB is stored, the correlation of the backscatter to AGB usually increases with decreasing frequency. CARABAS – an ultra-wideband airborne SAR system operating at VHF band (20-90 MHz) – has shown that the dynamic range of the scattering is significantly larger than at P-band (440 MHz), suggesting a greater sensitivity of the lower frequency [4].

3. Another approach takes advantage of the fact that tree height can be inferred using airborne single-pass Interferometric SAR (InSAR) dual frequency (X-and P-band) data, or alternatively, LIght Detection And Ranging (LIDAR) systems. AGB is subsequently retrieved by species using allometric equations. Moreover, by integrating into the

inference function the interferometric height and the P-band backscatter at different polarizations, it has been demonstrated that the well-known saturation level could be overcome [5]. On a still experimental basis, it has been shown that canopy height (CH) can be also retrieved by using airborne single-pass L-band polarimetric InSAR technique including a forest model involving a random volume of scatterers situated over a ground scattering model [6].

4. C-band backscatter is generally deemed as useless due to the weak sensitivity with respect to biophysical properties. Moreover, the strong sensitivity to the dielectric properties of the scattering objects makes C-band backscatter unreliable for monitoring forest. For this reason, low frequencies have been preferred. Santoro et al. [7] demonstrate that accurate estimates of forest GSV can be retrieved from C-band backscatter data under the requirement that a large stack of observations is available. The estimation of the GSV is carried out by means of an algorithm, which combines hyper-temporal C-band data stacks, the inversion of a water-cloud model relating the GSV to the forest backscatter, and a multi-temporal combination of GSV estimates from each image. Traditionally, model training is based on in situ measurements for unvegetated and dense forest areas and corresponding forest backscatter measurements. The novel aspect is that these are identified by means of the MODIS Vegetation Continuous Fields (VCF) product, where the corresponding measures are computed. The hyper-temporal combination exploits the different sensitivities of the forest backscatter to GSV, which can be retrieved from the estimates of the a priori unknown model parameters. Results show that the inferred RMSE is generally below 40% for full resolution data and below 20% for aggregated versions at reduced spatial resolution. The key outcome is that the retrieved GSV is never affected by saturation, with estimates in line with in situ data up to 300 m^3/ha.

5. Estimate forest biomass through regressions based on exponential function exclusively derived from single frequency single/dual polarizations, ensemble regression models at eco-regions level are developed by considering multi-sensor data and bio-and geo-physical gradient data (elevation, slope, aspect, canopy density, and land cover). Works performed in the United States and Chile have shown that ensemble regression models based on multi-temporal ALOS PALSAR-1 intensities and repeat-pass coherences at HH and HV polarization as well as Landsat data yield AGB and vegetation height accuracies with R^2 in the range of 0.7 to 0.85 [8].

In summary, looking at the thematic (area and disturbances) and bio-physical (AGB and CH in primis) information which can be extracted or inferred from remote sensing data, it can be stated that:

- Depending upon the eco-region, environmental conditions, and forest practices, remote sensing data should be accordingly selected – in particular considering the seasonality, i.e. vegetation phenology – and algorithms consequently adapted. For instance, due to the different practices, forest clear cuts in Amazon have a completely different response at L-band than in the tropical forest in Africa, where, typically, trunks are left on the ground for months, hence engendering a much stronger backscatter, compared to a cleared area.

- The availability of multi-/hyper-temporal SAR data considerably contributes, firstly, to enhance the data quality by significantly reducing the speckle. In this respect, it is worth mentioning that future hyper-temporal data stacks acquired from Sentinel-1A/B will play, in terms of signal processing, a relevant role for the provision of high quality data (i.e. high Equivalent Number of Looks) at highest level of detail. This will allow, at analysis level, the exploitation of a pixel based approach, simpler and less time consuming than a region based one. Secondly, the temporal component provides an additional source of information for the identification of land cover classes and the appropriate treatment of moisture related phonological influence on the backscatter variations not detectable in a single-date image. Thirdly, temporal data stacks, in particular at low wavelengths, allow the estimation of the interferometric SAR coherence, proven useful for thematic and bio-physical purposes. Also, multi-/hyper-temporal SAR data sets can additionally be used in fusion with optical data, as well as bio- and geo-physical gradient data, in order to develop specific ensemble regression models at eco-regions level for the retrieval of key biophysical parameters.
- Planned new SAR missions like ESA's Sentinel-1 C-Band SAR and BIOMASS P-band SAR, JAXA's ALOS-2 PALSAR, Argentina's SAOCOM L-Band SAR, and NASA/India's L-and S-Band NISAR will doubtlessly play a relevant role for the estimation of forest bio-physical parameters.

2. The ALOS PALSAR-1 mission

Between its launch in January 2006 and the end of the mission in April 2011, the ALOS PALSAR-1 system has acquired wall-to-wall global coverage on an annual basis, which has resulted in up to five acquisitions per year at a particular location [9]. The first-of-its-kind global observation strategy for the ALOS mission provided thus an unprecedented opportunity to take global snapshots of Earth's natural resources at very narrow time-steps and high resolution. Figure 1 shows a pan-tropical ALOS PALSAR-1 HH/HV mosaic: around 17, 000 ALOS PALSAR-1 single-look complex data frames (coverage 70x70 km per frame) were multi-look processed to a 4-look image corresponding to 15 m pixel spacing. Subsequently, the multi-looked images were speckle filtered, radiometric calibrated, normalized and terrain geocoded using the Shuttle Radar Topography Mission (SRTM) Digital Elevation Model (DEM). Geocoded frames where finally assembled to image mosaic tiles in resolutions of 15 m, 50 m, and 100 m. A similar procedure, but in this case starting from the K&C ALOS PALSAR-1 slant range amplitude data (16 looks in azimuth and 4 looks in range, corresponding to a pixel spacing of 37.5 m in range and 50.7 m in azimuth), has been exploited by De Grandi et al. [10], where the first African ALOS PALSAR-1 HH/HV mosaic has been generated. In the image mosaic illustrated in Figure 1, the HH information channel was assigned to red, HV was assigned to green, and the ratio between the two (HH/HV) was assigned to blue. With the applied color assignment, green and yellow tones correspond to instances where both HH and HV information channels have high energy returns e.g., over forested and urban areas. Blue and magenta colors are generally found in non-forested areas, where the HH polarized energy often exhibits a higher return from the surface than the HV polarized energy.

Figure 1. ALOS PALSAR-1 pan-tropical mosaic. Image Data © by JAXA/METI, Image Processing by WHRC/ASF/sarmap. Background image from MODIS by NASA/JPL.

L-band backscatter is correlated to increasing AGB or GSV, as, with increasing canopy density and height, the backscatter contribution from the forest floor declines and the volume scattering contribution from the canopy increases. The contribution of stem-ground interactions to the total backscatter is generally weak due to diffuse scattering at the rough forest floor and substantial attenuation of the signals in the vegetation layer unless the forest floor is flooded or the canopies are frozen. In order to capture the contribution of the environmental conditions to the measured forest backscatter, the retrieval of GSV or AGB with radar backscatter observations postulates a set of *in situ* measurements to tune models that relate the measured backscatter to the biophysical forest parameters to the prevailing conditions. However, extensive *in situ* data are often not available, either because of the vastness or remoteness of forests or because of restrictions on the use of existing measurements. Even if available, uncertainties connected to *in situ* measurements can be substantial. Two different approaches for model calibration that do require no or only very limited field data have been developed in recent years:

1. A number of investigators have assessed the possibility to calibrate models, relating radar observations to forest biophysical attributes, using LIDAR derived attribute estimates, which require only a limited set of in situ data for model calibration. The mapping of forest resources by means of fusion of LIDAR and SAR was tested, for instance, by Englhart et al. [11], Kellndorfer et al., [12], and Atwood et al. [14]. In [11], AGB estimates from a number of airborne LIDAR transects acquired over Kalimantan, Indonesia, were used to calibrate models, relating multi-temporal TerraSAR-X and ALOS PALSAR L-band data to AGB, and extrapolated to a 280, 000 ha area (RMSE of 79 t/ha, R^2 of 0.53). Kellndorfer et al. extrapolated airborne LIDAR derived estimates of CH for a 1,200 km^2 area in Maryland, USA, to an area of 110, 000 km^2 using SRTM, National Elevation Dataset (NED) and Landsat data as spatial predictor layers in an ensemble regression tree model. An RMSE of 4.4 m (Pearson correlation of 0.71) when independently validating against plot-level forest inventory data has been reported. Finally, in [14] for boreal forest, the RMSE of the AGB estimate was found to be 34.9 Mg/ha over a biomass range of 250 Mg/ha; only marginally less accurate than the 33.5 Mg/ha accuracy of the LIDAR technique.

2. Others have investigated the possibility of a fully automated algorithm that makes the retrieval based on radar backscatter data (mostly) independent of the availability of *in situ* data. Santoro et al. [7] presented a novel approach for the mapping of boreal GSV using multi-temporal ENVISAT ASAR ScanSAR C-band. A similar approach was presented in Cartus et al. [8] for ERS-1/2 Tandem coherence to map GSV classes in Northeast China. In both studies, the automation of the retrieval was accomplished with the aid of the MODIS VCF [13], which was used to calibrate semi-empirical models, relating the SAR/InSAR data to GSV.

In the following three sections, three case studies are presented. Particular emphasis in all these works is set 1) on the use of multi-temporal ALOS PALSAR-1 data; 2) on the data acquisition period; 3) on the wise integration with other data sets; 4) on the limitations of the ALOS PALSAR-1 data.

3. Fusion of ALOS PALSAR-1, Landsat ETM+ and ALS

Currently, small-footprint Airborne Laser Scanners (ALS) represent the most deployed type of LIDAR sensors. Numerous studies have illustrated the high performance of ALS for the estimation of forest biophysical attributes. Because of the scanning capability, ALS provide for the spatially explicit mapping of forests covered by transects of several hundred meters in width. However, wall-to-wall coverage of large forest areas with ALS is in most cases prohibitively expensive, which is why fusion with image data is required to generate wall-to-wall maps of forest attributes for larger areas. The goal of this study is to investigate robust methods for estimating CH and GSV by spatially extending ALS data using ALOS PALSAR-1 and Landsat ETM+ data, i.e. by calibrating models, relating the spaceborne data to CH and GSV, with the aid of ALS derived CH and GSV estimates. Landsat TM/ETM+ data have been considered as in several studies it was shown that a retrieval of forest biophysical parameters based on the fusion of SAR and optical data yielded higher retrieval accuracies [16, 17, 18].

The study area extended over three administrative regions: Maule, Biobio, and Araucania and covered parts of the coastal Cordillera and the Chilean Central Valley (Figure 2). The forest is dominated by even-aged plantations of *Pinus radiata* and to a lesser extent (<20% by area) *Eucalyptus globulus*. Stand-level forest inventory data for 437 stands that were collected in the timeframe of the LIDAR campaigns were provided by the ARAUCO timber company. The geospatial information company Digimapas Chile provided small-footprint airborne LIDAR data for an area of ~2.5 million ha. The data were acquired between 2006 and 2008. The used airborne platform consisted of a laser scanning system (Riegl LMS-Q560), two digital cameras (Applanix DSS 322) and navigation equipment (Applanix POS AV 401). The LIDAR data have a nominal range resolution of 2 cm and delivers an absolute vertical and horizontal accuracy of better than 15 and 25 cm, respectively. During the operation, the height and intensity of multiple discrete laser returns for each laser pulse were recorded. The laser point density on the ground varied between 1 and 3 hits/m^2 and the scan angles ranged up to 22.5°. Digimapas produced and delivered fully geocoded Digital Terrain Models (DTM) and Surface Models (DSM) with 1×1 m^2 pixel spacing.

Under a NASA/JAXA data agreement, the Alaska Satellite Facility (ASF) provided a multi-temporal ALOS PALSAR-1 dataset. In total, 189 Fine Beam Dual (FBD) polarization and 181 Fine Beam Single (FBS) polarization Single Look Complex images were available. The FBD data were acquired between June and December 2007 and the FBS data between January and June and November and December 2007. Ten FBD and 14 FBS images (from three different ALOS paths) covering the same area as the ALS data were used. The multi-annual ALOS PALSAR-1 acquisitions also allowed for the computation of the interferometric repeat-pass coherence, which describes the temporal stability of scattering between two images and generally decreases with increasing forest density and height. Despite the long repeat interval of 46 days – hence the increased risk of temporal decorrelation - spaceborne L-band repeat-pass coherence has shown some potential for the retrieval of forest biophysical parameters, in particular in combination with intensity measurements, when the imaging conditions were suitable. The interferometric coherence was computed for image pairs with temporal baselines of 46 or 92 days and all possible combinations of image modes (FBD-FBD, FBS-FBS, FBS-FBD). In total, coherence images for 11 acquisition date combinations were produced. The perpendicular baselines were between 40 and 900 m (Table 1). In addition, three Landsat 7 ETM+ images were obtained from the Global Land Cover Facility. Two images were acquired in December 2005 and one in January 2007 under cloud-free conditions over the study area. The L1T surface reflectance data were already calibrated and corrected for terrain as well as atmospheric effects [19].

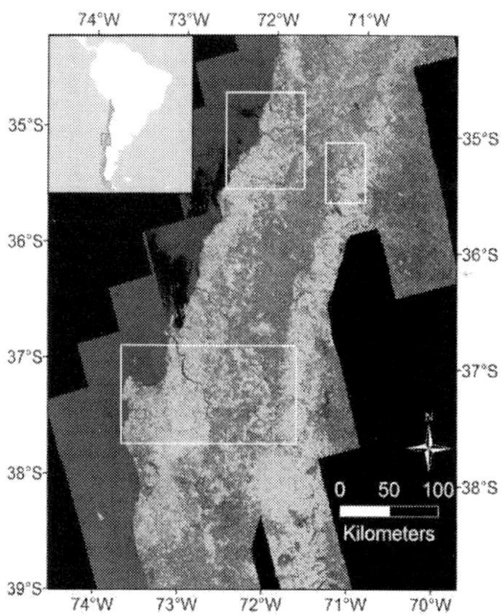

Figure 2. ALOS PALSAR-1 mosaic for Central Chile. RGB=HH, HV and HH/HV ratio. The white rectangles denote the areas with wall-to-wall ALS coverage and the red rectangles denote the areas used for testing the fusion of LIDAR, ALOS PALSAR-1 and Landsat ETM+.

A Canopy Height Model (CHM) with 1×1 m^2 pixel spacing was produced by subtracting the ALS DTM from the DSM. A suite of ALS canopy structure indices characterizing different aspects of the forests canopy structure were computed from the CHM for each stand in the inventory data. The indices comprised the percentiles of the height distribution of first returns in steps of 10%, the coefficient of variation, mean, kurtosis, skewness and several canopy density indices (i.e, the proportion of first returns from heights above different height thresholds). A Gaussian fit to the profile of first return heights was computed as an additional means of characterizing canopy vertical structure (i.e. with the number of Gaussians used to describe the profile). From the ALOS PALSAR-1 data, the average intensity and coherence were computed within each stand. From the Landsat ETM+ imagery, the average reflectance observed in bands 1 to 5 and 7 were computed.

3.1. Segmentation

Image segmentation was conducted on the ALS CHM using the multi-resolution segmentation algorithm implemented in the eCognition. Segmentation was used to delineate "stand-like" image object polygons across the entire study area (Figure 3). For the segmentation, the CHM was aggregated by means of simple block averaging from 1 to 4 m pixels as a tradeoff between preserving spatial detail and reducing the amount of data to a level that could be handled by the software in an acceptable amount of time. The segmentation parameters (i.e. scale, compactness, smoothness) were chosen so that segments had sizes comparable to the polygons in the inventory dataset (on average 8 ha) and smoothly followed stand boundaries visible in the CHM. For the segments, the same set of ALS, ALOS PALSAR-1 and Landsat ETM+ canopy structure indices were extracted as were for the inventory stands.

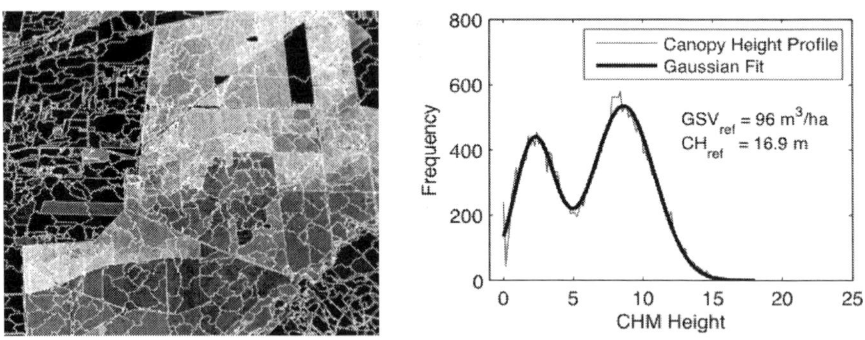

Figure 3. Segmented canopy height model (left) and height profile of ALS first returns for a radiata pine stand with a GSV of 96 m^3/ha and a CH of 17 m (right).

3.2. Modeling

For modeling the relationship between the suite of space- and airborne remote sensing data and the in situ measurements of CH and GSV we used randomForest [20], a data mining algorithm that has proven robust and computationally efficient and that has successfully been applied in several large-scale forest mapping efforts [17, 18]. In randomForest, a large number of regression trees are grown, each recursively partitioning the training data, considering at every node a random selection of predictors. The predictions from all regression trees are then averaged to obtain a single estimate. Each regression tree is grown using a random selection of samples. The rest of the samples, the so-called 'out-of-bag' cases (OOB), are estimated via the respective regression trees after training and the obtained OOB predictions for all trees are then averaged to obtain an unbiased estimate for the retrieval error.

3.3. Fusion experiments

In a two stage up-scaling approach, randomForest was used for the modeling of 1) the relationship between the ALS canopy structure indices and the in situ measurements of CH and GSV and 2) the relationship between the ALS-based estimates of GSV and CH and the ALOS PALSAR-1 / Landsat ETM+ datasets. The development of fusion models incorporating these data was performed within three 100 km^2 test sites. The test sites, as shown in Table 1, were selected so that 1) a wide range of stand growth stages were covered; 2) no management activities (e.g. thinning, logging, etc.) had occurred during the image acquisition timeframe; and 3) a cluster of inventory polygons (i.e. stands) was located within each site. Two of the selected test sites were located in the Cordillera along the Pacific coast and one in the Chilean Central Valley (Figure 2). In total, 105 inventory stands were located within the area of the three test sites. These 105 stands were used for validation purposes only (i.e. they were not used for the development of models, relating the ALS canopy structure indices to CH and GSV). At each of the test sites, ALS-derived estimates of CH and GSV for the segments in the ALS CHM were used as response variables in randomForest to develop new models, relating all available per-segment ALOS PALSAR-1 intensities and coherences as well as Landsat ETM+ reflectances to CH and GSV, respectively. The retrieval accuracy was assessed 1) by comparing the OOB estimates for CH and GSV to the per-segment ALS predictions of CH and GSV; and 2) by applying the developed models to the ALOS PALSAR-1 and Landsat ETM+ data extracted for the inventory polygons that were located within the area of the test sites and comparing the obtained estimates for CH and GSV to the respective *in situ* measurements.

3.4. Stand-level retrieval of canopy height and growing stock volume with ALS

Figure 4 illustrates the stand-level OOB estimates for GSV and CH when using all stand-level ALS canopy structure indices as predictors and the *in situ* measurements of CH and GSV obtained from 319 stands as response variables.

Test Site	Slope (Mean/SD)	CH and GSV (mean/SD)	PALSAR-1 Data		Coherence (B_n)
Date	Mode				
1	16/9°	CH_{FID}: 29/11 m CH_{ALS}: 25/9 m GSV_{FID}: 386/198 m³/ha GSV_{ALS}: 316/152 m³/ha	10 Jul. 10 Oct. 7 Jan. 25 Nov. 11 Dec.	FBD FBD FBS FBS FBS	10 Jul. & 10 Oct. (503 m) 10 Oct. & 25 Nov. (304 m)
2	2.6/5°	CH_{FID}: 19/4 m CH_{ALS}: 17/6 m GSV_{FID}: 172/70 m³/ha GSV_{ALS}: 154/98 m³/ha	5 Jul. 5 Oct. 17 Feb. 4 Apr. 20 Nov.	FBD FBD FBS FBS FBS	5 Jul. & 5 Oct. (597 m) 5 Oct. & 20 Nov. (214 m) 17 Feb. & 4 Apr. (567 m) 4 Apr. & 5 Jul. (320 m)
3	6.5/4°	CH_{FID}: 21/3 m CH_{ALS}: 16/6 m GSV_{FID}: 172/53 m³/ha GSV_{ALS}: 144/102 m³/ha	17 Jul. 1 Sep. 1 Mar. 2 Dec.	FBD FBD FBS FBS	17 Jul. & 1 Sep. (42 m) 1 Sep. & 2 Dec. (887 m)

Table 1. Properties of forest in inventory stands and segments derived from the ALS CHM (CH and GSV estimates from ALS) and the ALOS PALSAR-1 SAR/InSAR imagery available at three 100 km2 test sites.

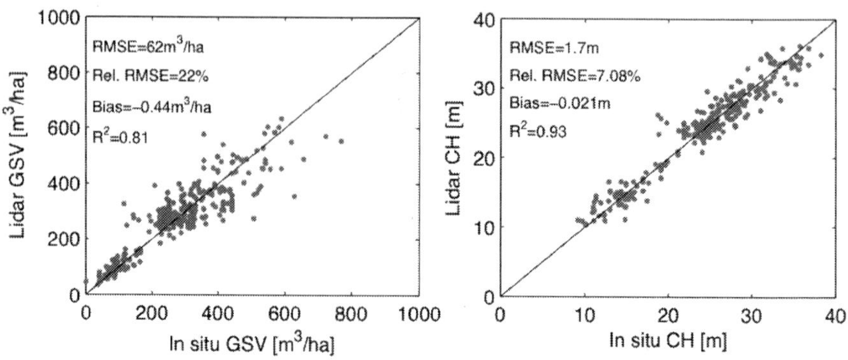

Figure 4. RandomForest out-of-bag GSV and CH estimates based on stand-level ALS canopy structure indices versus *in situ* GSV and CH.

The retrieval accuracy is given with the coefficient of determination (R^2), the root mean square error (RMSE), the relative RMSE (RMSEr) and the bias. The RMSEr represented the RMSE divided by the average GSV and CH in the *in situ* dataset and the bias was calculated from the difference between the average GSV and CH in the *in situ* dataset and the ALS predictions, respectively. In the case of the GSV retrieval, the R^2 was 0.81, the RMSE was 62 m^3/ha and the relative RMSE was 22 % when comparing the randomForest OOB predictions to the inventory data. In the case of CH, the R^2 was 0.93, the RMSE was 1.7 m and the RMSEr was 7.1 %. The bias was negligible in both cases. When using independent sets of training (67 %) and testing samples (33 %), the retrieval accuracies did not differ significantly (in the range of 1 %) from the OOB results.

3.5. CH and GSV retrieval through synergy of ALS, ALOS PALSAR-1 and Landsat ETM+

Figure 5 illustrates the performance of the stand-level CH and GSV at the three 100 km^2 test sites when using all available spaceborne predictor layers (i.e. up to 18 layers including multi-temporal ALOS PALSAR-1 HH/HV intensities, repeat-pass coherences and Landsat ETM+reflectances) and the ALS-based per-segment estimates for CH and GSV as response variables.

When comparing the OOB estimates of CH and GSV with the per-segment estimates from ALS, similar accuracies in terms of the R^2 and RMSEr were obtained at the three test sites. In the case of GSV, the R^2 was ~0.8 and the RMSEr was below 30% for all three test sites. In the case of CH, the R^2 was ~0.82–0.86 and the RMSEr was in the range of 15 to 17%. The RMSEs at test sites 2 and 3 were about 43 m^3/ha (GSV) and 2.5 m (CH), respectively. At test site 1, the RMSEs were higher but since the average GSV and CH values were also higher (Table 1), the RMSEr was comparable to that obtained at the other two sites. The bias was always close to zero. The result of the independent validation using 105 inventory stands was consistent with those obtained when comparing the ALOS PALSAR-1 / Landsat ETM+ OOB estimates for CH and GSV to the respective ALS derived estimates. In the case of GSV, the R^2 was between 0.72 and 0.87 and the RMSEr between 15 and 25%. In the case of CH, the R^2 was between 0.76 and 0.86 and the RMSEr between 8 and 13%. The GSV and CH estimates for the inventory polygons generally presented a somewhat larger bias of up to 20 m^3/ha and 1 m, respectively.

In order to evaluate the benefit of integrating multi-temporal ALOS PALSAR-1 FBD and FBS intensity images, repeat-pass coherence and Landsat ETM+ data, the CH retrieval has been repeated using different combinations of the spaceborne datasets. Eight different combinations of predictors were considered: (1) the best FBS intensity image (1×HH), (2) the best FBD intensity image (1×HH, 1×HV), (3) all FBD intensity images (2×HH, 2×HV), (4) all FBS/FBD intensity images (4–5×HH, 2×HV), (5) all FBD intensity (2×HH, 2×HV) and coherence images (1×HH, 1×HV), (6) all FBS/FBD intensity (4–5×HH, 2×HV) and coherence images (2–4×HH, 1×HV), (7) Landsat only, (8) Landsat and all FBS/FBD intensity (4–5×HH, 2×HV) and coherence images (2–4×HH, 1×HV). The retrieval accuracies that were achieved when using intensities from only one FBS or FBD acquisition were low with less than 50% of CH variance being explained and RMS errors in the range of 4 to 6 m at test sites 2 and 3 and 8 to 10 m at test site 1 (i.e. the test site with the highest average and maximum CH) when comparing the randomForest OOB against the ALS estimates (Figure 6).

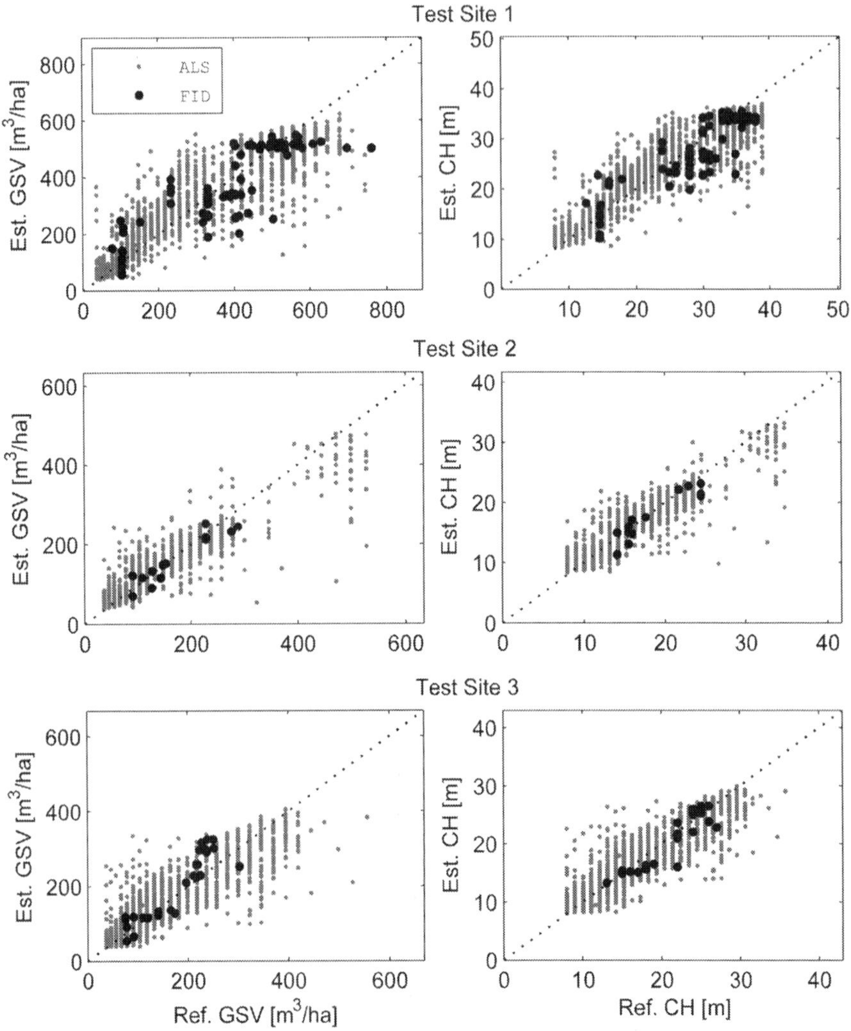

Figure 5. Estimates of GSV and CH obtained through fusion of multi-temporal dual polarization ALOS PALSAR-1 intensities and coherence and Landsat ETM+ data *versus* 1) LIDAR estimates of GSV and CH for segments derived from the CHM (grey dots); 2) *in situ* measurements of CH and GSV (black dots).

As was to be expected, the retrieval with the FBD images performed somewhat better than the retrieval with FBS since the FBD images included the HV intensity. The integration of multi-temporal intensity observations allowed substantial improvements of the retrieval performance. When combining all available FBD intensities, the R^2 and RMSE improved

http://dx.doi.org/10.5772/58220

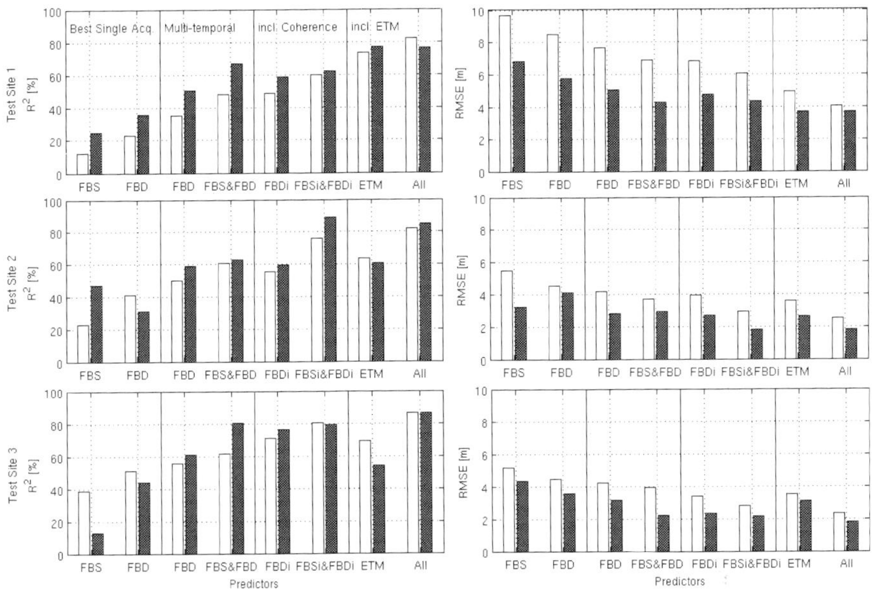

Figure 6. CH retrieval accuracy when using different combinations of the ALOS PALSAR-1 and Landsat ETM+ data as predictors. FBS, FBD and ETM stand for FBS/FBD intensity and the Landsat ETM+ data, respectively. FBDi and FBSi/FBDi denote the cases where intensities and coherences were used jointly. The white bars show the retrieval error when comparing the ALOS PALSAR-1 / Landsat ETM+ OOB against the ALS predictions. The grey bars refer to the comparison of the ALOS PALSAR-1/Landsat ETM+ predictions for the inventory polygons and the corresponding *in situ* measurements.

for about 5 to 12% and 0.25 to 0.8 m, respectively. The R^2 and RMSE improved further when adding the stack of FBS intensity images for 6.5 to 12% (R^2) and 0.3 to 0.7 m (RMSE). The integration of the coherence images resulted as well in higher retrieval accuracies. When using all available FBS/FBD intensities and coherences, the R^2 and RMSE were in the range of 0.75 to 0.8 and 3 m at test sites 2 and 3 and 0.60 and 6 m at test site 1, respectively. Finally, the R^2 and RMSE improved significantly for 6 to 22% and 0.4 to 2 m, respectively, when adding the Landsat ETM+ data to the stack of multi-temporal intensities and coherences. When testing the retrieval with only the Landsat ETM+ data, the retrieval accuracy was roughly comparable to that achieved with the multi-temporal ALOS PALSAR-1 intensities at test sites 2 and 3. At test site 1 (the test site with the highest average and maximum CH), the Landsat ETM+ image even outperformed the ALOS PALSAR-1 imagery with only minor improvements being achieved when combining the ALOS PALSAR-1 and Landsat ETM+ imagery. The improvement of the retrieval accuracy with the successive integration of multi-temporal intensities, coherences and Landsat ETM+ was generally confirmed when comparing the randomForest predictions for the inventory polygons to the corresponding *in situ* measurements.

4. Mapping AGB across the Northeastern US using multi-temporal ALOS PALSAR-1 data

Based on a multi-temporal ALOS PALSAR-1 dataset comprising 655 FBD dual polarization data for the Northeastern US (Figure 7), following two topics have investigated:

- the feasibility of an automated model training and inversion approach, similar to those presented in [7, 21] for ENVISAT ASAR intensity and ERS-1/2 InSAR data in the L-band case;
- the retrieval performance at different spatial scales, considering the influence of the imaging conditions and the benefit of having multi-temporal data stacks.

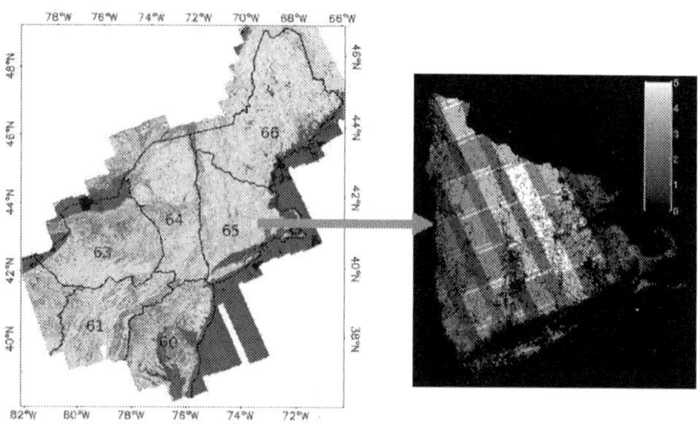

Figure 7. RGB mosaic of ALOS PALSAR-1 FBD data for the Northeastern US. The red channel shows the HH intensity, the green channel the HV intensity and the blue channel the HH/HV ratio. The black lines denote different mapping zones. The image on the right-hand side illustrates for one mapping zone the number of available FBD acquisitions.

4.1. Retrieval algorithm

Complex physically-based approaches for the modeling of backscatter as function of forest biophysical attributes have been developed that consider various aspects of the forest structure (e.g., the size and orientation of stem, branches and leaves) as well as scattering mechanisms. However, when aiming at retrieval the model formulation needs to be simple enough so that it can be inverted. A relatively simple physically-based Water-Cloud type of model [22] that has been tested extensively for retrieval with C- and L-band backscatter data models the backscatter from forest, σ^0_{for}, as a sum of three contributions [23]:

$$\sigma^0_{for} = (1-\eta)\sigma^0_{gr} + \eta\sigma^0_{gr}e^{-\alpha h} + \eta\sigma^0_{veg}\left(1-e^{-\alpha h}\right) \tag{1}$$

The first term describes the direct backscatter from the forest floor, σ^0_{gr}, through gaps in the canopy. The parameter η represents the percentage to which the ground is covered by the canopy. The second term describes the backscatter from the ground that was attenuated in the canopy. Herein, the exponential represents the two-way tree transmissivity, which is a function of the tree height, h, and the two-way signal attenuation, α. The last term describes the volume backscatter, σ^0_{veg}, from an opaque canopy without gaps. The model can be re-written in the following form:

$$\sigma^0_{for} = \sigma^0_{gr} T_{for} + \sigma^0_{veg}\left(1 - T_{for}\right) \tag{2}$$

where T_{for} represents the forest transmissivity:

$$T_{for} = \left(1 - \eta\right) + \eta e^{-\alpha h} \tag{3}$$

The model in Eq. (1) expresses the forest backscatter as a function of height and gap fraction. According to scatterometer experiments at X- and C-bands [24], T_{for} can also be expressed as function of growing stock volume, GSV [m^3/ha]:

$$T_{for} = e^{-\beta GSV} \tag{4}$$

with β being an empirical parameter. Since, in this study, AGB was the observable and GSV is commonly considered a proxy for AGB (i.e. AGB can be estimated from GSV using age-dependent expansion factors, e.g., IPCC, 2006), we replaced the volume with biomass, B [t/ha]:

$$\sigma^0_{for} = \sigma^0_{gr} e^{-\delta B} + \sigma^0_{veg}\left(1 - e^{-\delta B}\right) \tag{5}$$

In Eq. (5), β has been replaced with δ to underline that the forest transmissivity is now expressed as function of biomass.

Two of the three unknowns in the model in Eq. (5) are related to the backscatter from open ground not covered by vegetation (σ^0_{gr}) and to what is considered the backscatter from opaque forest canopies with infinite biomass (σ^0_{veg}). In [7, 21], it was shown for C-band that the backscatter properties of open ground and dense forest canopies, and their temporal and spatial variations, could be identified with the aid of MODIS VCF by masking the intensity images for areas with low and high VCF canopy cover and taking the mean or median of the measured intensities in the masked areas, respectively. In the case of open ground with low canopy cover, ancillary datasets have to be used to exclude land cover classes (settlements, industrial areas, water surfaces, etc.) for which the backscatter can differ substantially from that of open ground. In the case of the intensity observed over dense forests, denoted as σ^0_{df}, an additional compensation for residual backscatter contributions from the ground has to be carried out. The compensation of σ^0_{df} for residual ground contributions can be accomplished with:

$$\sigma^0_{veg}\left(B_{df}\right) = \frac{\sigma^0_{df} - \sigma^0_{gr} e^{-\delta B_{df}}}{1 - e^{-\delta B_{df}}} \quad (6)$$

where B_{df} represents the AGB of dense forest. The estimation of σ^0_{veg} requires knowledge of δ and B_{df}.

Once the parameters have been estimated, the model can be inverted to estimate the biomass from the SAR data. When inverting the model, some intensity measurements might in fact exceed the range of modeled intensities between σ^0_{gr} and σ^0_{veg}. Inversion for intensities lower than σ^0_{gr} would result in negative biomass estimates, which is why a biomass of 0 t/ha has to be assigned. In the case of high intensity values exceeding σ^0_{veg}, an inversion is not possible and for intensities slightly lower than σ^0_{veg}, the inversion could result in biomass estimates far exceeding the range of realistic biomass values, which is why a maximum biomass level, B_{max}, has to be defined up to which inversion is carried out. In case of multi-temporal datasets, a weighted combination of the biomass estimates from each image covering a particular pixel location, B_i, can be computed to obtain new multi-temporal estimates, B_{mt}. Weights can be calculated using the difference between σ^0_{veg} and σ^0_{gr}, referred to hereafter as the dynamic range. The larger the dynamic range, the more weight is given to the particular biomass estimate.

4.2. AGB retrieval

Model training and inversion were carried out for each of the 1310 intensity images (655 HH and 655 HV PALSAR FBD scenes, off-nadir angle of 34.3°) available for the Northeastern United States. The Landsat-based canopy density maps and land cover maps from the National Land Cover Database [25] were used to identify sparse and dense forests in each backscatter image and to mask settlements, industrial areas, agricultural land and water surfaces, respectively. The forest transmissivity parameter δ, which describes the backscatter trend with increasing biomass, could be expected to depend on 1) the imaging conditions, and 2) the forest structural characteristics [26]. Model simulations indicated, however, that in the case of the ALOS PALSAR-1 data available for the Northeastern United States, the potential variations in the forest transmissivity parameter as function of the forest type and imaging conditions could be expected to be minor and that the use of a fixed value for δ (0.008 ha/t) for the biomass retrieval represented a justifiable compromise. For a detailed discussion of the forest transmissivity as function of AGB refer to [15]. The AGB of dense forests, B_{df} (Eq. 6), was determined via the plot data from the Forest Inventory and Analysis (FIA) plot network of the US Forest Service [27] with the 90th percentile of the regional plot biomass distribution [7]. A fixed biomass offset, ΔB, was then used with respect to B_{df} to define the maximum retrievable biomass, B_{max} ($B_{max}=B_{df}+\Delta B$), when inverting the models to estimate the biomass for all pixels in the intensity images for which the NLCD land cover map reported forest. B_{max} affects the retrieval results in the higher biomass ranges and therefore (primarily) derived population statistics such as county totals. That is why ΔB was adjusted so that the differences between the total biomass estimates per county and FIA county-level estimates were minimized; this was the case when setting ΔB to 30 t/ha (see below). The multi-temporal combination and mosaicing of the single image biomass maps were carried out in a single step for each of six mapping zones in the Northeastern United States. Both, HH- as well as HV-intensity based biomass estimates were considered for the multi-temporal combination. The resulting AGB map is shown in Figure 8.

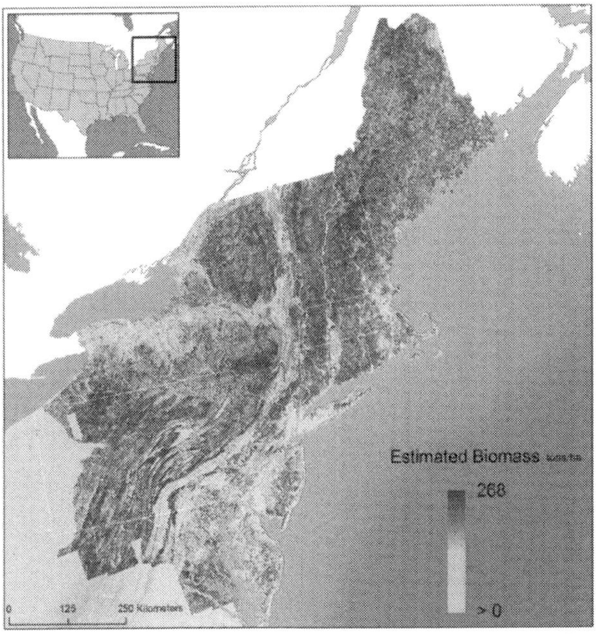

Figure 8. Forest AGB map for the Northeastern United States produced from 655 ALOS PALSAR-1 dual polarization intensity images acquired in 2007/08.

Since the ALOS PALSAR-1 data could not be compared directly to the plot data of the FIA database (note that the exact FIA plot locations are not publicly available), we compared the ALOS PALSAR-1 biomass maps to the biomass maps from the National Biomass and Carbon Dataset 2000 (NBCD) that were produced through the fusion of SRTM and NED elevation, Landsat TM and the FIA forest inventory data [17, 28]. The algorithm performance was assessed with the root mean square difference (RMSD) between the ALOS PALSAR-1 and NBCD maps calculated separately for different mapping zones (see Figure 7), which were adapted from the National Land Cover Database (NLCD) [25].

At full resolution, the RMSDs were large (>80 t/ha). This might on one side have been a consequence of noise in the ALOS PALSAR-1 data (residual speckle, small scale environmental effects) but it also has to be considered that at the level of 30 m pixels, the NBCD map contains considerable uncertainty as well; note that the NBCD map was validated at segment (i.e. hectare) level. The comparison of the maps at 30 m pixel size was therefore of limited explanatory value. Hence, the comparison at different pixel aggregation scales between 150 m and 6 km have been repeated. When aggregating the maps by means of simple block-averaging, the agreement increased substantially (Figure 9, right). The largest improvement could be observed when aggregating up to ~1 km pixel size, for which the RMSD was in a range of 20 to 25 t/ha. In Figure 9 (left), the ALOS PALSAR-1 biomass estimates for two mapping zones have been plotted against the NBCD biomass estimates for pixel sizes of 150 and 600 m. The comparison of the ALOS PALSAR-1 and NBCD biomass maps revealed a good agreement along the 1:1 line. The spread along the 1:1 line reduced substantially with increasing pixel size, which can be seen in Figure 9 (left) from the decreasing length of the vertical bars. The ALOS PALSAR-1 biomass estimates, however, tended to be lower than the NBCD estimates when approaching a biomass of 200 t/ha, indicating saturation effects in the L-band data.

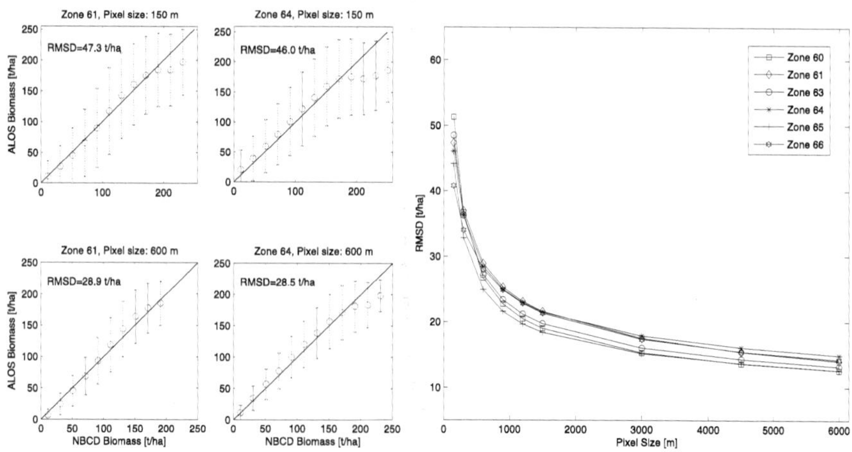

Figure 9. Left: Average (circles) and standard deviation (vertical bars) of the multi-temporal ALOS PALSAR-1 biomass estimates for intervals of NBCD biomass at zones 61 and 64. Right: Root Mean Square Difference (RMSD) between the ALOS PALSAR-1 and NBCD biomass maps as a function of the pixel aggregation scale.

4.3. Importance of multi-temporal acquisitions

To evaluate the importance of having multi-temporal stacks of L-band intensity, the retrieval performance for single images is discussed first. Figure 10 shows for the L-band intensity images covering one of the mapping zones the RMSD between the biomass estimates from single intensity images and NBCD as function of the dynamic range (at 150 m pixel size).

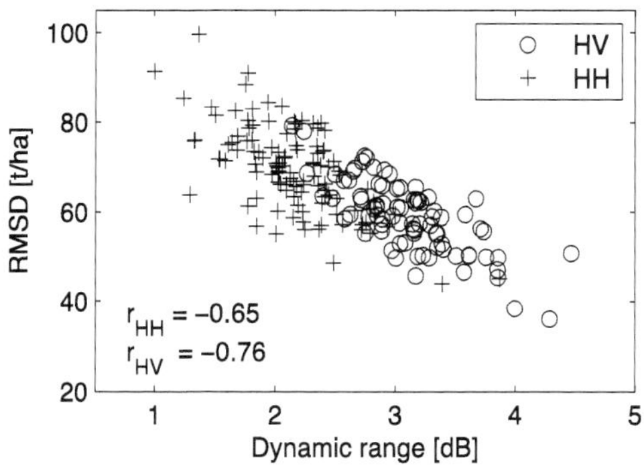

Figure 10. RMSD between the biomass estimates from single HH (+) and HV (o) intensity images and NBCD as function of the dynamic range (zone 64, 150 m pixel size).

The figure clearly shows that the dynamic range can be considered one of the main factors influencing the retrieval performance. The figure also shows that the RMSDs for HV intensity images tended to be lower (40-80 t/ha) than for HH intensity images (50-100 t/ha). For a given FBD HH and HV image pair, the RMSD was always lower for the HV image (5 to 10 t/ha). The differences in the dynamic ranges were most likely a consequence of differing imaging conditions. For the images covering New York State, we compared the dynamic ranges with the weather conditions at the time of the sensor overpasses. The comparison with the weather data revealed no correlation with temperature; note that the temperature was consistently above the freezing point for all images so that no major temperature related fluctuations of the dielectric properties of the trees were to be expected. Weak negative correlations were observed when relating the dynamic range to the total amount of rain in the days prior to the sensor overpasses. In both polarizations, there was a trend towards lower dynamic ranges with increasing amounts of precipitation (i.e. with increasing wetness of the soils and vegetation). The Pearson correlation coefficients were between -0.3 and -0.5 depending on which timeframe prior to the sensor overpasses was considered. This result is consistent with previous finding [29].

Figure 11 demonstrates the benefit of having multi-temporal data for an area where five FBD images were available. The dashed line shows the RMSD for each intensity image (HV: 55-59 t/ha, HH: 58-71 t/ha, 150 m pixel size), the solid line shows the change in RMSD when successively integrating the particular images into the multi-temporal retrieval.

The multi-temporal combination resulted in a clear improvement of the RMSD when combining the available 5 HV images for about 10 t/ha (compared to the best HV image). When, in addition, integrating the corresponding HH images, only slight additional improvements of the RMSD for about 3 t/ha were achieved. These results confirmed that the multi-temporal combination allowed for significant improvements of the biomass estimates. However, it has to be noted that the multi-temporal coverage acquired by ALOS PALSAR-1 was not consistent

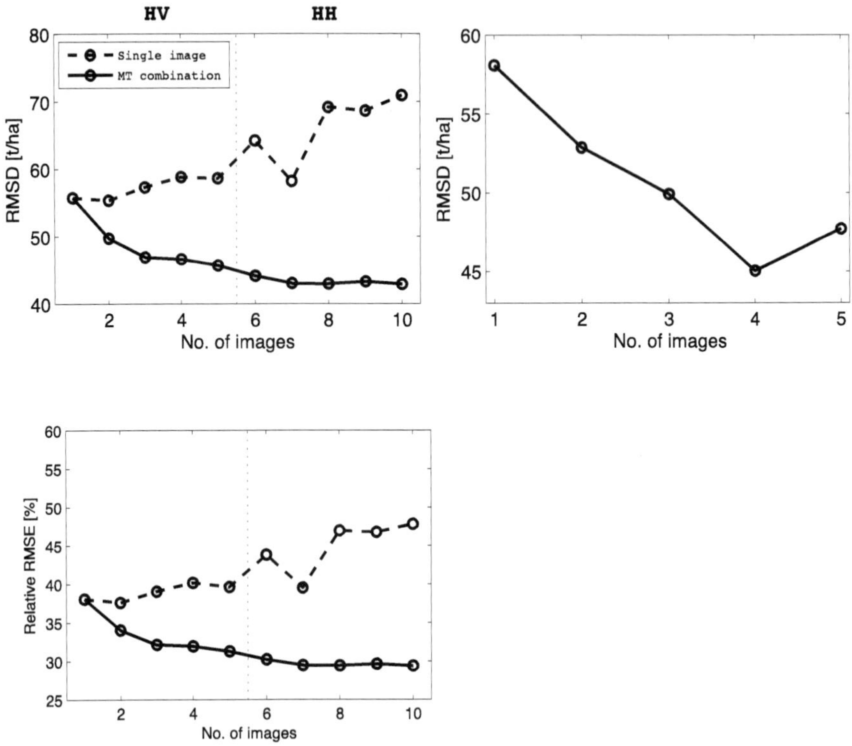

Figure 11. Effect of the multi-temporal combination on the agreement of the ALOS and NBCD biomass maps (at 150 m pixel size). The circles connected by the dashed line denote the RMSD for each image and those connected by the solid line show how the RMSD changed when successively integrating the single image estimates into the multi-temporal retrieval.

across larger areas as regionally between one and five images were acquired per year. As a result of the varying multi-temporal coverage (see Figure 7), the accuracy of the map in Figure 8 was not consistent across the entire study area. The comparison with NBCD confirmed that locally, the performance of the retrieval depended strongly on the multi-temporal coverage (i.e. the number of images) as well as the weather conditions under which the particular set of images have been acquired. The results therefore stress the need for consistent multi-temporal acquisition strategies for upcoming spaceborne L-band missions.

4.4. Accuracy at county scale

The comparison of the ALOS PALSAR-1 biomass maps with NBCD indicated that, at least at aggregated scales, the spatial distribution of biomass could be depicted with the retrieval approach presented. To further assess the performance of the retrieval algorithm, ALOS

PALSAR-1 biomass maps have been compared to FIA county-level total AGB statistics for 143 counties in Maine, New Hampshire, Vermont, Massachusetts, Rhode Island, Connecticut, New York and New Jersey. The county statistics were obtained via the EVALIDator online inventory tool of the US Forest Service. For the comparison of the ALOS PALSAR-1 and FIA estimates, the ALOS PALSAR-1 per-pixel (30x30 m^2) biomass estimates were summed per county. The comparison of the ALOS PALSAR-1 and FIA county-level estimates of total AGB resulted in a coefficient of determination (R^2) of 0.98 and a root mean square error (RMSE) of 2.75 10^6 t. When calculating the average biomass per county (i.e. the total biomass divided by the county size in hectares), the RMSE was 12.9 t/ha and the R^2 was 0.86 (Figure 12).

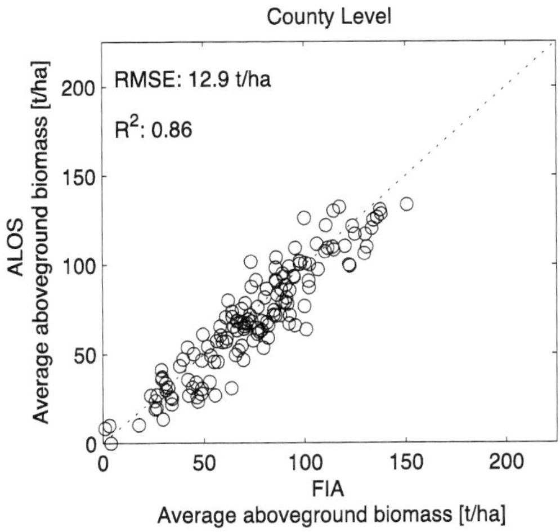

Figure 12. Average AGB according to the ALOS PALSAR-1 biomass map and FIA biomass statistics for 143 counties in the Northeastern United States.

The FIA county statistics included information about the sampling error, which could be used to approximate the confidence intervals of the FIA estimates [27]. The sampling error was between 2 % and 110 % for the largest counties (with the largest number of sample plots) and for the smallest counties, respectively; the size of the counties ranged from 64 to 17, 686 km^2. The ALOS PALSAR-1 total AGB estimates were within the 95 % confidence intervals of the FIA estimates for most (92 %) of the 143 counties.

5. Synergetic use of multi-temporal ALOS PALSAR-1 and ENVISAT ASAR data for forest and agricultural mapping at national scale in Africa

The forest cover in Malawi is approximately 32, 000 sqkm corresponding to 34% of country surface [30]. Natural forests represent the remainder of the Miombo (Swahili word for

Brachystegia) forests that once covered almost the whole country. The area of natural forests (primary forest 29%, other naturally regenerated forest 60%) over the years has remained unchanged – annual change rate between 2005 and 2010 is less than 1% – with the exception of forest reserves that have continued to grow in number. Characteristically, the trees shed their leaves for a short period in the dry season (June to October) to reduce water loss, and produce a flush of new leaves just before the onset of the rainy season (November-December to March-April). Planted forests consist of softwood (mainly *Pinus patula*) and hardwood species (mainly *Eucalyptus*).

Thanks to the availability of a multi-year ALOS PALSAR-1 FBD data set acquired in a systematic way between 2006 and 2011 [9] and a seasonal ENVISAT ASAR Alternating Polarization (AP) data stack planned and regularly acquired during the wet season, the possibility is explored to generate countrywide a forest map and a cultivated area map during the crop wet season. In synthesis, the method is based on the synergetic use of multi-temporal data by considering the different data characteristics and given acquisition modes, the acquisition periods, and the vegetation phenology during the acquisitions. Moreover, since the ultimate goal is to provide this type of information over large areas, i.e. at least national scale, and to upscale the proposed method to other regions [31], the processing chain has been developed in such a way that the products are generated automatically. Finally, it is worth mentioning (cf. *Fusion of interferometric SAR and photogrammetric Elevation Data*) that using the same ALOS PALSAR-1 repeat-pass InSAR data, a DEM with higher quality than the SRTM one can be provided in those nearly equatorial regions characterizes by non-dense forest condition.

5.1. Method

The data processing flow can be divided into two distinct steps. The first one converts the multi-temporal SLC data into terrain geocoded backscattering coefficient ($\sigma°$). In addition, for the 46-days ALOS-PALSAR-1 and 1-day Cosmo-SkyMed Stripmap interferometric data, coherence (γ) is computed and terrain geocoded. Noteworthy, but not discussed here, are the ionospheric effects in the equatorial region observed in the L-band data [35]. In this case, less than 10% of the ALOS PALSAR-1 images have been omitted from the processing due to severe (several dB) and non-systematic striping along the azimuth. However, in other cases [31] around one fourth of the ALOS PALSAR-1 data could not be used. In the second step, the forest and the cultivated area map are generated.

1. Sigma nought ($\sigma°$) is derived as follows:

   - Co-registration – Images acquired with the same observation geometry and mode are co-registered in slant range geometry. This step is mandatory to allow time-series speckle filtering.

   - Time-series speckle filtering – Within the multi-temporal filtering an optimum weighting filter is introduced to balance differences in reflectivity between images at different times [32]: this allows enhancing significantly the radiometric resolution and preserve the spatial one. Multi-temporal filtering is based on the assumption that the same

http://dx.doi.org/10.5772/58220

resolution element on the ground is illuminated by the radar beam in the same way, and corresponds to the same slant range coordinates in all images of the time series. The reflectivity can change from one time to the next due to a change in the dielectric and geometrical properties of the elementary scatters, but should not change due to a different position of the resolution element with respect to the radar.

- Terrain geocoding, radiometric calibration – A backward solution by considering a DEM is used to convert the positions of the backscatter elements into slant range image coordinates. The three dimensional object coordinates given in a cartographic reference system into the two-dimensional row and column coordinates of the slant range image are transformed using a range-Doppler approach. Radiometric calibration is performed at the same time by means of the radar equation, where scattering area [33], antenna gain patterns and range spread loss are considered.

- Radiometric normalization – In order to compensate for the range dependency and topographic effects, the backscattering coefficient is normalized according to a modified cosine law by considering the relationship between the local incidence angle and the backscattering coefficient of forest [14] and agriculture, respectively.

- Anisotropic Non-Linear Diffusion Filtering – This filter significantly smoothes homogeneous targets, whilst also enhancing the difference between neighbouring areas. The filter uses the diffusion equation, where the diffusion coefficient, instead of being a constant scalar, is a function of image position and assumes a tensor value [34]. In this way, it is locally adapted to be anisotropic close to linear structures such as edges or lines.

- Removal of atmospheric attenuation – Although microwave signals have the ability to penetrate through the clouds, in case of severe storms, as it often occurs in Africa during the raining season period, it may occur at C-band that locally the backscattering coefficient is attenuated by water vapor in the range of several dB. The temporal signature of the backscatter coefficient can be affected in two ways: i) the thick layer of water vapor generates a strong decrease of backscattering coefficient, followed by a strong increase; ii) the strong rainfall generates a strong increase of the backscattering coefficient, followed by a strong decrease. These effects are corrected in the processing chain by analyzing the temporal signature: anomalous peaks are identified and the backscattering coefficient values are corrected by means of an interpolator. The correct application of this process relies strongly on a priori knowledge of the land cover type and the weather conditions when the image was acquired.

2. Forest area – Given the regular country-wide ALOS PALSAR-1 FBD acquisitions between 2006 and 2011, the high acquisition rate between June and October (dry season), and that forest extent in Malawi is relatively constant (annual change rate between 2005 and 2010 is less than 1%), a pseudo-seasonal time-series with a relatively high temporal occurrence by combining all years of ALOS PALSAR-1 FBD observations into one year has been developed. The seasonal HH-HV forest signatures – except some rare outliers representing clear cuts – are typically characterized by a constant high σ°, while other cover types show noticeable lower (bare soil, water, dry or low vegetation) or higher (settlements) values. It has to be pointed out that this net discrimination between the different land covers is given by the combination of two factors. First, the selected dry season period, which, due to the very limited vegetation development, considerably reduces the confusion between low forest biomass and others vegetation types, agriculture (i.e. maize) in particular. Second, the long wavelength, on one hand tends to smooth out the radar backscatter from land covers characterized by limited surface roughness, on the other hand it guarantees a relatively high backscatter of those land covers, like forest, where double bounce and volume scattering prevail. Note that this separability is not obtainable at shorter wavelengths, due to similarities in the volume scattering contributions.

3. Cultivated area – Pre-requisite for the generation of this product is to obtain a seasonal data set as far as possible regularly acquired along the whole crop season and according to the crop practices. This allows i) to reduce the confusion between cropped areas and the surrounding vegetated (non crop) areas; and ii) to monitor the crop development [36, 37]. The specific sensitivity of active microwave short wavelength sensors (C- and X-band) to soil properties, such as roughness and moisture content, enables the possibility to detect these changes already at the earliest stage during the field preparation, i.e. ploughing (high backscatter) and harrowing (low backscatter). During the second phase (namely from flowering to plant drying stage), the dielectric and structure properties of the plant are the key factors determining the high reflectivity at C-band HH polarization. Finally, the lower reflectivity during the plant drying is caused by the loss of plant moisture. As proposed in [36, 37], an efficient way to quantitatively describe the σ^o temporal crop signature is to derive appropriate temporal features, i.e. the relative minimum and maximum including corresponding dates; their difference; the minimum and maximum ratio between two subsequent acquisitions. These features are used to generate the product according to:

   a. The crop start of season, which is identified when there is a relative minimum followed by a maximum increment between two subsequent acquisitions;

   b. The crop peak of season, which is identified when there is a relative maximum followed by a minimum increment between two subsequent acquisitions;

   c. The pixel, which is classified as crop if:

      - conditions 1 and 2 are satisfied;
      - the range between relative minimum and maximum attains a minimum value;
      - the temporal duration between 1 and 2 is within a given duration.

5.2. Forest and cultivated area products

To cover the entire country, 65 ALOS PALSAR-1 FBD/FBS frames distributed over 5 adjacent tracks are necessary. Three to four coverages per dry season per year during four years have been used, resulting in the around 900 SLC scenes. A coherence-intensity mosaic using 46-days interferometric ALOS PALSAR-1 FBS data acquired during the wet season (January-February

2008) has been additionally generated. A total of 225 ENVISAT ASAR AP images acquired five times from October 2007 to April 2008 are used to monitor the crop growth along the whole wet season. The step to transform the SLC data into geo-referenced $\sigma°$ and γ products is doubtless the most time consuming one. For this reason, the data processing is performed by means of a PC based cluster solution. Noteworthy, the algorithms have been written in a such way to fully exploit the characteristics of the processors. This setting allows to carry out the processing without any intervention of the operator and to process the data sets within few days.

Figure 13 and 14 illustrate four multi-temporal mosaics: the multi-year ALOS PALSAR-1 FBD mosaic acquired during the dry season (Figure 13, left); the seasonal (October to April) ENVISAT ASAR HH mosaic (Figure 13 right); an ENVISAT ASAR HH mosaic (October and January, the months showing the most significant radiometric changes) combined with the ALOS PALSAR-1 HV July one (Figure 14 left); the ALOS PALSAR-1 HH coherence-intensity mosaic (Figure 14 centre).

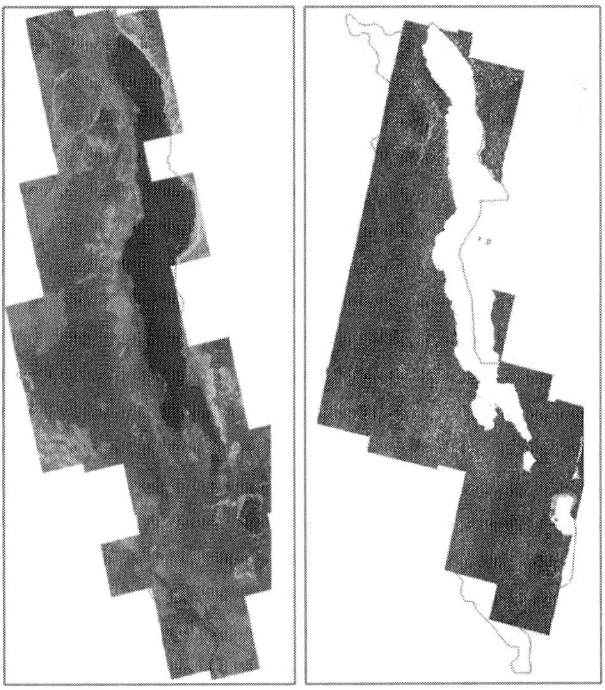

Figure 13. (left) Multi-year ALOS PALSAR-1 FBD mosaic, 15m acquired during the dry season (mean HH=red, mean HV=green, mean HH / mean HV=blue); (right) Seasonal ENVISAT ASAR HH mosaic, 15m (ASAR HH October=red, ASAR HH December=green, ASAR HH January=blue).

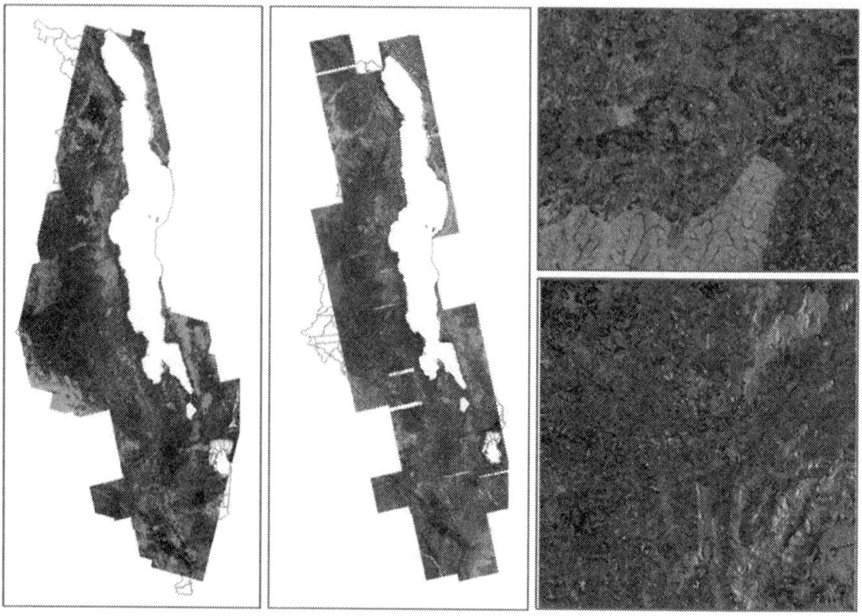

Figure 14. (left) ENVISAT ASAR HH-ALOS PALSAR-1 HV mosaic, 15m (ASAR HH December=red, PALSAR-1 HV July=green, ASAR HH January=blue); (centre) ALOS PALSAR-1 HH (FBS) coherence-intensity mosaic, 10m (coherence January-February=red, mean intensity=green, intensity difference=blue); (top right) detail of figure left; (bottom right) detail of figure centre.

These examples clearly show that depending upon the selected sensor, the acquisition mode and time, and a smart data integration, different types of information (i.e. products) can be derived. It is worth mentioning that the purpose of data synergy is not exclusively to obtain higher accuracies or more information, but it is also intended to simplify and automatize the products generation. *Conditio sine qua non* is, on one hand, to recognize the sensor capabilities and limitations – including the processing techniques – on the other hand, to understand the object, its characteristics, and the environmental surrounding conditions.

The multi-year ALOS PALSAR-1 FBD mosaic (Figure 13 left) acquired during the dry season undoubtedly shows a clear distinction between forest (green colour) and other cover types (blue tonalities). Note that this net separation is the outcome of i) the excellent quality of the multi-temporal speckle filtering, and ii) the averaging (omitting some rare outliers) of the HH intensities, and the HV respectively over the dry season period and years (total of 14 images). Both operations contribute to considerably improve the signal-to-noise ratio and to enhance the level of detail. Concerning the temporal averaging, it has to be pointed out that this procedure is more than reasonable, because the SAR data have been exclusively selected in the dry period, where the vegetation phenology is stable, in terms of roughness and dielectric properties, and, the forest extent variations are negligible. In order to demonstrate the usefulness of this approach, for a selected area, a 1-day interferometric Cosmo-SkyMed Stripmap image pair (3m resolution) acquired in September has been acquired and processed. Figure 15 shows, for comparison purposes, the single-date and multi-year ALOS PALSAR-1 products, and the 1-day interferometric Cosmo-Skymed Stripmap pair (3m resolution). The high level of detail of the multi-year ALOS PALSAR-1 product (bottom left) is particularly appreciable by comparing the speckle effect in the forest in the two ALOS-PALSAR-1 products, and the better feature delineations (for instance in the riparian forest) in the multi-year one. Furthermore, by visually inspecting the single trees in the multi-year ALOS PALSAR-1 product and in the Cosmo-SkyMed one, it is noticeable that almost all single trees identifiable in the 3m image are easily recognisable in the 15m multi-year scene, hardly in 15m single-date one. With respect to the use of X-band data for forest applications, it should be shortly remarked that, in general, the 1-day Cosmo-SkyMed coherence or the bistatic TerraSAR-X-Tandem one are essential, because the limited σ^o dynamic range at this frequency significantly reduces the discrimination capabilities in vegetated areas, if exclusively intensity is used. For details refer to Holecz et al. [38].

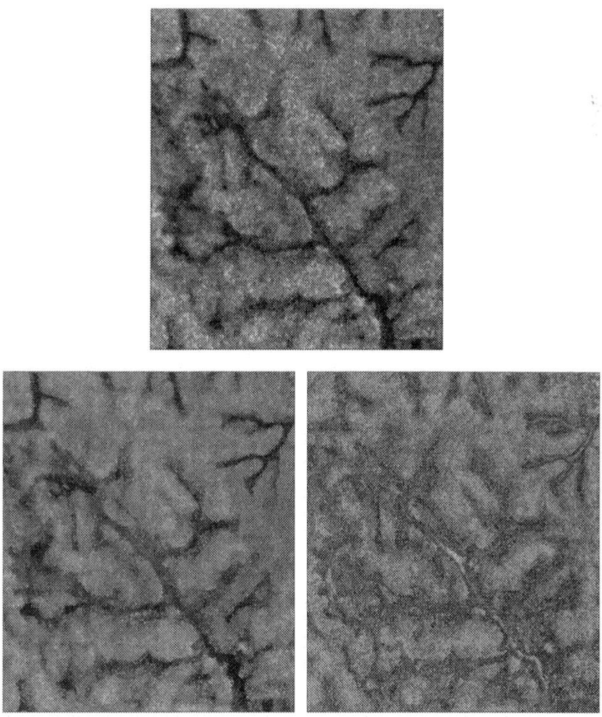

Figure 15. (top left) Single-date ALOS PALSAR-1 FBD mosaic, 15m (mean HH=red, mean HV=green, mean HH / mean HV=blue); (bottom left) Multi-year ALOS PALSAR-1 FBD mosaic, 15m; (bottom right) Cosmo-SkyMed Stripmap coherence-intensity, 3m (1 day coherence=red, mean intensity=green, intensity difference=blue).

As expected, due to the shorter wavelength, forest is quite less distinguishable in the seasonal ENVISAT ASAR mosaic (Figure 13 right) confirming the conclusions presented in the comparative study carried out by Mitchell et al. [39]. The predominant colours in this image are the cyan (southward) and blue-violet (northward), corresponding to the crop growth after the start of the rain (around November). This is reflected in the December (green) and January (blue) acquisition, where the C-band backscattering coefficient almost attains its highest values due to the rapid crop growth (i.e. increase in the surface and volume scattering) and the high crop moisture content (increase in the dielectric constant). It is interesting to note the correspondence between the blue areas in the ALOS PALSAR-1 FBD mosaic (bare soil during the dry season) becoming brightly coloured in the ENVISAT ASAR mosaic (cultivated area during the wet season). This is becoming obvious in Figure 14 left (and detail, top right), where the ALOS PALSAR-1 HV July mosaic (green) has been merged with the ENVISAT ASAR HH mosaic of December (red) and January (blue). This colour composite demonstrates, in a qualitative but evident way, that data synergy is undoubtedly beneficial, if exploited in a wise manner.

Coherence is unquestionably a complementary and valuable source of information. However, in practice, useful interferometric correlation is often not easy to obtain due to unfavourable baseline conditions and unsystematic atmospheric effects. Moreover, the temporal decorrelation sometimes causes interpretation uncertainties. Figure 14 centre (and detail, bottom right) shows a 46-days repeat-pass ALOS PALSAR-1 HH coherence-intensity mosaic generated from an FBS image pair acquired in January and February 2008. As expected, forest has a low correlation and an average high intensity (green), meaning that the forest didn't vary in this gap of time; blue (large intensity difference), particularly visible in the detail bottom right, represents the growing fields; the red tonalities correspond to those areas where the cover changes were minimal, i.e. meaning primarily rough bare soil areas. Nonetheless, it should be considered that tiny crops at L-band are almost transparent, hence resulting into a relatively medium coherence, as the rough bare soil. It turns out that the cultivated area is underestimated in favour of bare soil. In synthesis, a forest map could be generated even if, due to the long repeat-pass and the different baselines, the uncertainties can be relevant; these ambiguities are significantly higher for the cultivated area also considering that for this product long acquisition time intervals are unsuitable [37]. A final example on coherence is illustrated in Figure 16, which it has been obtained from a July-August FBD image pair.

Observing the coherence, the Miombo forest is not distinguishable in the HH polarization (average γ is 0.6), hardly detectable in the HV (average γ is 0.5). In the multi-year ALOS PALSAR-1 FBD intensity colour composite, as extensively presented and discussed above, forest is clearly separable from the surrounding land covers. It is anticipated that the perpendicular baseline of the interferometric pair is 280m, which for this frequency, is thereby appropriate for thematic analysis. The Miombo forest – bare in the dry season – on average has a tree height significantly less than 10m and a diameter breast height ranging from 10 to 20 cm corresponding to a relatively low biomass (often considerably less than 100 tons/ha). This means that the main scattering contribution is the volume, primarily induced by the tree branches; hence the HH radar backscatter is markedly attenuated. This is reflected in both, HH

Figure 16. (left) ALOS PALSAR-1 46-days HH July-August coherence; (centre) HV July-August coherence; (right) multi-year ALOS PALSAR-1 FBD image, 15m resolution (mean HH=red, mean HV=green, mean HH / mean HV=blue).

coherence (Figure 16 left) and HH intensity (Figure 16 right). At HV intensity, forest is well recognizable in green (Figure 16 right). This distinction is mainly given by the temporal averaging over the dry seasons and four years, which strongly enhances the separability of the various land types. On the opposite, for the HV coherence (Figure 16 centre), the discrimination between forest and the surrounding area (both having a relatively high coherence) is very limited: this is due, on one hand, to 46-days temporal decorrelation, on the other hand, to the impossibility to perform a temporal averaging. In summary, coherence is doubtless a valuable source of information, however, it should be used with care.

Based on the above considerations and evaluations, the most suitable solution is to use the multi-year ALOS PALSAR FBD data set for the forest area product, and the seasonal ENVISAT ASAR one for the cultivated area. Furthermore, in order to understand the contribution of the seasonal ENVISAT ASAR for the forest area (Figure 17 left), the obtained cultivated area (Figure 17 right) is merged in IF condition with the forest area one.

The two maps are performed using a hierarchical prior knowledge-based classifier. For the forest area the input data were the mean HH intensity and the corresponding mean HV intensity of the multi-year ALOS PALSAR FBD data set. Concerning the cultivated area, the temporal features of the seasonal ENVISAT ASAR data set have been used.

5.3. Accuracy at national scale

Validation involves the collection of ground reference data for the validation of remote sensing based products. Usually it is carried out by sampling units, i.e. points unambiguously identified by co-ordinates. As shown in Figure 18, systematic grids are used with randomly selected starting corner co-ordinates in order to ensure a representative and spatially well-distributed sample.

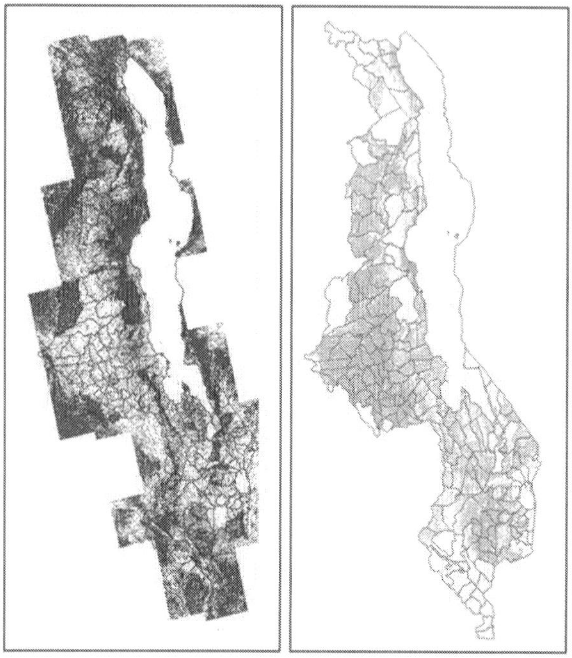

Figure 17. (left) Forest area generated from multi-year ALOS PALSAR-1 FBD data; (right) Cultivated area generated from seasonal ENVISAT ASAR data.

From an operational perspective, the grouping of sample points in clusters is encouraged. Although this approach may introduce some degree of statistical bias, by significantly reducing travelling time between sample locations, it leaves more resources available for data collection in the field. The relevant parameters of this systematic cluster approach are: 1) distance between clusters; 2) number of points per cluster; 3) distance of points within cluster. The values can be fixed considering several criteria and constraints: 1) available budget; 2) size and shape of the area of interest; 3) resolution of the remote sensing images; 4) logistics; 5) average dimension of the area to be classified. In this case, it has been opted for 1km distance between the clusters, 16 points per cluster, and 250m distance between the points within the cluster. A total of 868 valid points have been collected.

Table 2 shows the obtained confusion matrices for the forest product exclusively based on the multi-year ALOS PALSAR-1 FBD (top) and for the forest product generated by combining the forest and cultivated area product (bottom). In general, the obtained accuracies are high. The main difference between the two tables is in the reduction of the omission errors of the classes sugar cane and crop, and urban. Sugar cane is a tall crop with a long crop season (typically one year). This fully explains the large omission error. By merging the cultivated area map, this error could be reduced by 30% only, because the cultivated area product, as defined here, exclusively considers the crop planted at the start of the rainy season. The same explanation is valid for the class crop. In essence, in order to almost completely remove this omission error, an annual (and not just seasonal) ENVISAT ASAR monitoring should be carried out. Concerning the class urban (mainly small cabins in the rural areas), the omission error has been reduced by 40%, leading to a 10% error. In this specific case, the radar backscatter is often random, therefore the combination of the two frequencies strongly contributes to better detect this land cover type.

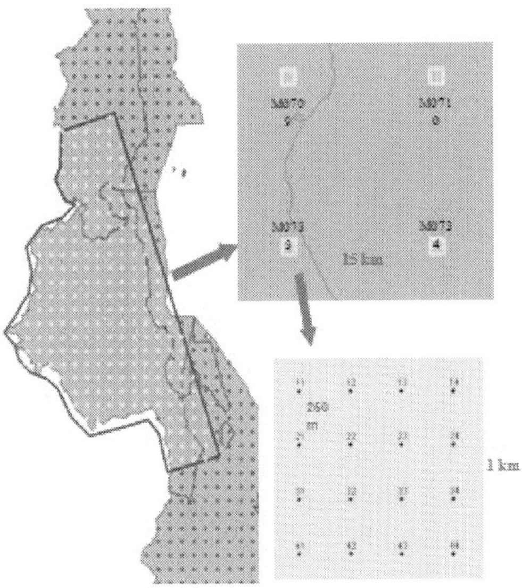

Figure 18. Validation scheme.

With respect to the cultivated area, the obtained overall accuracy is 80%. For details refer to Holecz et al. [36, 40]. In summary, in that work, it was recognized that the limiting factor for cultivated area estimation in small plot agriculture in Africa is the spatial resolution. A possible way to overcome this limitation is on the synergetic use of sensors with different spatial resolutions and characteristics, therefore by optimizing the spatial and temporal resolution in a way that both dynamics are taken into account (Cf. *Estimation of cultivated areas using multi-temporal SAR data*).

	forest	sparse veg	other	Total	Omission error (%)
Urban	10	0	10	20	50
Sugarcane	19	3	7	29	76
Crop	42	0	347	389	11
Forest	365	0	37	402	9
Other	1	0	27	28	4
Total	437	3	428	868	**K-coeff 0.75**
Commission error (%)	16	0	9	**overall accuracy 87%**	

	forest	sparse veg	other	Total	Omission error (%)
Urban	2	0	18	20	10
Sugarcane	10	3	16	29	45
Crop	12	0	377	389	3
Forest	357	0	45	402	11
Other	1	0	27	28	4
Total	382	3	483	868	**K-coeff 0.84**
Commission error (%)	7	0	9	**overall accuracy 92%**	

Table 2. Confusion matrices forest area product – (top) Multi-year ALOS PALSAR-1 FBD; (bottom) Multi-year ALOS PALSAR-1 FBD and seasonal ENVISAT ASAR.

Acknowledgements

We are grateful to the Japan Aerospace Exploration Agency, the European Space Agency, NASA, and the Italian Space Agency (ASI) for the provision of the ALOS PALSAR-1, ENVISAT ASAR, and Cosmo-SkyMed data. Parts of the work included in this article have been carried out within the JAXA Kyoto and Carbon initiative and the ESA Global Monitoring for Forest Security project. Support is acknowledged from the Gordon and Betty Moore Foundation, The David and Lucile Packard Foundation, and the Google.org Foundation.

Author details

Josef Kellndorfer[1*], Oliver Cartus[1], Jesse Bishop[1], Wayne Walker[1] and Francesco Holecz[2]

*Address all correspondence to: josefk@whrc.org

1 Woods Hole Research Center, MA, USA

2 Sarmap, Purasca, Switzerland

References

[1] Dobson M.C., Ulaby F.T., Le Toan T., Beaudoin A., Kasischke E., Dependence of radar backscatter on coniferous forest biomass, IEEE Transactions on Geoscience and Remote Sensing, Vol. 30(2), 1992.

[2] Imhoff M.L., A theoretical analysis of the effect of forest structure on SAR backscatter and the remote sensing of biomass, IEEE Transactions on Geoscience and Remote Sensing, Vol. 33(2), 1995.

[3] Ferrazzoli P., Guerriero L., Solomini D., Comparison between predicted performances of bistatic and monostatic radar in vegetation monitoring, Proceedings IGARSS Symposium, 1994.

[4] Smith G., Ulander L., Forest biomass retrieval using VHF SAR, ESA/ESTEC, 1998.

[5] Santos J., Neeff T., Dutra L., Araujo L., Gama F., Elmiro M., Tropical forest biomass from dual frequency SAR interferometry (X-and P-band), Proceedings of ISPRS Conference, 2004.

[6] Papathanassiou K.P., Cloude S.R., Single-baseline polarimetric SAR interferometry, IEEE Transactions on Geoscience and Remote Sensing, Vol. 39(11), 2001.

[7] Santoro M., Beer C., Cartus O., Schmullius C., Shvidenko A., McCallum I., Wegmüller U., Wiesmann A., Retrieval of growing stock volume in boreal forest using hyper-temporal series of ENVISAT ASAR ScanSAR backscatter measurements, Remote Sensing of Environment 115, 2011.

[8] Cartus O., Kellendorf J., Rombach M., Walker W., Mapping canopy height and growing stock volume using airborne LIDAR, ALOS PALSAR and Landsat ETM+, Remote Sensing, 4, 2012.

[9] Rosenqvist A., Shimada M., Ito N., Watanabe M., ALOS PALSAR: A Pathfinder Mission for Global-Scale Monitoring of the Environment, IEEE Transactions on Geoscience and Remote Sensing, 45(11), 2007.

[10] De Grandi G., Bouvet A., Lucas R., Shimada M., Monaco S., Rosenqvist A., The K&C PALSAR mosaic of the African continent: processing issues and first thematic results, IEEE Transactions on Geoscience and Remote Sensing, Vol. 49(10), 2011.

[11] Englhart S., Keuck V., Siegert F., Aboveground biomass retrieval in tropical forests—The potential of combined X-and L-band SAR data use, Remote Sensing of Environment, *115*, 2011.

[12] Kellndorfer J., Walker W., LaPoint E., Kirsch K; Bishop J., Fiske G., Statistical fusion of LIDAR, InSAR, and optical remote sensing data for forest stand height characterization: A regional-scale method based on LVIS, SRTM, Landsat ETM+, and ancillary data sets, Journal of Geophysical Research, 115, G00E08, 2010.

[13] Hansen M.C., Potapov P.V., Moore R., Hancher M., Turubanova S.A., Tyukavina A., Thau D., Stehman S.V., Goetz S.J., Loveland T.R., Komardeey A., Egorov A., Chini L., Justice C.O., Townshend J.R.G., High-Resolution Global Maps of 21st-Century Forest Cover Change, Science 342(6160):850-853. doi: 10.1126/science.1244693, 2013.

[14] Atwood D., Andersen H.E., Matthiss B., Holecz F., Impact of topographic correction on estimation of above ground boreal biomass using multi-temporal, L-band backscatter, IEEE Journal of Selected Topics in Applied Earth Observations and *Remote Sensing* (J-STARS), paper accepted to be published in 2014.

[15] Cartus O., Santoro M., Kellndorfer J., Mapping Forest Aboveground Biomass in the Northeastern United States with ALOS PALSAR Dual-Polarization L-Band, Remote Sensing of Environment, 124, 2012.

[16] Moghaddam M., Dungan J., Acker S., Forest variable estimation from fusion of SAR and multispectral optical data, IEEE Transactions on Geoscience and Remote Sensing 40, 2002.

[17] Walker W., Kellndorfer J., Lapoint E., Hoppus M., Westfall J., An empirical InSAR-optical fusion approach to mapping vegetation canopy height, Remote Sensing of Environment, 109(4), 2007.

[18] Kellndorfer J., Walker W., LaPoint E., Kirsch K., Bishop J., Fiske G., Statistical fusion of LIDAR, InSAR, and optical remote sensing data for forest stand height characterization: A regional-scale method based on LVIS, SRTM, Landsat ETM+, and ancillary data sets, Journal of Geophysical Research, *115*, G00E08, 1–10, 2010.

[19] Masek J.G., Vermote E.F., Saleous N.E., Wolfe R., Hall F.G., Huemmrich K.F., Feng G., Kutler J, Teng-Kui L., A Landsat surface reflectance dataset for North America, 1990–2000, IEEE Geoscience Remote Sensing Letters, 3, 2006.

[20] Breiman L., Random forests, *Machine Learning, 45*, 2001.

[21] Cartus O., Santoro M., Schmullius C., Li Z, Large area forest stem volume mapping in the boreal zone using synergy of ERS-1/2 tandem coherence and MODIS vegetation continuous fields, Remote Sensing of Environment, 115, 2011.

[22] Attema E.P.W., Ulaby F.T., Vegetation modeled as a water cloud, Radio Science, 13, 1978.

[23] Askne J.I.H., Dammert P.B.G., Ulander L.M.H., Smith G., C-band repeat-pass interferometric SAR observations of the forest, IEEE Transactions on Geoscience and Remote Sensing, 35(1), 1997.

[24] Pulliainen J.T., Heiska K., Hyyppä J.M., Hallikainen M.T., Backscattering Properties of Boreal Forests at C-and X-bands, IEEE Transactions on Geoscience and Remote Sensing, 32(5), 1994.

[25] Homer C., Dewitz J., Fry J., Coan M., Hossain N., Larson, C., Completion of the 2001 National Land Cover Database for the Conterminous United States, Photogrammetric Engineering & Remote Sensing, 2007.

[26] Dobson M.C., Ulaby F.T., Pierce L.E., Sharik T., Bergen K., Kellndorfer J.M., Estimation of forest biophysical characteristics in Northern Michigan with SIR-C/X-SAR, IEEE Transactions on Geoscience and Remote Sensing, 33(4), 1995.

[27] Bechtold W.A., Patterson P.L. (Eds), The Enhanced Forest Inventory and Analysis Program-National Sampling Design and Estimation Procedures, General Technical Report GTR-SRS-080, Asheville, NC: U.S. Department of Agriculture, Forest Service, Southern Research Station, 2005.

[28] Kellndorfer J., Walker W., La Point E., Bishop J., Cormier T., Fiske G., Hoppus M., Kirsch K., Westfall J., NACP Aboveground Biomass and Carbon Baseline Data (NBCD 2000), U.S.A., 2000, Data set, 2012. Available on-line at http://daac.ornl.gov from ORNL DAAC, Oak Ridge, Tennessee, U.S.A. http://dx.doi.org/10.3334/ORNLDAAC/1081.

[29] Lucas R.M., Armston J., Fairfax R., Fensham R., Accad A., Carreiras J., An Evaluation of the ALOS PALSAR L-Band Backscatter – Above Ground Biomass Relationship Queensland, Australia : Impacts of Surface Moisture Condition and Vegetation Structure, IEEE Journal of Selected Topics in Applied Earth Observations and Remote Sensing, 3(4), 2010.

[30] ftp://ftp.fao.org/docrep/fao/004/ab585e/ab585e00.pdf.

[31] Holecz F., Collivignarelli F., Gatti L., Barbieri M., Earth Observation information services for UN-IFAD, Final report to ESA, January, 2014.

[32] De Grandi F., Leysen M., Lee J., Schuler D., Radar reflectivity estimation using multiplicative SAR scenes of the same target: technique and applications, Geoscience and Remote Sensing Symposium, IGARSS'97, 1997.

[33] D. Small, Flattening Gamma: Radiometric Terrain Correction for SAR Imagery, IEEE Transactions of Geoscience and Remote Sensing, 49 (8), 2011.

[34] Aspert F., Bach-Cuadra M., Cantone A., Holecz F., Thiran J-P, Time-varying segmentation for mapping of land cover changes, ESA ENVISAT Symposium, 2007.

[35] Shimada M., Muraki Y., Otsuka Y., Discovery of anomalous stripes over the Amazon by the PALSAR onboard ALOS satellite, Geoscience and Remote Sensing Symposium, IGARSS'08, 2008.

[36] Holecz F., Barbieri M., Cantone A., Pasquali P., Monaco S., Synergetic use of multitemporal ALOS PALSAR-1 and ENVISAT ASAR data for topographic/land cover mapping and monitoring at national scale in Africa, Geoscience and Remote Sensing Symposium, IGARSS'09, 2009.

[37] Holecz F., Collivignarelli F., Barbieri M., Estimation of cultivated area in small plot agriculture in Africa for food security purposes, ESA Living Planet Symposium, 2013.

[38] Holecz F., Barbieri M., Eyre C., Mönnig N., Forest Management – Mapping, monitoring, and inference of biophysical parameters using ALOS PALSAR and Cosmo-SkyMed data, JAXA Kyoto and Carbon Initiative, Tokyo, 2010.

[39] Mitchell A.L, Williams M., Tapley I., Milne A.K., Interoperability of multi-frequency SAR data for forest information extraction in support of national MRV systems, Geoscience and Remote Sensing Symposium, IGARSS'12, 2012.

[40] Holecz F., Barbieri M., Global Monitoring for Food Security, Malawi Service Validation Report, Report to ESA/ESRIN, 2009.

Chapter 3

Estimation of Cultivated Areas Using Multi-Temporal SAR Data

Nada Milisavljević, Francesco Collivignarelli and Francesco Holecz

Additional information is available at the end of the chapter

http://dx.doi.org/10.5772/58282

1. Introduction

Estimation of cultivated areas in small plot agriculture is an important issue for food security purposes in Africa. One way to obtain this information is through classical field surveys and aerial photography, which are both time and resource consuming. Multi-temporal high resolution Synthetic Aperture Radar (SAR) systems, as sources of reliable and overall information [1], [2], [3], [4], are an alternative solution, satisfying also the demand of continuous monitoring. Namely, for food security purposes, large scale agricultural products are requested at different times throughout the rain-fed crop season. Concerning the cultivated area, typically, a first product is required after the fields preparation; a second one prior to the harvesting time. In this respect, it is worth mentioning that i) the cultivated area product at start of crop season – today not available in food security services – is an excellent indicator to quantify the overall situation of the upcoming rain-fed crop season; ii) SAR systems – on the contrary to optical sensors – are suitable to map these areas due to their sensitivity to the soil roughness, a typical characteristic of the fields at this stage.

Based on these considerations, a three-step approach for estimation of cultivated area in small plot agriculture in Malawi is envisaged and presented in this chapter. The first step of this approach is the estimation of crop extent prior to the crop season. The estimation of the potential area at start of the crop season is the second step, while the third step consists in determining the crop growth extent during the rain-fed crop season. Taking into account that various vegetation types grow during the rainy season, the key issue is to know what is really cultivated and not simply vegetated. The final result is crucial when dealing with food security and agriculture in developing countries, where available land-cover map is either inaccurate, out of date or it does not even exist. Once derived, this global information, which should give a basis for deciding where to perform more detailed analysis, should be relevant for a longer period of time in normal situations, so it should not need to be updated annually.

As shown in [5], [6], [7], [8], radar-only approaches possess an important operational advantage w.r.t. optical-only or optical-radar approaches, especially in cloud-prone regions, due to their all-weather working ability, high spatial resolution and sensitivity to biomass and moisture. In addition, the penetration depth into crop canopy depends on SAR frequency (the longer the wavelength, the longer the penetration depth) so using multi-frequency data (L-, X- and C-bands) in function of the phenological stages of crops should bring improvements in the estimation of cultivated areas [9], [10]. Finally, since the growth periods of different crops are not equal, in order to differentiate cultivated areas in small-plot agriculture from other areas with different plants and crops that are out of our interest, multi-temporal data are needed.

The main goal of the work presented here on deriving the potential crop extent prior to the start of the rain-fed season is to provide a first information layer regarding the extent of the bare soil area where crop will potentially grow. Multi-temporal L-band SAR data having resolution in the order of 15 m should be sufficient to extract this type of information. In a next step, where potential cultivated area at start of season should be looked for, a very high resolution sensor is needed. The acquisition coverage of this very high resolution sensor (such as X-band Cosmo-SkyMed with the resolution of 3 m) can be limited thanks to the output of the first step. In the final step, monitoring of the crop growth, multi-temporal ASAR (C-band, 15 m resolution) are used, starting from the period before December (in order to have the reference bare soil) as well as covering the period from December to April (the completion of crop season). Finally, several voting strategies are tested for the combination of the outputs of each of the three sensors.

The key aspect of the proposed approach is the generation of three independent and complementary products – each one with a clear meaning within agriculture and food security – derived from different spaceborne SAR sensors, which in turn are fused, by yielding the cultivated area product. Moreover, it is also intended to demonstrate the usefulness of SAR data for the targeted application. Note that in this context, the meaning of cultivated area is the effective cropped land (i.e., cultivated and not fellow land) during the rain-fed season. Commercial and irrigated fields, typically cultivated during the dry season, are out of our interest.

The chapter is organized as follows. In the following section, we briefly describe the food security situation in Malawi. Then we present the three steps of our method, one by one. For each of them, we describe the data used, the procedure, we show the obtained results and their validation using ground-truth information. After that, we present the final combination by describing our fusion methods, obtained results and their validation. Finally, we derive some conclusions.

http://dx.doi.org/10.5772/58282

2. Food security situation in Malawi

According to [11], [12], [13], [14], [15], Malawi is one of the poorest and least developed countries in Sub-Saharan Africa and a huge challenge facing Malawian agriculture is producing more food for a growing population. Many rain-fed smallholder farmers in the country have been shifting to farming systems that are increasing food crop yields and household food security. Urban poverty is increasing in Malawi as well and a pragmatic solution is seen in urban agriculture (i.e., "food production conducted in or around urban regions" [16]). Nevertheless, it is important that the amount available to small plot agriculture, especially in urban conditions, is extended with an understanding of local environmental conditions and that a careful assessment of cultivated areas in small plot agriculture is performed. As a result, the real extent of the food insecurity conditions can be estimated in order to facilitate an adequate humanitarian response if necessary and diminish the country's vulnerability to hunger.

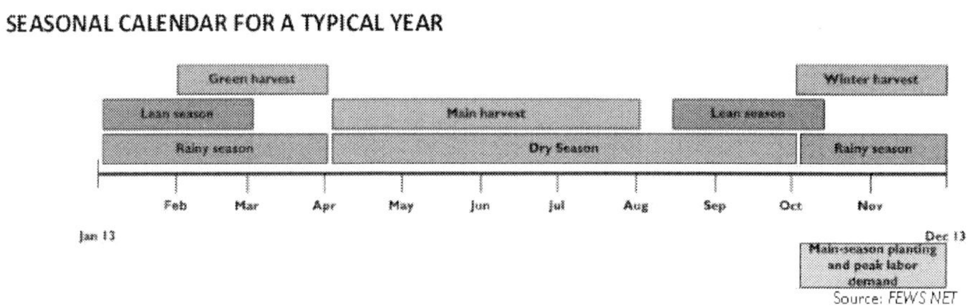

Figure 1. Seasonal calendar and critical events timeline in Malawi [17]

In Figure 1, the seasonal calendar for a typical year and critical events timeline in Malawi are shown. This country consists in three agro-ecological zones: 1) high altitude areas (more than 1300 m above sea level, cool temperatures, agricultural areas are rain-fed, with wheat and beans as main crops), 2) low altitude areas (less than 600 m above sea level, irrigated, rice, maize and beans are main crops) and 3) medium altitude areas (600-1300 m above sea level). In this chapter, the region around Lilongwe is analyzed. It belongs to medium altitude areas of Malawi that comprise about 60% of the total cultivated surface of the country. These areas are characterized by moderate temperature and a fairly long rainy season (December to February/March) and their major agriculture practice is maize. The crop calendar for maize in these areas is given in Figure 2.

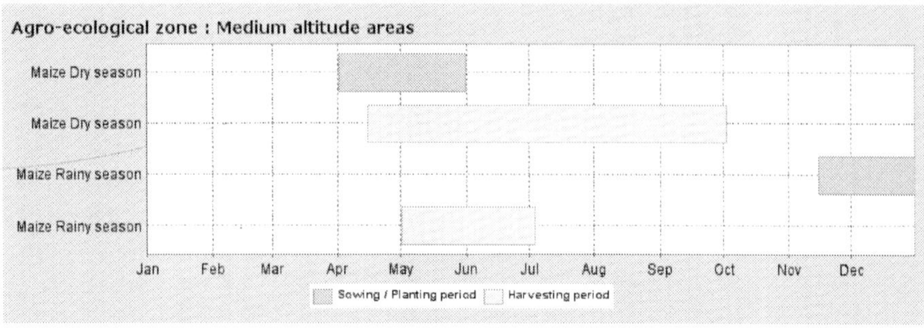

Figure 2. Generic maize crop calendar for medium altitude areas of Malawi [18]

3. First step

3.1. Multi-year L-band data

The aim of this step is to define the potential cultivable extent of the region prior to the start of the rain-fed season, in order to: 1) provide a first layer of information with the potential crop extent in bare soil conditions, i.e. the bare soil area where crop will potentially grow; 2) limit the acquisition coverage of the very high resolution sensor. The generation of this product is relevant when the available land cover map does not exist, is not updated, is inaccurate or the spatial scale is not appropriate (i.e. small scale), which are typical problems when dealing with food security and agriculture in developing countries.

Multi-temporal L-band SAR data having resolution in the order of 15 m should be sufficient to extract this type of information. Namely, in dry conditions, L-band HH/HV data have a potential of distinguishing between bare soil and other land cover classes (sparse to strong vegetation, forest, settlement, bush, wetlands, water).

Therefore, in order to estimate crop extent, multi-annual ALOS PALSAR-1 [19] acquired during the dry season are chosen, since we are interested in an average bare soil area, and not in small changes. The archive of these data, acquired by the Japanese Space Agency (JAXA), is consistent and can be processed in a multi-temporal way, which enables the speckle reduction and allows the generation of a more accurate map (based on a multi-temporal classifier) than the one obtained using a single date, because not relevant temporal variations are filtered out. Optionally, ENVISAT ASAR AP, Landsat TM-5 or SPOT-4/5 data can be used; however, in general, the latter options are not suitable, since less performing for the targeted product during the selected period. Concerning the failure of ALOS PALSAR-1 system in March 2011, it should be noticed that it does not represent a major problem, because i) the available archived data is sufficient for the generation of this intermediate product; ii) these archived data can be used even in the years to come, since the crop pattern are usually not rapidly changing; and iii) the launch of several L-band sensors is planned in the next 1-2 years (PALSAR-2, SAOCOM-1/2).

Taking into account that the dry season prior to the start of the rainy season in Malawi is from April/May to October (Figure 2), the multi-annual PALSAR data that cover that period of the year are selected.

3.2. Procedure

As proposed in [20], preprocessed multi-temporal PALSAR data in HH and HV polarization are the input data for estimating the crop extent. The preprocessing phase consists in: multi-looking, orbital correction, co-registration, multi-temporal speckle filtering, terrain geo-coding, radiometric calibration and normalization, and anisotropic non-linear diffusion filtering. Such a preprocessing allows us to work at pixel level.

It is possible to classify these preprocessed multi-temporal SAR data using various approaches. Taking into account that our goal is to develop an approach that could be reused in other regions, where we might not have any reliable information about the existing land-cover classes, we opt for an unsupervised classification method. This means that we can either perform classification on each image separately and then combine the classification results or analyze the multi-temporal signatures of pixels or regions and perform classification based on the similarity of signatures. We choose the former in order to cover situations where we have only a few multi-temporal images or where the data are not radiometrically calibrated. The key issue at this step is to determine which pixels change in time (as a potential bare soil) and which pixels remains the same (so they can be used as a mask that covers the regions that are not interesting for further steps of our three-step method). Therefore, to each image of the multi-temporal data set, we need to apply such an unsupervised classification algorithm that preserves the grayscale information so that the classes from one image can be compared with the classes of another image and that the decision whether the class changed or not is meaningful. Based on this, our final choice is an algorithm using the Principal Components Transform (PCT) and median cut [21]. This algorithm looks for the most discriminative information based on which it divides the image into a preset number of classes, taking into account the colour (or grayscale) values.

At this step of the global approach, we have to make sure that none of the pixels that might change in time is excluded from further steps (i.e., masked as stable), while misclassifying stable pixels as changing ones is not critical here and will be further corrected in the two following steps. Due to that, we proceed in the following way. Once all the pixels of each image are classified into n classes, labeled 1, 2, …, n, they are compared and as soon as a pixel changes its label, it is marked as 0, i.e., a potential bare soil. Only if the pixel preserves its class label in all images, that is its final label too.

Taking into account that we have both HH and HV data sets, the above procedure is applied to each of the sets and the two outputs are analyzed using ground-truth information. In a final phase of the first step, these two outputs are combined and the result is compared with the

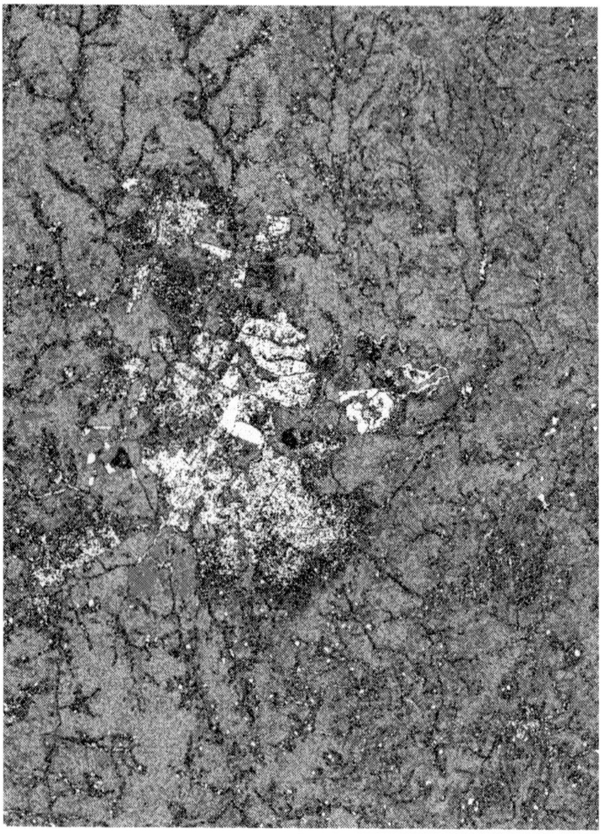

Figure 3. An example of the PALSAR HV image classification result

ground-truth as well, in order to verify the usefulness of fusing the HH/HV information at this level.

3.3. Results and validation

The test site is a relatively flat region of Malawi, around its capital, Lilongwe. We have ten ALOS PALSAR FBD (Fine Beam Double Polarisation) intensity data (so, HH and HV image pairs from ten different dates) acquired from 2007 to 2010 (from April to October each year, Figure 2). Based on the information from field, we classify each image in eight classes. (Note that we have repeated the whole procedure for ten and for twelve classes, and there was no significant change in the final result.) An example of the classified image is shown in Figure 3. The result of comparing the classification results for the HV data set is given in Figure 4, and for the HH data set, it is presented in Figure 5. Label 0 is in gray in all the images.

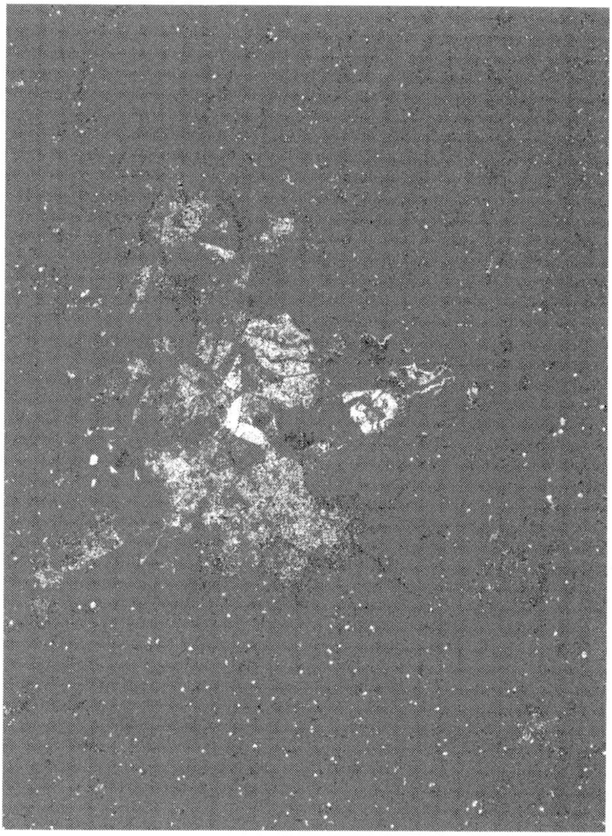

Figure 4. Comparison of the PALSAR HV classification results

Finally, a combination of the results of the HV and HH data sets is shown in Figure 6. This combination is based on the idea that if the HH and HV results label differently the pixel, its neighborhood is analyzed in HH and HV results and of the two labels, the one that is more present is chosen. If HH and HV results label equally the pixel, that label is preserved in the combined result.

As a validation, we use the ground-truth information consisting of 422 points that correspond to regions that do not change in time (buildings, water ...) and 330 points that belong to arable land. For HV result (Figure 4), 202 points that do not change in time are correctly classified (having label other than 0), while 330 points that change in time are correctly labeled as 0. In the HH case (Figure 5), 91 points that belong to regions that do not change in time are well classified, and 320 points are correctly classified in case of arable land. Finally, the combination result (Figure 6) classified well 242 points that do not change in time and all 330 points of arable land.

The final output of this step, i.e. the resulting PALSAR mask, used in further steps, is given in Figure 7. It is obtained simply by changing the color of the gray pixels from Figure 6 into white (representing potential bare soil) and assigning black colour to all other classes/colours of Figure 6 (since they are out of interest, thus masked).

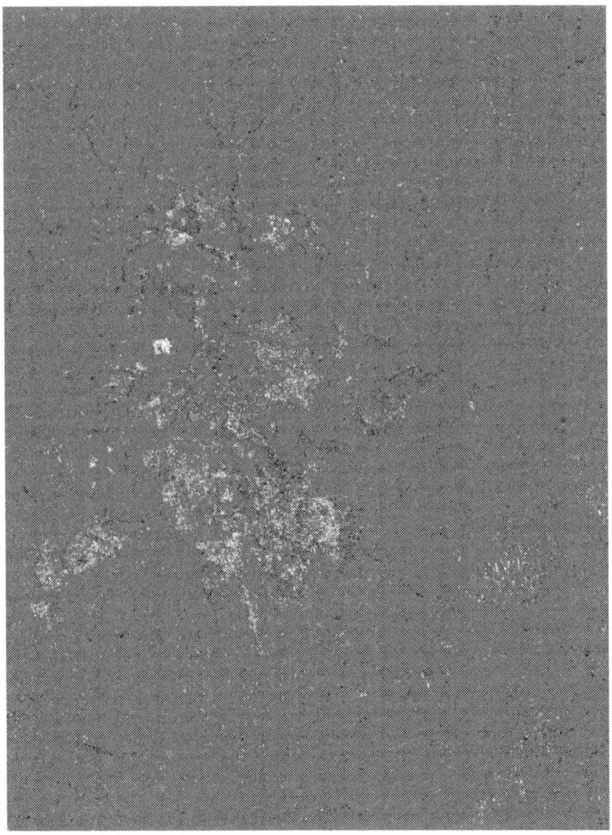

Figure 5. Comparison of the HH classification results

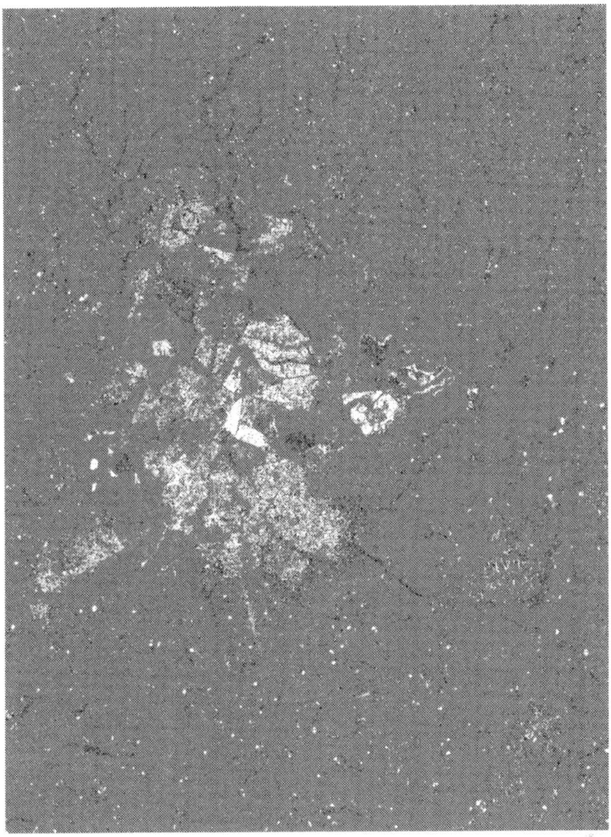

Figure 6. Combination of the HH and HV results

4. Second step

4.1. One-day interferometric X-band data

The aim of this step is to define the potential cultivated area, in particular to delineate the ensemble of fields, where crops will later grow.

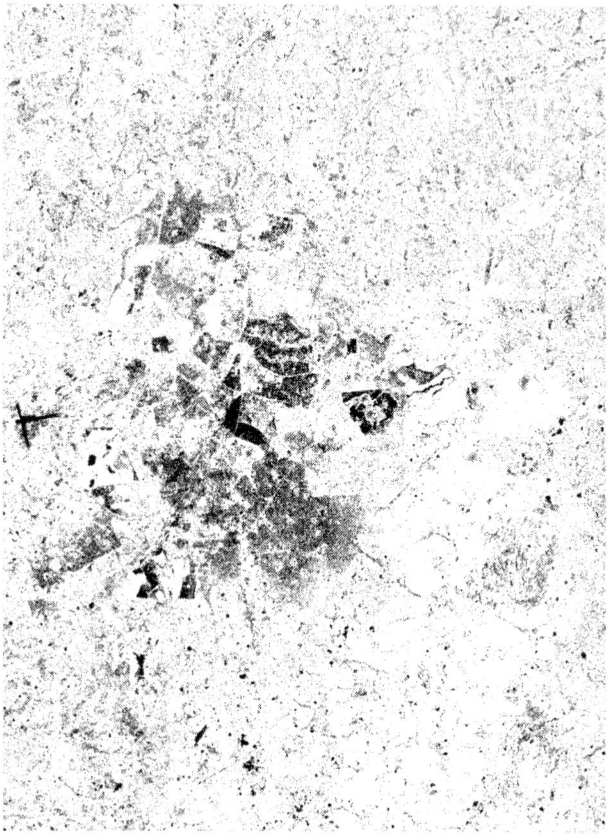

Figure 7. PALSAR mask: white – potential bare soil, black – masked.

X-band radars have the least penetration depth of the three bands used in this work, which also makes them more vulnerable to atmospheric effects. The COSMO-SkyMed (CSK) X-band SAR system is capable of acquiring data twice a day, at a high spatial resolution, which allows for very short-term analysis such as one-day correlation. Taking into account its high sensitivity and poor penetration capabilities, this system provides excellent means for analyzing short-term (lack of) changes at the very start of the crop season, when the crops are only being planted. Thus, potential cultivated area at start of season is derived from one-day interferometric CSK pair (3 m) acquired during the field preparation period.

In this respect, it is worth mentioning that this step does not have to be performed on a yearly base, if the area and the pattern of the fields remain stable: it should be exclusively updated when changes occur.

4.2. Procedure

Potential cultivated area at start of season is derived from one-day interferometric CSK pair acquired during the field preparation period (December, Figure 2). Very high resolution optical data would not be useful at this stage since fields are not yet covered by vegetation, so it would be very difficult to discriminate them from the surrounding bare soil area using optical sensors. On the contrary, because of the rough nature and dry conditions of the fields, short wavelength SAR data acquired in an interferometric mode (1-day interval) provide useful information at this second step.

This step should provide information on the status of the fields at the beginning of the crop season in terms of vegetated or bare soil condition. Since the purpose of the overall cultivated area product here is to map the effective crop growth from the start to the end of the crop season, only those fields with bare soil conditions are considered. This means, in terms of interferometric X-band data, that these areas are defined by a medium to high coherence (absence of human activities) and a medium to high backscattering coefficient (rough bare soil).

Image 1 Image 2	L	M	H
L	L	ML	C
M	ML	M	MH
H	C	MH	H

Table 1 Combination of two CSK amplitude images taken with one-day interval.

Taking into account that we have one one-day interferometric data set and two corresponding amplitude images, we should first combine the two amplitude images into one. Using the PCT and median cut algorithm mentioned in Subsection 3.2, we split each of the two amplitude images into three classes (L - low, M - medium and H - high amplitude). We combine them as given in Table 1, where: L – low, M – medium, H – high, C – (significant) change, ML - medium low, MH – medium high.

Amplitude Coherence	L	C	ML	M	MH	H
L	O	O	O	O	O	O
M	O	O	M1	M3	M4	H1
H	O	O	M2	M4	H1	H2

Table 2 Combination of the combined CSK amplitude image and the corresponding coherence image.

Note that the situations in which there is a strong change in the backscattering coefficient in such a short time most possibly refer to on-going works (therefore, they are most possibly not among the areas of interest).

Figure 8. CSK amplitude image taken on December 9, 2010

Once the combined amplitude image is obtained, we combine this information with the corresponding coherence image, split in three classes (L, M and H) using the PCT and median cut algorithm (Subsection 3.2). The combination is based on the idea that, at this period of the year that we consider, neither low coherence nor low amplitude correspond to potential cultivated area. Thus, we combine these two data sets as indicated in Table 2, where: O – out of interest, M1 – medium 1 (one amplitude image has a low value, the other and the coherence have medium values), M2 – medium 2 (one amplitude value is low, the other is medium, coherence value is high), M3 – medium 3 (all images have medium values), M4 – medium 4 (two out of these three values are medium and one is high), H1 – high 1 (two out of these three values are high and one is medium), H2 – high 2 (all images have high values).

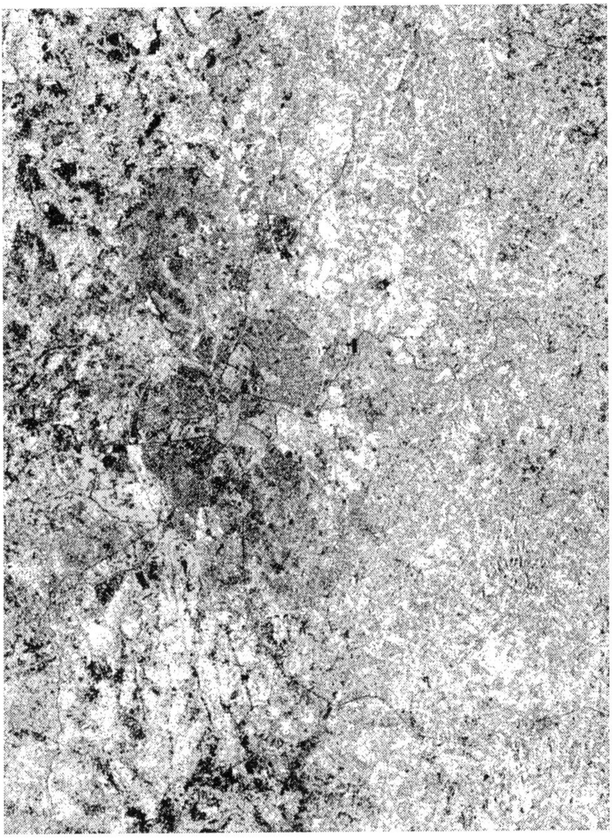

Figure 9. Result of combining the classification results of the two amplitude images from Figure 8 according to Table 1, where: L – black, C – blue, ML – green, M – yellow, MH – red, H – white.

In such a way, the order O-M1-M2-M3-M4-H1-H2 corresponds to the increasing probability that the pixel belongs to a cultivated area (bare soil).

4.3. Results and validation

Two CSK amplitude images of Lilongwe, taken on December 9 and 10, 2010 are used (the one of December 9 is shown in Figure 8, as an illustration). After splitting each of the two images in three classes (L, M and H) and combining them as indicated in Table 1, we obtain the result presented in Figure 9. Once the coherence image is split in three classes (L, M and H) and combined with the combination of the two amplitude images following the rules of Table 2, we get the image in Figure 10.

Figure 10. Result of combining the combined amplitude images from Figure 9 with the coherence image according to Table 2, where: O – black, M1 – blue, M2 – green, M3 – yellow, M4 - orange, H1 – red, H2 – white.

As far as validation is concerned, we use the same ground-truth information as described in Subsection 3.3. However, the question here is how to treat different degrees of possibility of belonging to cultivated areas, taking into account the variety of output values (O-M1-M2-M3-M4-H1-H2), i.e. where to put the threshold level between not cultivated and cultivated. If we decide to assign everything that is not O to cultivated areas, we obtain 293 (out of 330) correct classifications of bare soil and 184 (out of 422) correct classifications of the rest. If we assign O and M1 to not-cultivated areas, then we get 289 (out of 330) correct classifications of bare soil and 212 (out of 422) correct classifications for not-cultivated areas. Finally, if we assign O, M1, M2 and M3 to not-cultivated areas, we obtain

280 (out of 330) correct classifications of bare soil and 294 (out of 422) correct classifications for not-cultivated areas. In other words, as we move the threshold, the number of correct classifications for cultivated areas slowly decreases while the number of correct classifications for not-cultivated areas significantly increases.

Figure 11. CSK mask: white – potentially cultivated, black – masked

The final output of our approach is the combination of PALSAR (that performs very good for bare soil and moderately good for the rest, as shown in Subsection 3.3), CSK and ASAR. Thus, we should keep the threshold level for CSK in such a way that the classification of not-cultivated areas is the best possible (thus, between M3 and M4) and pay attention during the final combination so that complementarities of the three sensors are exploited in a best possible way. Figure 11 contains the image from Figure 10 where LO, M1, M2 and M3 are labeled as black (masked) and M4, H1 and H2 as white (potentially cultivated).

5. Third step

5.1. Seasonal C-band data

At this step, crop growth extent is derived from multi-temporal ENVISAT Advanced Synthetic Aperture Radar (ASAR) data [22] acquired during the crop season. Optionally, Radarsat-1/2 FB can be used at this step. Optical data such as Landsat TM-5, SPOT-4/5, Ikonos, and QuickBird, can also provide information of the crop growth. But, due to the persistent cloud coverage during this period, their use is very difficult, even more if large areas, such as national coverage, are targeted.

Multi-temporal ENVISAT ASAR (15 m) are C-band data proven to be useful in monitoring agricultural activities on a regular basis, i.e. seasonal land cover changes [1], [23], [24], [25]. Generally speaking, C-band SAR data are not hindered by atmospheric effects; their penetration capability with respect to vegetation canopies is restricted to the top layers.

5.2. Procedure

The goal of the third step is to monitor the crop growth during the crop season, which explains the necessity of having multi-temporal data acquired regularly all along the crop season. In such a way, the confusion between cropped and the surrounding vegetated, non-crop, areas should be reduced, and at the same time, the crop development could be monitored.

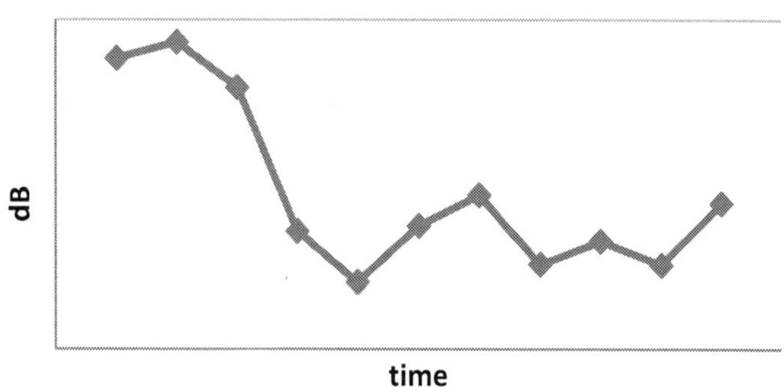

Figure 12. An illustration of the typical behavior of the C-band maize signature (intensity in function of time) from the beginning to the end of the season.

Our procedure for analyzing the ASAR data is as follows. Firstly, we perform an unsupervised classification based on multi-temporal signatures (so, grouping together the pixels having a similar multi-temporal behavior) in a preset number of classes. As a result, we have an output image where the pixels with similar signature have the same label. As this is an unsupervised classification, we are unable to determine which of these signatures is similar to the one of maize. Thus, in the following level of our analysis, we introduce our knowledge regarding the maize signature at C-band (note that, depending on the area, season and the type of the crop

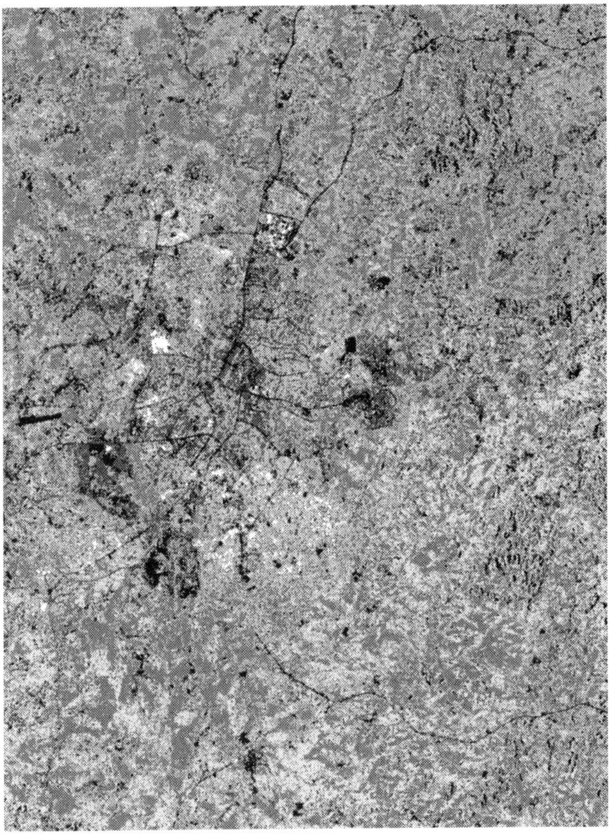

Figure 13. Result of an unsupervised classification based on multi-temporal pixel behavior in eleven ASAR images

we want to distinguish, this step can be easily modified for other applications): at the very beginning of the season, the intensity of the signature is low, and then it grows, reaching rapidly its maximum value (which corresponds to ploughing), and then dropping to its first minimum (sowing). Then the second phase begins, i.e. from flowering to plant drying stage, where the intensity of the signature raises, reaches another maximum, possibly drops a bit and raises again, in function of the plant moisture and the surface scattering at the top of the plant. This behavior is similar to the one illustrated in Figure 12. Since the later stages can vary from season to season, the key indicators we use to select which of the signatures from the unsupervised classification output resemble to the one of maize are the starting raise of the signature, its sudden drop followed by its next raise. Thus, we analyze the moments when the first maximum, the first minimum and the second maximum occur and label the class(es) with the corresponding tendencies as the one(s) of maize.

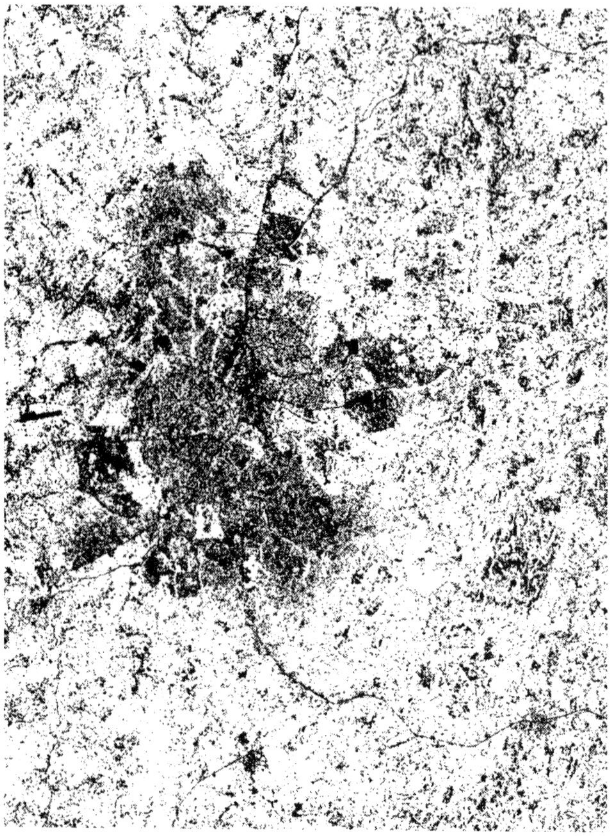

Figure 14. Grouping of the classes from Figure 13: white – crop growing extent, black – the rest (masked)

5.3. Results and validation

Once a multi-temporal unsupervised classification of multi-temporal pixel signatures in twelve classes is performed, using eleven ASAR intensity images covering the period from September 2010 to March 2011, we obtain the result given in Figure 13. After analyzing the multi-temporal signatures of each of twelve classes, we select the classes having tendencies similar to the ones of maize (Figure 12), and we mask the rest. The obtained result is given in Figure 14. Note that the classification has been also performed in eight and ten classes and that there was no significant change in the final result.

The validation results are as follows: 303 (out of 330) pixels from cultivated areas are well classified, as well as 235 (out of 422) pixels from non-cultivated areas.

http://dx.doi.org/10.5772/58282

6. Final combination

6.1. Procedure

There are various methods of combining the results obtained from each of the three sensors, depending on the quality of each of the sensors, requested computation speed, application, etc. In our case, looking for a simple, fast yet reliable method, we test several voting combination rules:

- ANDc (a pixel is declared as cultivated if all sensors labeled it as such; it is masked otherwise);
- ANDm (for the masked class: a pixel is declared as masked if all sensors labeled it as such; otherwise, it is declared as potentially cultivated);
- MAJ (for class labeled as potentially cultivated: if at least two out of three sensors labeled it as potentially cultivated, the pixel is labeled at the combination output as cultivated; otherwise, it is masked), and
- OR (for class labeled as potentially cultivated: a pixel is declared as masked only if all sensors labeled it as such; otherwise, i.e., if at least one of the sensors labeled a pixel as potentially cultivated, it will keep that label at the combination output).

Note that the difference between ANDc and ANDm is that in the former case, AND voting is applied to the cultivated class while in the latter case, AND voting is applied to the masked class. This makes the two AND voting approaches complementary and it depends on the application which one is more useful than the other. Finally, the difference between MAJ and OR is that the former labels a pixel as potentially cultivated if at least two of the three sensors have given it that label and it is masked otherwise, while the latter labels a pixel as potentially cultivated if any of the three sensors labeled it as such.

6.2. Results and validation

The four voting strategies are applied to the outputs of the three sensors given in Figures 7, 11 and 14. As an illustration, Figure 15 contains the result of ANDc voting, while the result of MAJ voting is shown in Figure 16.

We validate these results using the same validation set as in the previous steps and obtain the following results:

- ANDc voting: 225 out of 330 pixels from cultivated fields well classified; 377 out of 422 pixels belonging to non-cultivated fields well classified;
- ANDm voting: all 330 pixels belonging to cultivated fields well classified, as well as 78 out of 422 pixels from non-cultivated fields;
- MAJ voting: all 330 pixels from cultivated fields well classified, as well as 323 out of 422 pixels from non-cultivated fields;

- OR voting: all 330 pixels from cultivated fields well classified, as well as 59 out of 422 non-cultivated field pixels.

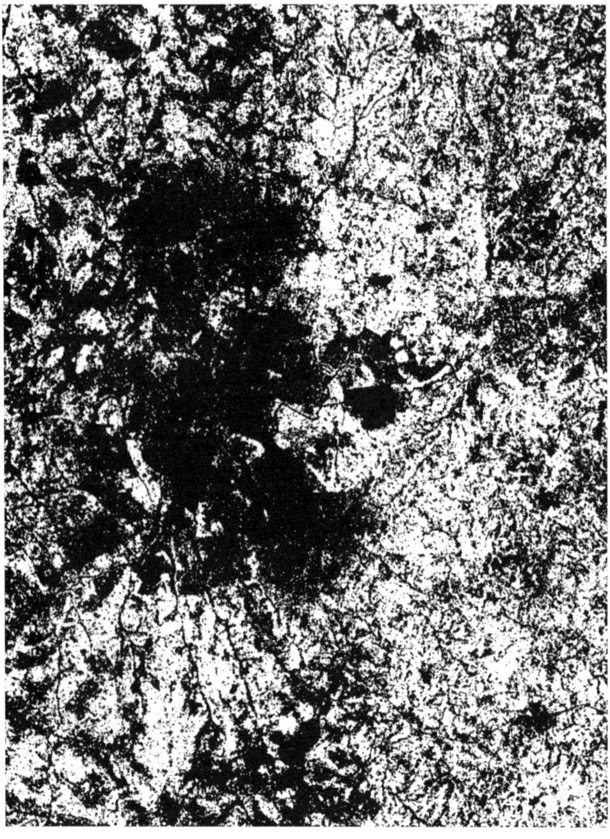

Figure 15. Result of ANDc voting: white – crop growing extent, black – masked.

These validation results are in accordance with our expectations. Namely, if we keep as cultivated only those pixels where all three sensors claim that it is cultivated indeed, and label all the rest as masked (non-cultivated), we can expect to have a high detection of non-cultivated fields and only modest results for cultivated fields (ANDc voting). With the inverse logic, we preserve all cultivated fields, but also label many non-cultivated as being cultivated (ANDm voting, as well as OR). Finally, with the majority voting (MAJ), it can be expected to obtain results that optimize the two extremes, and benefit from the complementarities of the sensors.

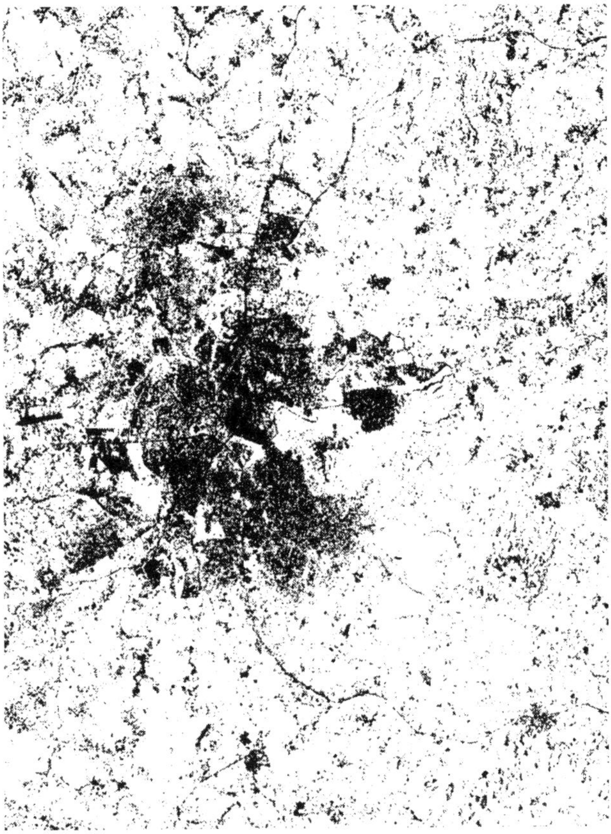

Figure 16. Result of MAJ voting: white – crop growing extent, black – masked.

The same conclusion can be derived from Table 3, which represents the summary of the validation results, for each of the steps and each of the combination approaches tested. From the three steps, step 1 has the best correct classification of pixels belonging to cultivated fields, but 42.64% of pixels belonging to non-cultivated fields are also classified as cultivated. On the other hand, step 2 has the best correct classification of pixels from non-cultivated fields, but 15.15% of cultivated field pixels are also classified as non-cultivated. Regarding the four fusion strategies used, we can conclude that with the MAJ voting, we preserve the maximum number of correct classifications as the one of PALSAR in step 1, while we increase the correct classification of non-cultivated fields as well, thanks to the output of CSK and ASAR. Although the priority is not to miss cultivated fields, an improvement in correct classification of non-cultivated fields is also an important issue since it leads to a better estimation of the extent of the problem.

	cultivated fields well classified (out of 330)	non-cropped fields well classified (out of 422)
step 1	100%	57.35%
step 2	84.85%	69.67%
step 3	91.82%	55.69%
final ANDc	68.18%	89.34%
final ANDm	100%	18.48%
final MAJ	100%	76.54%
final OR	100%	13.98%

Table 3 Correct classifications: summary of the validation results (in percents)

Note that the validation is performed on the pixel level. If done on the region level, the validation results would certainly have been even better since a region would have been declared as correctly classified if majority of its pixels (and not all of them, as here) were correctly classified.

The achieved accuracy confirms the validity of the methodology, in particular that: 1) the use of very high resolution data is an indispensable condition for the identification of small agricultural plots; 2) the differentiation between cultivated areas (i.e. growing vegetation during the crop season) and other land cover classes is first and foremost possible if multi-temporal data are used; 3) the combination of various SAR sensors (bands) improves the final results.

7. Conclusion

Agriculture is the land cover type showing the largest spatial and temporal dynamics during a relatively short period. Therefore, pre-requisite for the generation of an accurate cultivated area product is to combine very high resolution data with multi-temporal high resolution data acquired throughout the whole crop season. This approach has been tested through a three-step procedure for estimation of cultivated area in small plot agriculture in Malawi and the obtained results have proven its validity.

The first step of this procedure is the estimation of crop extent prior to the crop season using multi-temporal L-band PALSAR data. The estimation of the potential area at start of the crop season using X-band COSMO-SkyMed is the second step, while the third step consists in determining the crop growth extent during the rain-fed crop season, with the help of multi-temporal C-band ASAR data. The final result is crucial when dealing with food security and agriculture in developing countries, where available land-cover map is either inaccurate, out of date or it does not even exist. Once derived, this global information, which should give a basis for deciding where to perform more detailed analysis, should be relevant for a longer period of time in normal situations, so it should not need to be updated annually.

At each step, the obtained results are validated using ground-truth information.

Four voting combination strategies are tested in the final combination of the three sensors, based on "majority", "or" and "and" logic (two versions of it - one prioritizing cultivated, and the other non-cultivated fields). For our application, the majority voting gives the most interesting results of the four, while for some other applications (such as mined area reduction, for example), one of the other strategies might be useful.

As demonstrated here, the spatial resolution of existing space-borne remote sensing systems and the wise integration of different remote sensing sources enable the achievement of a high level of detail and accuracy, as long as the data are understood, processed and used in the right way. The proposed solution is attractive, less time consuming and less expensive compared to area regression estimators exclusively based on field survey. Furthermore, the remote sensing solution intrinsically provides a monitoring component (as agricultural area can vary during a season): this is often (or fully) not taken into account in the area regression estimator approach, simply because it is too time consuming to frequently repeat the field survey.

In our future work, other combination approaches will be tested, in order to optimize the exploitation of the complementarities of the three SAR sensors.

Acknowledgements

The European, Japanese, and Italian Space Agency are acknowledged for the provision of the ENVISAT ASAR, ALOS PALSAR-1, and Cosmo-SkyMed data. EFTAS, C-ITA, and the Ministry of Agriculture and Food Security of Malawi are acknowledged for the collection and provision of the ground survey data in Malawi.

This work has been done in the scope of the SARLAT project launched by the Belgian Ministry of Defense.

Author details

Nada Milisavljević[1], Francesco Collivignarelli[2] and Francesco Holecz[2]

1 Department of Communication, Information Systems & Sensors (CISS), Royal Military Academy, Brussels, Belgium

2 Sarmap, Purasca, Switzerland

References

[1] F. Holecz, M. Barbieri, A. Cantone, P. Pasquali and S. Monaco, "Synergetic Use of Multi-temporal ALOS PALSAR and ENVISAT ASAR Data for Topographic/land Cover Mapping and Monitoring at National Scale in Africa," in *IEEE International Geoscience & Remote Sensing Symposium (IGARSS 2009)*, Cape Town, South Africa, 2009.

[2] V. Krylov and J. Zerubia, "High resolution SAR image classification," Technical report 7108, INRIA Sophia Antipolis, France, 2009.

[3] N.-W. Park, "Accounting for temporal contextual information in land-cover classification with multi-sensor SAR data," *International Journal of Remote Sensing 31(2)*, pp. 281-298, 2010.

[4] D. Bargiel and S. Herrmann, "Multi-Temporal Land-Cover Classification of Agricultural Areas in Two European Regions with High Resolution Spotlight TerraSAR-X Data," *Remote Sensing 3(5)*, pp. 859-877, 2011.

[5] H. McNairn, J. Ellis, J. J. van der Sanden, T. Hirose and R. J. Brown, "Providing crop information using RADARSAT-1 and satellite optical imagery," *International Journal of Remote Sensing 23(5)*, pp. 851-870, 2002.

[6] X. Blaes, L. Vanhalle and P. Defourny, "Efficiency of crop identification based on optical and SAR image time series," *Remote Sensing of Environment 96(3)*, pp. 352-365, 2005.

[7] J. Shang, H. McNairn, C. Champagne and X. Jiao, "Application of Multi-Frequency Synthetic Aperture Radar (SAR) in Crop Classification," in *Advances in Geoscience and Remote Sensing*, InTech, DOI: 10.5772/8321, 2009.

[8] X. Wang, L. Ge and X. Li, "Pasture Monitoring Using SAR with COSMO-SkyMed, ENVISAT ASAR, and ALOS PALSAR in Otway, Australia," *Remote Sensing 5*, pp. 3611-3636, 2013.

[9] X. Jiao, H. McNairn, J. Shang and J. Liu, "The Sensitivity of Multi-Frequency (X, C and L-Band) Radar Backscatter Signatures to Bio-Physical Variables (LAI) over Corn and Soybean Fields," in *ISPRS TC VII Symposium—100 Years ISPRS*, Vienna, Austria, 2010.

[10] R. Fieuzal, F. Baup and C. Marais-Sicre, "Sensitivity of TerraSAR-X, RADARSAT-2 and ALOS satellite radar data to crop variables," in *IEEE International Geoscience and Remote Sensing Symposium (IGARSS 2012)*, Munich, Germany, 2012.

[11] D. P. Garrity, F. K. Akinnifesi, O. C. Ajayi, S. G. Weldesemayat, J. G. Mowo, A. Kalinganire, M. Larwanou and J. Bayala, "Evergreen Agriculture: a robust approach to sustainable food security in Africa," *Food Security 2(3)*, pp. 197-214, 2010.

[12] D. D. Mkwambisi, E. D. G. Fraser and A. J. Dougill, "Urban Agriculture and Poverty Reduction: Evaluating how Food Production in Cities Contributes to Food Security,

Employment and Income in Malawi," *Journal of International Development 23(2),* pp. 181-203, 2011.

[13] M. Douillet, "La relance de la production agricole au Malawi: succès et limites," Fondation pour l'agriculture et la ruralité dans le monde (FARM), Montrouge, France, 2012.

[14] F. Ellis and E. Manda, "Seasonal Food Crises and Policy Responses: A Narrative Account of Three Food Security Crises in Malawi," *World Development 40(7),* p. 1407–1417, 2012.

[15] M. Tafirenyika, "What went wrong? Lessons from Malawi's food crisis," *Africa Renewal 26(3),* p. 9, 2013.

[16] L. J. A. Mougeot, "Urban agriculture: Definition, presence, potential and risks," in *Growing Cities, Growing Food: Urban Agriculture on the Policy Agenda, a Reader on Urban Agriculture,* Feldafing, Germany, German Foundation for International Development (DSE), 2000, pp. 1-42.

[17] FEWSNet, "Malawi Food Security Outlook," July 2013. [Online]. Available: http://www.fews.net/docs/Publications/Malawi_OL_07_2013.pdf. [Accessed 17 September 2013].

[18] Food and Agriculture Organization of the United Nations (FAO), "Crop calendar" [Online]. Available: http://www.fao.org/agriculture/seed/cropcalendar/searchbycountry.do. [Accessed 24 11 2013].

[19] A. Rosenqvist, M. Shimada and M. Watanabe, "ALOS PALSAR: Technical outline and mission concepts," in *Proceedings of 4th International Symposium on Retrieval of Bio- and Geopyhisical Parameters from SAR Data for Land Applications,* Innsbruck, Austria, 2004.

[20] N. Milisavljević, F. Holecz, I. Bloch, D. Closson and F. Collivignarelli, "Estimation of crop extent using multi-temporal PALSAR data," in *IEEE International Geoscience and Remote Sensing Symposium (IGARSS 2012),* Munich, Germany, 2012.

[21] S. E. Umbaugh, Digital Image Processing and Analysis: Human and Computer Vision Applications with CVIPtools, Boca Raton, USA: CRC Press, Taylor & Francis Group, 2011.

[22] Y. L. Desnos, C. Buck, J. Guijarro, J.-L. Suchail, R. Torres and E. Attema, "ASAR – Envisat's Advanced Synthetic Aperture Radar," *ESA Bulletin, No. 102,* pp. 91-100, May 2000.

[23] S. M. Tavakkoli, P. Lohmann and U. Soergel, "Multi-temporal Segment-based Classification of ASAR Images of an Agricultural Area," in *Global Change Issues in Developing and Emerging Countries: Proceedings of the 2nd Göttingen GIS and Remote Sensing Days 2006,* Göttingen, Germany, Universitätsverlag Göttingen, 2007, pp. 175-188.

[24] S. M. Tavakkoli Sabour, P. Lohmann and U. Soergel, "Monitoring Agricultural Activities using Multi-temporal ASAR ENVISAT Data," in *Proceedings of XXIst ISPRS Congress "Silk Road for Information from Imagery", Vol.XXXVII, ISSN 1682-1750,* Beijing, China, 2008.

[25] A. Bouvet and T. L. Toan, "Use of ENVISAT/ASAR wide-swath data for timely rice fields mapping in the Mekong River Delta," *Remote Sensing of Environment 115(4),* pp. 1090-1101, 2011.

Chapter 4

Combining Moderate-Resolution Time-Series RS Data from SAR and Optical Sources for Rice Crop Characterisation: Examples from Bangladesh

Andrew Nelson, Mirco Boschetti, Giacinto Manfron,
Franceco Holecz, Francesco Collivignarelli,
Luca Gatti, Massimo Barbieri, Lorena Villano,
Parvesh Chandna and Tri Setiyono

Additional information is available at the end of the chapter

http://dx.doi.org/10.5772/57443

1. Introduction

Operational monitoring of agricultural extent and production is part of several national and international efforts to provide transparent, rapid and accurate information related to food security and food markets. Initiatives such as the G20 Agricultural Market Information System (AMIS), the European Commission's Monitoring Agricultural Resources mission (MARS), the Global Agricultural Monitoring (GEOGLAM) component of GEO, and the United States Department of Agriculture's Global Agricultural Monitoring Foreign Agricultural Service (GLAMFAS), are just some examples of operational services that do, or will require remote-sensing–based information on crop status in almost any part of the world.

Of the thousands of edible plants, just three—rice, wheat, and maize —provide 60% of the global population's food energy intake, and the top 15 crops amount to 90% [1]. Seasonal or monthly estimates of production and availability of these staples form a part of many agricultural bulletins and agricultural outlook reports. These reports are used for decision-making and policies on imports, exports, subsidies, and investments, which, in turn, affect prices. Food security is fundamentally about availability and price; such reports, and the responses to these reports, affect both. International events, such as the food price crisis of 2008, are the unintended outcome of national and international policy decisions that can detrimentally affect millions of people. More accurate and timely information to better inform policymakers is one way to reduce the likelihood of similar events in the future.

Crop bulletins and other regular assessments of crop area and production are a combination of statistics, surveys, reports from agricultural offices, Met Office forecasts, expert opinion, and remote-sensing information on vegetation, soil moisture, and rainfall. However, there is scope to increase the role of remote sensing in these reports if the information can be provided at the appropriate time, scale, and, wherever possible, with a focus on specific crops. Highly anticipated programmes such as the European Space Agency's (ESA) Sentinel constellation can provide exactly this type of information, although the demands for sustainably acquiring, processing, validating, and delivering the information are considerable.

This chapter provides a proof of concept for how a key staple crop can be monitored on a national scale using existing remote-sensing (RS) products that will soon be complemented and superseded by forthcoming sensors. Our exemplar crop is rice and our test country is Bangladesh (Figure 1). This region is highly suitable for a demonstration because of the triple remote-sensing challenge of pervasive cloud cover, small field size, and complex cropping patterns, which are typical of the vast and important agricultural areas of Asia, Africa, and Latin America. It is these areas where future gains in productivity must and will be made, not the agricultural areas of Europe, the US, and other developed regions where crop monitoring is substantially easier.

We first briefly describe the rice environments of Bangladesh, and then demonstrate how a combination of hypertemporal synthetic-aperture radar (SAR) and optical RS data can be combined to generate both baseline map information and near–real-time monitoring information on crop extent and crop seasonality.

2. Rice in Bangladesh

Agriculture is one of the most important sectors in Bangladesh's economy, accounting for approximately 20% of the gross domestic product (GDP) in 2010. It also accounts for the employment of over 60% of the country's population [2]. Rice is far and away the most important crop in the agricultural sector. Almost 60% of the land area is planted to rice and it provides food, employment, and income for much of the rural population. Bangladesh is one of the most populous countries in Asia, with almost 150 million people, according to the 2011 census. With over 1,000 people per square kilometre, it is one of the most densely populated countries in the world [2]. The country's milled rice consumption per capita is the highest in the world at 173 kg per year or almost 500 g of uncooked rice per day [1].

Rice can be cultivated in any of three seasons in Bangladesh in a myriad of cropping systems such as rice/pulse, rice/maize, rice/wheat, rice/vegetables, rice/shrimp and double-or triple-rice monoculture. The dry *boro* season, which runs from November to April, is largely dependent on irrigation and is of growing importance to Bangladesh's rice output. This is followed by the *aus* season (sometimes referred to as early *kharif*), which runs from March to

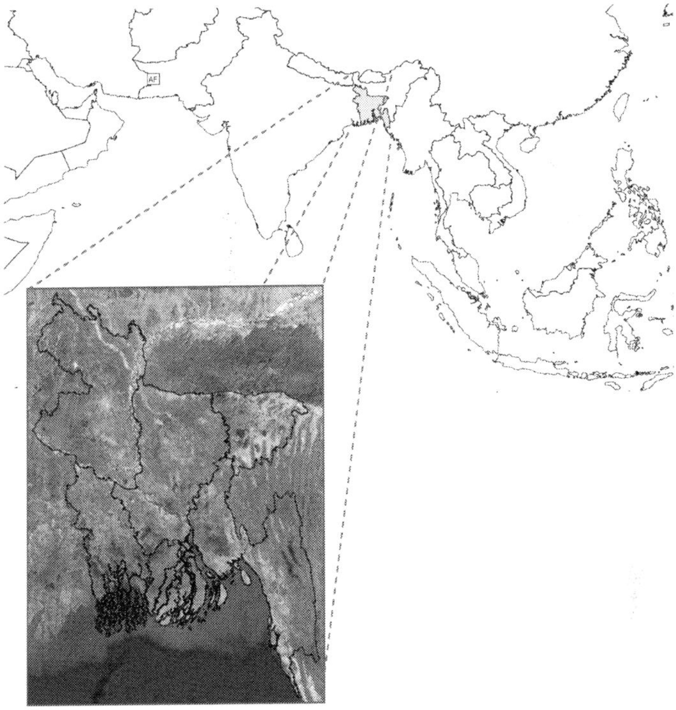

Figure 1. Location of Bangladesh in Asia.

August and relies on irrigation to establish the crop and early monsoon rains during the rest of the season. The *aman* season from June to December relies almost entirely on monsoon rains and is another major source of rice production. Deepwater aman rice grown in flooded conditions using tall varieties or floating rice starts in April or May, but is harvested in November/December. Figure 2 shows how these seasons overlap due to the large variation in (trans)planting and harvesting windows across the country. In general, the boro season starts in the east and south and moves northward; whereas, the rainfed aman season starts in the northernmost reaches of the delta and moves from north to south, although, local conditions, cropping patterns, and other management decisions can affect this trend.

Figure 3 shows the trends in rice cropped area per season for 1999-2012 and Figure 4 shows the reported area for each of the 64 districts in 2011 based on BBS annual reports [2-3]. Figure 3 shows the general stability in total rice area with extremes of 10.4 (2004) and 11.5 (2011) million hectares; the recent increases are mainly due to the expansion of the area in the *boro* season. Figure 4 shows the general spatial distribution of rice per season where the boro crop is mainly in the north and east, the much smaller aus crop area is in various clusters across the country, and the aman crop is dominant in both north and southern coastal regions.

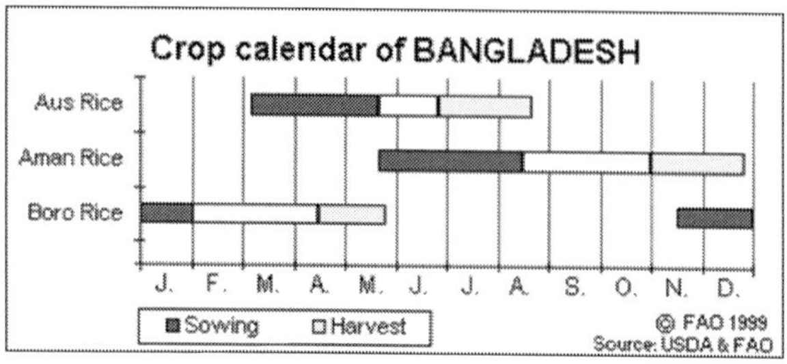

Figure 2. Generic rice crop calendar for Bangladesh (from FAO).

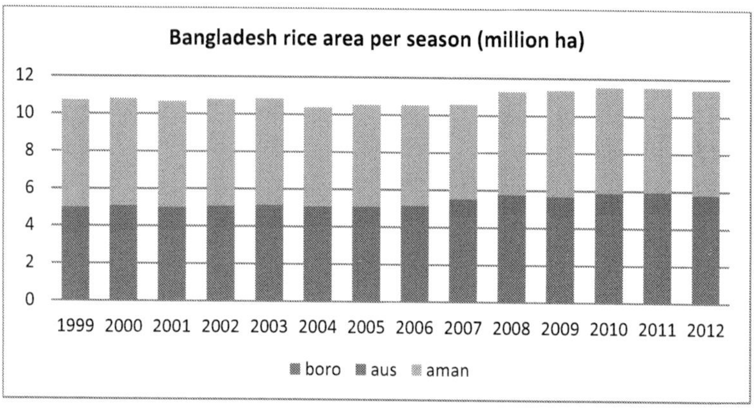

Figure 3. Bangladesh rice area by season 1999-2012 (Source: [2-3]).

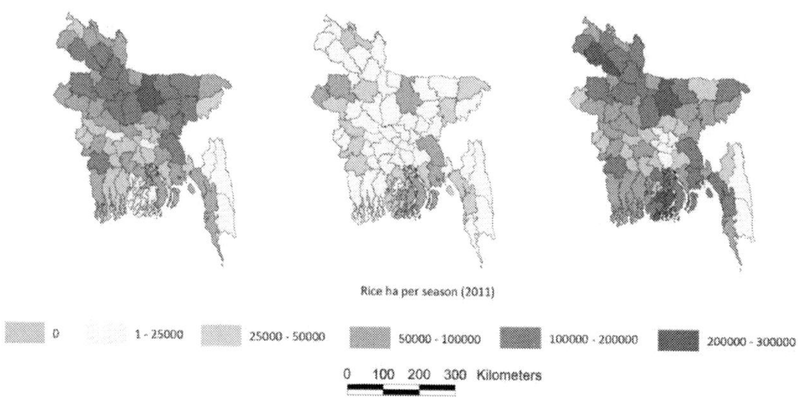

Figure 4. Bangladesh rice area per district in 2011 (l to r: boro, aus, aman) (Source: [3]).

Whilst production has increased steadily over the past 10 years, this trend faces multiple challenges. Flooding from excessive rainfall and the massive inflow from the Ganges (called Padma in Bangladesh), Brahmaputra (called Jamuna in Bangladesh), and Meghna rivers is estimated to severely affect 1.32 million ha and moderately affect a further 5.05 million ha. Crop losses from such massive inundation can be enormous. At the same time, drought brought on by a shortage of rainfall affects both the rainfed aman crop and the dry season crops, with a total drought-affected area estimated at 3.52 million ha. Salinity and tidal surges from cyclones affect a further 1 million ha of cultivated land, and any given area can suffer from one or more of these stresses during the year [4]. Severe yield losses are also attributed to pests, diseases, weeds, and other biotic yield reducers.

The possible impacts, both positive and negative, of changes in climate, society, economy and technology are hard to quantify, but the fundamental message is that Bangladesh faces multiple challenges to food production which are the focus of ongoing research and development across the country, such as stress tolerant varieties, innovative cropping systems, investments in infrastructure, extension and training. Bangladesh has been identified as one the countries with the highest vulnerability to anticipated climate change and there is continued need to deliver the best possible crop status information to policy makers as a contribution to sustainable and resilient agricultural production.

In summary, rice can be cultivated during any month and it faces different pressures in each of the three seasons. Rice is cultivated in varying amounts in every district with differing spatial patterns each season. Thus any RS-based monitoring requires frequent observations throughout the year across the entire country. For this reason, hypertemporal imagery with a large footprint, such as MODIS and ENVISAT WS products, are good choices for baseline rice extent mapping and rice seasonality monitoring. They also serve as a proof of concept of what can be achieved at higher spatial and temporal resolutions with future satellite programmes in an operational monitoring context.

3. Methods and data

3.1. Multi-temporal SAR and optical data for rice mapping and monitoring

Field conditions and crop evolution follow well understood seasonal changes, which can be observed by multi-temporal remote-sensing data. Combined with a priori knowledge of the crop calendar and land practices, this multi-temporal remote-sensing data can be correctly interpreted to deliver valuable information. For rice, at the start of the season, remote-sensing time-series can determine when and where fields are prepared and irrigated. The same time-series information can also capture the various crop stages, starting from seeding or transplanting and then through the vegetative, reproductive, and ripening stages until harvest. In synthesis, assuming that images are regularly acquired, the key source of information about crop presence and status is the temporal signature [5-9]. Systematic acquisitions are essential; one image or randomly acquired scenes, even with very high resolution, are of limited use for crop monitoring.

SAR systems have a proven ability to detect irrigated and lowland rice through the unique backscatter temporal signature. In the past three decades, a significant number of publications have been dedicated to the rice signature and to its detection and monitoring [10-17]. In summary, these studies have shown that:

- Lower frequencies (L-and C-band) penetrate deeper into the rice plant than higher frequencies, whereas higher frequencies (X-band) interact better with rice grains.
- Double bounce dominates at L-band, whereas volume scattering prevails at X-band.
- The correlation between the backscattering coefficient and rice bio-physical parameters shows that, lower frequencies are more related to the total fresh weight, leaf area index, and plant height; whereas higher frequencies better correlate with grain weight and grain water content.
- The date of the maximum backscattering coefficient at X-band precedes those of C-and L-band.
- The VV backscattering coefficient increases only during the vegetative stage; it is quite stable at the reproductive stage, and it decreases at the ripening stage due to the canopy attenuation.
- The HH backscattering coefficient increases at the reproductive stage, and it is quite stable at the ripening stage. The temporal trend of the HV backscattering coefficient is similar to the HH one.
- The VV/HH polarization ratio, at C-and L-band, significantly decreases throughout the course of the crop season, hence proving its sensitivity to rice plant age.
- The frequency ratios for HH and VV (C-VV/L-VV and C-HH/L-HH) are significantly lower in the latter part of the rice season, when thick vegetation canopy hampers wave penetration.

High-resolution optical images in the tropics are strongly limited for crop monitoring purposes, especially rice. Most rice is grown in the rainy or monsoon season and images often suffer from persistent and widespread cloud contamination. High-resolution images are rarely available with high temporal frequency, so gap-filling and smoothing options are limited. On the contrary, moderate-resolution systems with daily revisiting cycles, such as MODIS, have the appropriate spectral bands and temporal resolution needed for crop identification and phenological monitoring at regional scales.

Although sometimes neglected and rarely the subject of discussion in remote-sensing literature, the impact of meteorological conditions and crop practices on rice plant growth is often significant. Some prior knowledge of the rice variety, the crop calendar, varietal maturity, crop management practices for water and inputs, and meteorological conditions is prerequisite for the correct interpretation of the temporal signature and for the generation of an accurate 'rice map', whatever the specific information (i.e., rice area, start of season date, etc.) in the map. This agronomic and meteorological information must be considered in the development of a rice detection algorithm.

http://dx.doi.org/10.5772/57443

Thus, there is an opportunity to use both SAR and optical time-series information for rice crop detection and monitoring, provided that there is sufficient a priori knowledge of the rice system under observation. The following approach is based on specific peculiarities and complementarities of high temporal resolution and moderate spatial resolution SAR and optical data; however, the approach can also use higher resolution imagery if available.

Given the higher spatial resolution but limited revisiting time (ranging from 16 to 24 days), SAR data are primarily used to estimate the spatial extent (resolution 100 m) of the rice crop. Medium-resolution optical data with a quasi-daily revisiting cycle are used to derive the land surface phenology as a means to identify key stages of the rice crop season. There are two distinct processing chains to develop rice monitoring products on an operational basis: one for the rice extent/area estimation from SAR, another for rice seasonal monitoring within that extent/area from MODIS. The main advantages of this approach are:

- It overcomes the spatial-temporal problem, hence, assuring an appropriate temporal repetition at an adequate scale (i.e., spatial resolution) even over large areas;
- It provides an operational monitoring system, which is based on multi-sensor data redundancy.

Figure 5 illustrates the overall approach. Depending upon data availability, rice products based on archive and actual remote-sensing data at different spatial scales can be generated. Here, we will discuss the use of archive ENVISAT ASAR Wide Swath data (100 m) and MODIS (500 m), leading to a multi-year rice extent and phenological monitoring product.

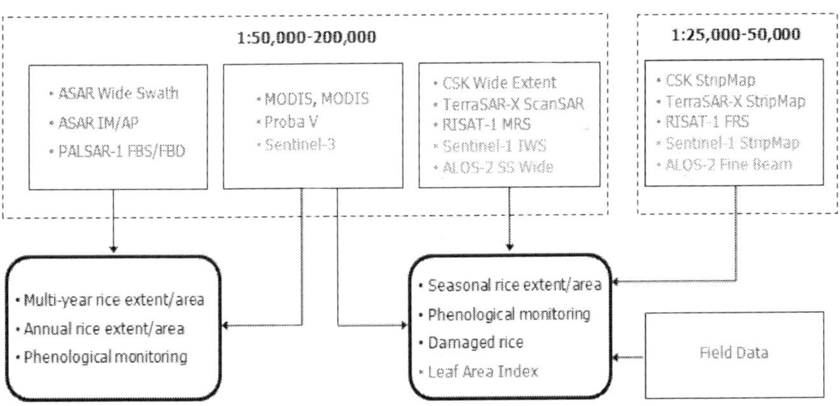

Figure 5. Products and supported sensors (past, current, near future) including acquisition modes.

3.2. Multi-year rice extent from archive ENVISAT ASAR Wide Swath data

The C-band Advanced Synthetic Aperture Radar (ASAR) on board of the ENVISAT satellite ensured continuity of the ERS-1/2 SAR systems between 2002 and 2012. It featured enhanced capability in terms of coverage, range of incidence angles, polarisation, and modes of operation. There were three ASAR acquisition modes with a 35-day repeat cycle:

- Wide Swath mode-WS (400 by 400 km, 100 m resolution, HH or VV),
- Image mode – IM (100 by 100 km, 15-25 m resolution, HH or VV), and,
- Alternating Polarization – AP (100 by 100 km, 15-25 m resolution, HH/VV or HH/HV or VV/VH).

The 10-year ENVISAT ASAR WS data archive is mainly based on background mission acquisitions, hence, it was not specifically planned and not optimised for agricultural applications. Nevertheless, the large archive can be exploited for rice mapping with appropriate processing and interpretation. A fully automated processing chain has been developed to convert the images into terrain geocoded backscattering coefficient (σ°).

1. Co-registration – Images acquired with the same observation geometry and mode are co-registered in slant range geometry. This step is mandatory to allow time-series speckle filtering.

2. Time-series speckle filtering – An optimum multi-temporal filtering is used to balance differences in reflectivity between images at different times [18]. Multi-temporal filtering is based on the assumption that the same resolution element on the ground is illuminated by the radar beam in the same way, and corresponds to the same slant range coordinates in all images of the time series. The reflectivity can change from one time to the next due to a change in the dielectric and geometrical properties of the elementary scatters, but should not change due to a different position of the resolution element with respect to the radar.

3. Terrain geocoding, radiometric calibration, and normalization – A backward solution by considering a Digital Elevation Model is used to convert the positions of the backscatter elements into slant range image coordinates. A range-Doppler approach is used to tranform the three-dimensional object coordinates in a cartographic reference system into the two-dimensional row and column coordinates of the slant range image. During this step, the radiometric calibration is performed by means of the radar equation, where scattering area, antenna gain patterns, and range spread loss are considered. Finally, the backscattering coefficient is normalized according to the cosine law of the incidence angle in order to compensate for the range dependency.

4. Anisotropic Non-Linear Diffusion Filtering – This filter significantly smoothes homogeneous targets, whilst also enhancing the difference between neighbouring areas. The filter uses the diffusion equation, where the diffusion coefficient, instead of being a constant scalar, is a function of image position and assumes a tensor value [19]. In this way, it is locally adapted to be anisotropic close to linear structures such as edges or lines.

5. Removal of atmospheric attenuation – Although microwave signals have the ability to penetrate through the clouds, it is possible, particularly at short (X-and C-band) wavelengths, that severe localised storms can affect the backscattering coefficient in the range of several dB. The temporal signature of the backscatter coefficient can be affected in two ways: (1) the thick layer of water vapor generates a strong decrease of backscattering coefficient, followed by a strong increase; (2) the strong rainfall generates a strong increase of the backscattering coefficient, followed by a strong decrease. These effects are corrected in the processing chain by analyzing the temporal signature—anomalous peaks or troughs are identified and the backscattering coefficient values are corrected by means of an interpolator. This correct application of this process relies strongly on a priori knowledge of the rice system and the weather conditions when the image was acquired.

Once the WS archive has been processed, it is ready for interpretation. Rice detection relies on interpretation of the temporal signature from regular acquisitions within a given year or season. The 2002-2012 WS archive does not have regular acquisitions over any one area and so, we made the following assumption. Given that rice extent in Asia is relatively constant (see Figure 3 for evidence of this in Bangladesh) and cropping calendars are relatively stable, we can then develop a pseudo-annual time-series with a relatively high temporal occurrence by combining all years of WS observations into one year. This is illustrated in Figure 6 where the multi-year SAR data for a selected time frame (e.g., weekly, bi-monthly) is temporally averaged leading to an annual signature.

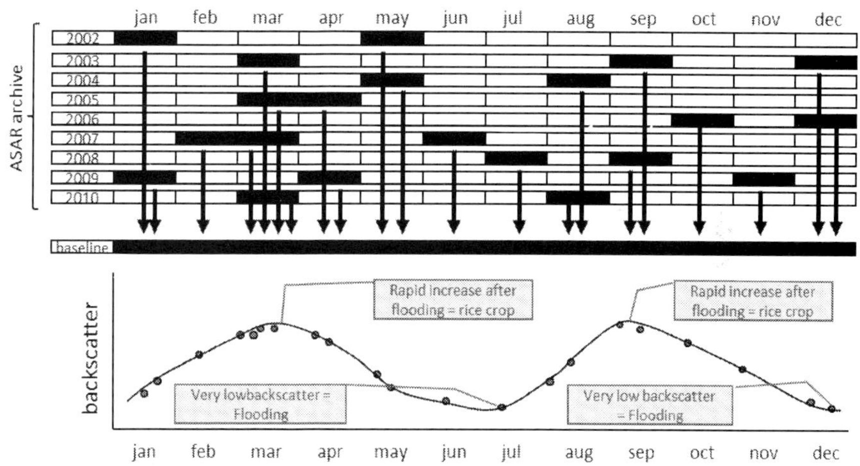

Figure 6. Example of the pseudo-annual time series for a single pixel based on multi-year ASAR Wide Swath data (2002-2010).

This temporal signature can be interpreted in two ways:

1. The temporal evolution of the backscattering coefficient is analyzed from an agronomic perspective. This assumes that the data have been regularly acquired in temporal terms, and that a priori knowledge of rice type, calendar and duration, crop practice, and meteorological conditions during the whole season is available. This is not the case for the WS archive, so we move to option two.

2. From the backscattering coefficient time series, representative temporal features (or descriptors) can be derived. Even if this approach cannot derive specific information on the rice phenology [9], it can provide valuable information on the rice location and, depending upon the temporal frequency of the SAR acquisitions, the various rice seasons. We use this approach for interpreting the signature from the WS archive.

The most representative temporal features for rice have been found to be the relative minimum and maximum, their difference, and minimum and maximum increment between two subsequent acquisitions. The selected five temporal features are used to generate rice extent and rice seasons according to:

1. The start of the rice season is identified when there is a relative minimum followed by a maximum increment between two subsequent acquisitions;

2. The peak of the rice season is identified when there is a relative maximum followed by a minimum increment between two subsequent acquisitions;

3. The pixel is classified as rice if:

   - conditions 1 and 2 are satisfied;
   - the range between relative minimum and maximum attains a minimum value;
   - the temporal duration between 1 and 2 is within a given duration;

4. Rice is further distinguished in single, double, triple, etc., according the number of valid minima and maxima (condition 3).

3.3. Phenological monitoring with MODIS data

3.3.1. MODIS data for land surface monitoring

With the launch of the Moderate Resolution Imaging Spectroradiometer (MODIS), onboard the NASA TERRA (EOS AM-1) satellite, a new era in multi-spectral satellite remote sensing began. MODIS sensors permit continuous monitoring of the environment by measuring spectral bands from the blue to the thermal infrared. The 36-band MODIS spectrometer provides a global data set every 1-2 days. The swath dimensions of MODIS are 2,330 x 2,330 km and the spatial resolution (pixel size at nadir) is 250 m for channels 1 and 2 (0.6 μm-0.9 μm), 500 m for channels 3 to 7 (0.4 μm-2.1 μm) and 1,000 m for channels 8 to 36 (0.4 μm-14.4 μm).

There are many standard MODIS data products that are provided to users in near–real-time and easy-to-use formats for no cost. The MOD09A1 product (a full technical description is available online at https://lpdaac.usgs.gov/products/modis_products_table/mod09a1) provides 8-day composite reflectance data across seven spectral bands (red-0.6 μm, NIR-0.9 μm, blue-0.4 μm, green-0.5μm, NIR-1.2 μm, SWIR-1.6 μm, SWIR-2.1 μm) at 500 m spatial resolution, as well as pixel-specific quality control flags and observation dates. This product type is derived from a multi-step process that considers atmospheric, cloud, and aerosol corrections, and records the best reflectance data registered during the time composite window for each pixel. MODIS product data are provided in Hierarchical Data Format (HDF) on a tile system with the Sinusoidal projection grid. Each tile of the grid covers an area of 1,200 × 1,200 km, more or less 10° of latitude and longitude (Figure 7A). Despite drawbacks related to spatial and spectral resolution, it is the zero cost, ease of access, clarity of product description, and ready-to-use nature of MODIS products that have contributed to their widespread use as a source of remote-sensing information for monitoring.

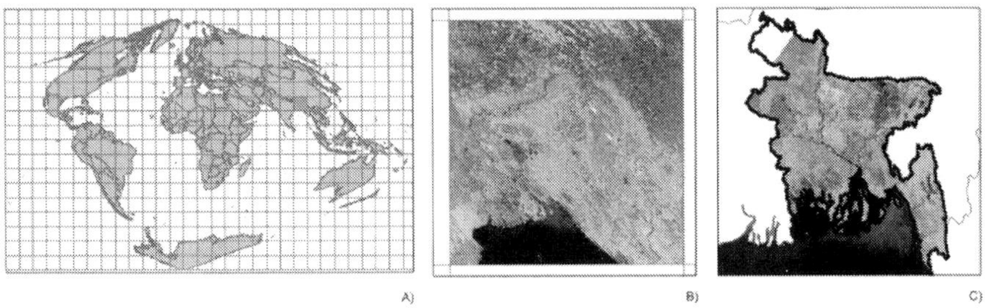

Figure 7. The study area: The location of MODIS tile H26V06 in the Global Sinusoidal projection map (A), tile H26V06 which covers most of Bangladesh (B), and, Bangladesh in geographic coordinates showing the extent of H26V06, which covers all the country except the far the north west (C).

3.3.2. PhenoRice algorithm

MOD09A1 data for 2011 from the tile H26V06 (Figure 7B) were used to demonstrate seasonal rice monitoring in Bangladesh (Figure 7C) using a rule-based automatic algorithm called "PhenoRice." The aim of the PhenoRice approach is to detect paddy rice by analysing multi-spectral data with quasi-daily revisiting cycles (e.g., TERRA/AQUA-MODIS, SPOT-VGT, etc.) in a consistent and flexible way that minimises dependency on local threshold adaptation.

The basis for this approach is described in previous publications by CNR-IREA [7,20,21]. Before any seasonal phenological monitoring can be performed, PhenoRice needs an estimate of the rice-growing area. If there is no rice extent map available, then the PhenoRice process relies on the work of [6,22-23] which identifies irrigated or lowland rainfed rice when a clear and unambiguous agronomic flood is detected and is shortly followed by a rapid increase in vegetation growth. However, the process can also use available land cover maps that show rice extent, or use multi-year rice extent maps such as in this case study. The seasonal phenological monitoring part of PhenoRice is performed by analysing the temporal signatures of various vegetation indices [20].

Figure 8 shows a schematic diagram illustrating the concept and steps of PhenoRice algorithm. The algorithm involves three fundamental processing steps.

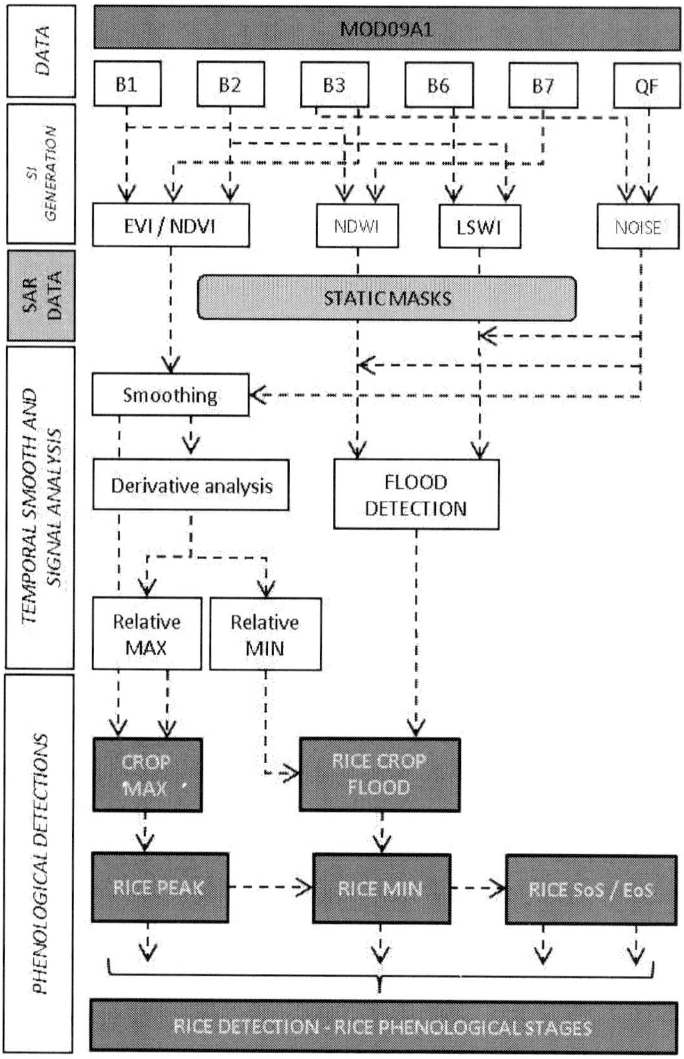

Figure 8. PhenoRice flow chart.

1. Pre-processing of MODIS composite data

  - **a.** The identification of noise in the data by analysing the MODIS Science Data Set (SDS) cloud and snow quality flags and blue band reflectance values. Quality flag values are extracted from HDF layer 11 (500 m State Flag): bits 0-1, 3-5, and 15 are analysed to identify cloud, land, and snow noises, respectively. This information, together with a threshold on the blue band, is then combined to evaluate noise in the spectral data and error in the computed spectral indices.

  - **b.** The computation of time series of spectral indices (SI) to perform (1) flood detection, using moisture [Land Water Surface Index, LWSI [22]] and water [Normalized Difference Water Index, NDWI [24]] indices, and (2) vegetation growth monitoring using the Enhanced Vegetation Index, EVI [25]. The red (layer 1), NIR (layer 2), blue (layer 3), and SWIR (layer 7) bands are extracted from each HDF file to compute the EVI, LSWI, and NDWI indices.

2. Smoothing and temporal signal analysis

  - **a.** The EVI time-series is smoothed in order to reduce contamination that still affects the MODIS data after the temporal compositing procedure. In the smoothing process, the raw EVI profile is analysed to detect and clean outlier values, i.e., anomalous peaks and irregular small drops. Finally, a temporal smoothing on the EVI data is applied using a local polynomial function that weighs observation in relation to cloud contamination (output of step 1a), which is based on a Savitzky-Golay filter [26]. This filter is specifically recommended for the interpolation of vegetation index time-series.

  - **b.** Derivative analysis: The smoothed signal is then analysed in order to calculate the first derivative. These derivatives are used to automatically identify all points of local (relative) minima (where the first derivative changes from negative to positive) and local (relative) maxima (where the first derivative changes from positive to negative).

3. Rice and phenological detection

  - **a.** Rice transplanting/seeding (MIN point) is identified when a local EVI minima occurs at the same time as a flood (NDWI, LSWI), and a series of positive EVI derivative values (indicating plant growth) occur shortly after.

  - **b.** Rice heading/flowering (MAX point) is detected when there is a local absolute maxima in the EVI time-series.

  - **c.** Finally, following [7], the smoothed EVI signal is analysed to extract rice emergence [start of season (SoS)] and maturity [end of season (EoS)]. These metrics are identified for each rice pixel, when EVI values match pixel-specific relative thresholds.

All the PhenoRice outputs refer to a temporal window of 12 months from 1 January to 31 December with a temporal step of 8 days corresponding to a MODIS composite granularity. For the year under analysis, referred to as Current Year (CY), the method is able to provide information on the different rice crop seasons (up to three) that occur in the period. It is assumed that a rice season belongs to the given CY, if at least the maximum (i.e., heading) of the crop occurs in this period (from January to December).

In order to provide a flexible method able to detect rice seasons in different environmental and climatic conditions with a specific crop calendar, we assume that the start of the season (flooding and seeding/transplanting) can occur in the previous calendar year (PY) up to six months before the start of CY. This results in an 18-month analysis period, which corresponds to 24 8-day MOD09A1 composites (192 days) in the PY and 46 8-day composites in the CY under analysis.

Finally, phenological metrics maps are produced for each quarter of the year. A rice detection is therefore attributed to the quarter, in which heading occurs. Consequently, the associated MIN, SoS, and EoS detection are referred to the same quarter, but as separate maps. In the case study of Bangladesh, output maps have been post-processed and synthesized for the specific cropping seasons of the country. Following the knowledge of the local crop calendar, all the detection that occurs between February and late April are attributed to boro, the ones between May and late July to aus, and others in the period from August to late November to aman.

4. Results

4.1. Multi-year rice extent from the ENVISAT WS archive

The 2002-2010 ASAR Wide Swath data stack consists of around 340 frames. Four scenes (in two different orbits) were needed to completely cover the country. Hence, on average, there are 85 frames across the 8-year period, corresponding to around 10 images per month in the pseudo-annual time-series approach (Figure 6). In practice, the amount of images per month in the pseudo-annual stack varies between 3 and 16, meaning that the archive data is not suitable for annual or regular monitoring purposes. Figure 9A illustrates the country coverage mosaic of three temporal features: the relative minimum, the relative maximum, and the maximum variation.

The three temporal features, illustrated as a mosaic in Figure 9A, are exclusively carried out for visualization purposes. Indeed, the applied gradient algorithm, which is used to produce a seamless image, may significantly change the radiometric content in the overlapping areas; hence, it is not suitable for data analysis purposes. Looking at the temporal features at country level (Figure 9A), three main colours can be identified: blue, yellow, and light and dark orange to brown. The blue colour represents the minimum, i.e., a constant low radar backscatter during the whole year. Yellow, identified as rice and primarily located in the northeast of Bangladesh, is the combination of a strong temporal variation (red) and a high maximum (green). Light and dark orange to brown, distributed over the whole country, classified as rice, is essentially a combination of a strong temporal variation (red) with a medium maximum (green).

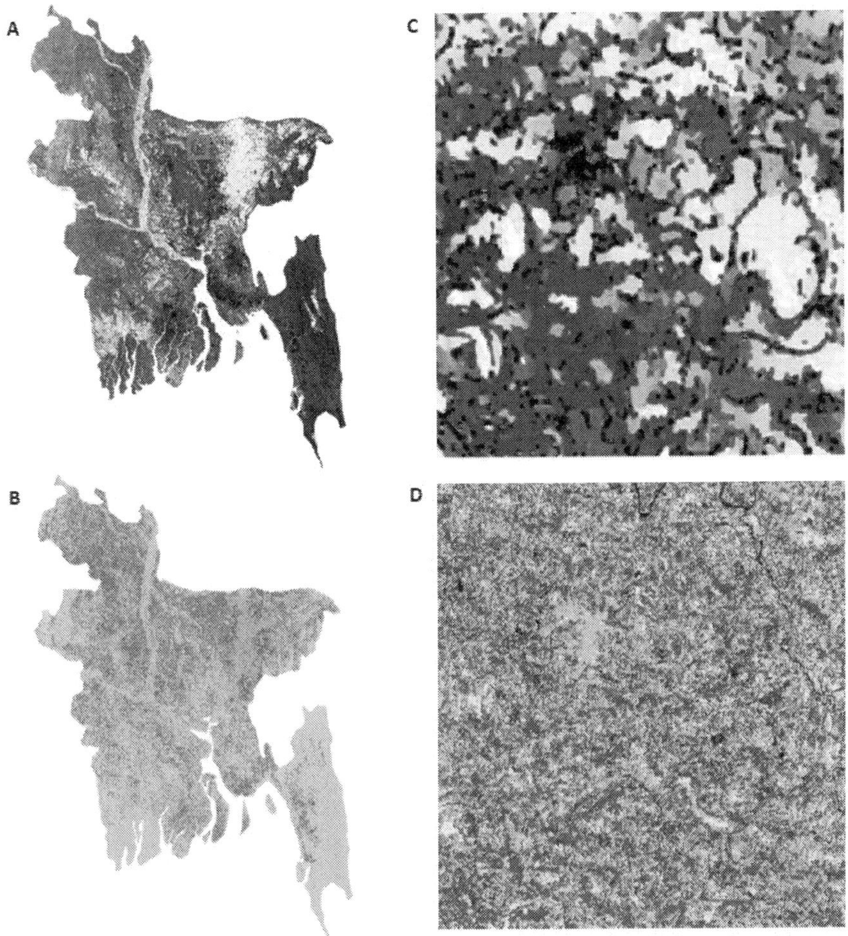

Figure 9. (A) Mosaic (100 m) of temporal features based on ENVISAT ASAR Wide Swath data acquired from 2002 to 2010. Red is the maximum variation, green is the relative maximum, and blue is the relative minimum. (B) Rice extent based on temporal features of archive ENVISAT ASAR Wide Swath data. Light green is single rice, dark green is double rice, aquamarine is rice grown after a long flooding duration (up to six months), and orange is rice mixed with other crops. (C) Enlargement of A (red box). (D) Temporal features (15 m) based on ENVISAT ASAR Alternating Polarization data acquired from June 2011 to March 2012 on the same area shown in C. Red is the maximum variation, green is the relative maximum, and blue is the relative minimum.

Due to the different acquisition times and geometry between the two orbits, the five temporal features – derived from the monthly averaged backscattering coefficient – are computed for each orbit separately. The rice extent is generated according to the rules described in section 3.1 and, subsequently, products belonging to the two orbits are shown as a mosaic at semantic level (Figure 9B). In this specific case, rice extent additionally includes the various rice crop seasons, which could be identified with this data set because of the good temporal frequency. Further distinction between single rice, double rice, and rice cultivated after prolonged flooding is provided by the time component—the single rice crop has a short to medium (90-120 days) duration; double rice usually means two short duration crops; and boro rice, after prolonged flooding in the northeast, is usually medium duration. This is a coarse characterisation of rice seasons, due to the fact that temporal features, derived from monthly averaged backscattering coefficient acquired irregularly, are not sufficient to follow the rapid spatial-temporal changes of agriculture. Nevertheless, this pseudo-annual data set enables the generation of a detailed rice–non-rice extent at a one-hectare scale.

With regard to the consistency of the proposed approach and to the impact of spatial resolution, temporal features derived from ASAR Wide Swath (Figure 9C) are compared to those derived from ASAR Alternating Polarization data (Figure 9D). This 15-m resolution data set has been acquired from June 2011 to March 2012 every 35 days, according to the repeat cycle of ENVISAT ASAR. The visual comparison shows that the land cover patterns, particularly of rice, are similar. It means that the multi-year (or pseudo-annual) approach is conceptually correct and viable in this case.

Another interesting observation can be made by comparing the two colour composites—spatial resolution plays an important role. Large and homogeneous regions are clearly identifiable in both acquisition modes, while small features tend to be smoothed out and to disappear. Rice covers a substantial part of the country, and although plots are small, the rice areas in much of the country are large and connected, especially in the rainy season, where rice is often the only viable crop. From analysis of SAR data at a range of resolutions (3 to 100 m), we estimate that an accurate area estimation using this approach would need a pixel size at least five times smaller than the field size. Medium and lower spatial resolution systems should be therefore exclusively used for the provision either of extent information — a proxy of area — or, in case of agricultural applications, of phenological crop monitoring.

The ASAR Alternating Polarization HH/HV data were used to further compare the pseudo-annual ASAR Wide Swath data to the quasi-annual HH/HV data, particularly with respect to the contribution of the cross-polarization for rice detection. Figure 10 (left) illustrates the temporal signatures of the ASAR WS Swath monthly averaged HH backscattering coefficient from January to December and those at HH and HV acquired in Alternating Polarization mode. First, this reaffirms that the use of multi-temporal data is essential, particularly at HH polarization. Second, the correspondence between the two HH radar backscatter signatures is evident, confirming the validity of the proposed approach. Third, and confirming the studies of [12,16], the HV polarization contributes in the discrimination of rice from other land covers. Typically, the HH/VV ratio is preferred because it shows a variation up to 7 dB from the beginning of the season to the plant maturity phase [11].

Finally, the quasi-annual HH/HV time-series is used to determine two key moments of the rice growth — start of season and peak of season (Figure 10, right). The first happens when the relative minimum is detected (September, blue colour), whereas the second is detected when the relative maximum is reached (November, green colour). A comparison with the generic

http://dx.doi.org/10.5772/57443

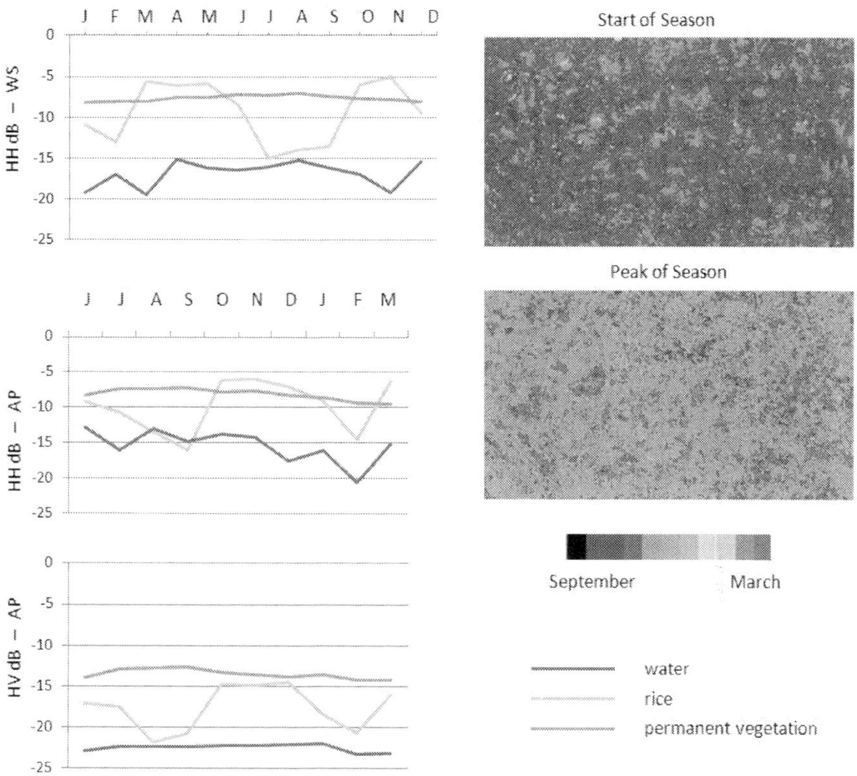

Figure 10. Top left: Temporal signatures derived from ENVISAT ASAR Wide Swath HH data acquired from 2002 to 2010. Middle left: Temporal signatures derived from ENVISAT ASAR Alternating Polarization HH data acquired from June 2011 to March 2012. Bottom left: Temporal signatures derived from ENVISAT ASAR Alternating Polarization HV data acquired from June 2011 to March 2012. Right: Detected start of season and peak of season based on ENVISAT ASAR Alternating Polarization HH data acquired from June 2011 to March 2012.

crop calendar of Figure 6 shows a good correspondence with the identified dates. However, it should be noted that a 35-day repeat cycle is not sufficient to obtain an accurate estimate of these two key moments. For this reason, the use of remote-sensing data acquired with higher temporal frequency is of advantage.

4.2. 2011 seasonal rice monitoring with MODIS

4.2.1. Rice season map

The number of rice seasons detected is reported in Figure 11A. For the year 2011, the analysis of optical data time-series revealed that most areas have one rice season. Two seasons were identified in the northern regions, while up to three seasons (red area in Figure 11A) were found in the northwest part of the country in Rajshahi Division. Areas with one or two rice seasons per year may cultivate a second or third non-rice crop, but those non-rice cropping seasons are not captured in the PhenoRice algorithm, nor is it represented in the map.

The same analysis also produces rice extent maps for the boro, aus, and aman seasons. The main rice cultivation (~% 49) occurs in the boro period (Figure 11B), followed by aman (Figure 11D) during the wet season (~% 45). The aus crop is very small (6%) in comparison to the other two, and occurs mainly in the central and western part of the country (Figure 11C).

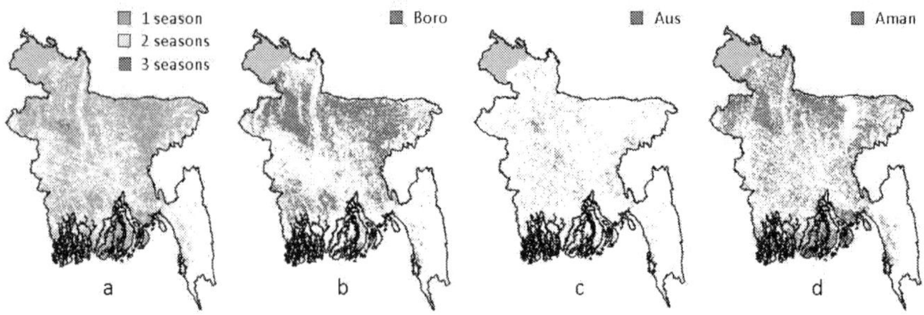

Figure 11. Total detected rice seasons (A) and cultivated area for boro (B), aus (C) and aman (D) seasons.

Figure 12 reports some examples, extracted from single pixels, of temporal analysis performed by PhenoRice. The panels provide examples for single rice in boro and aman (A, B), double rice on both boro and aman (C), and triple rice in all three seasons (D). As previously described, the algorithm analyses the temporal signature of three spectral indices — EVI (in green), LSWI (blue), and NDWI (light blue)—for vegetation and moisture/water. The figures also report the noise level for each composite data, mainly because of cloud contamination and are represented with grey dots, and the occurrence of the detected phenological stages. The symbols in red (square, triangle, diamond, and circle) represent agronomic flooding (MIN – crop transplanting/seeding),; the plants' emergence (SoS-start of the vegetative period of rapid growth); the crop heading/flowering (MAX – start of the reproductive period); and finally, the estimated crop maturity after senescence (EoS).

Figure 12A (pixel, 24°17′54.00″N – 90°59′48.46″E in the Sylhet basin in northeastern Bangladesh) shows the presence of a single rice crop in the boro season; the peculiarity of this location is (1) the strong presence of surface water for nearly six months and (2) a medium crop vegetative phase of about 120 days (period between EoS and SoS occurrence). The identification of these characteristics leads to the assumption that cultivation occurs after prolonged flooding. This observation perfectly matches the ones from the ASAR map and represents an area of low cropping intensity because of excess surface water.

In particular, Figure 12B (pixel, 22°1′30.00″N – 90°6′50.37″E in Barisal Division in coastal Bangladesh) reveals a single long-duration (150 days) rice crop in the rainfed aman season

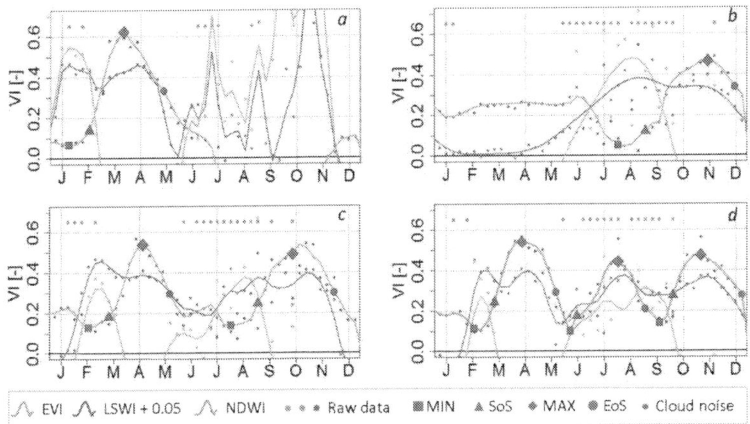

Figure 12. Example of the temporal dynamics of the vegetation indices — EVI (green line), LSWI+0.05 (blue line), and NDWI (light-blue line), and phenological detection for MIN, SoS, MAX, and EoS. Panels A, B, C, and D show a single season for boro rice, a single season for aman rice, a double season (boro-aman), and a triple season (boro-aus-aman), respectively.

with (1) crop heading (MAX) occurring in November in the fourth quarter of the analysed season, (2) a transplanting period (MIN) in June, (3) a green-up (SoS) in July, and (4) a crop maturity (EoS) in December. This area is highly saline in the dry boro season and no cultivation is possible so the land is left fallow for much of the year, until freshwater availability from rivers or canals increase after the monsoon rains. Thus, this is another example of environmental conditions limiting cropping intensity. The agronomic flooding from May to August that precedes the crop greening is particularly clear from the interpretation of LWSI and NDWI indices trends. The former, as described by Xiao et al. (2005), shows values far greater than EVI, whereas the latter presents values greater than zero.

Figure 12C (pixel, 25°10′60.00″N – 89°14′42.64″E in Ranjpur Division, northwestern Bangladesh) shows the detection of two rice crops in 2011. The first boro crop occurs during the first and second quarters, from February to June, with crop heading in late March; whereas, the second one, aman rice, takes place in the third and fourth quarters, starting in July and finishing in late October. The boro crop is a short-duration one, 90 to 100 days, whereas the aman rainfed crop is much longer, 130-150 days.

Finally, Figure 12D (pixel 24°36′45.00″N – 89°24′6.25″E, Rajshahi Division in central Bangladesh) illustrates the case of three rice seasons, heading occurrences are in February (boro), July (aus), and October (aman), respectively. In this case, it is possible to appreciate that the crop durations are much shorter than the ones in the previous examples — they are all 90 to 110 days — and could represent short-duration, high-yielding varieties that have done much to increase rice productivity in Bangladesh since the Green Revolution.

4.2.2. Phenological detection maps

Figure 13 shows the phenological detection results for the year 2011. The data represent a spatial analysis conducted at sub-district or *thana* level, taking into account the median value of the different phenological data occurring within each sub-district. Pixel-level interpretation and visualisation is challenging because of the inability of the algorithm to detect every occurrence, the well-known low-resolution bias of MODIS data when observing fields, and the heterogeneity of the cropping systems [27]. Thus, given that PhenoRice is a very conservative approach to detection (in order to minimize false positive error) and, if we accept that PhenoRice can only capture a proportion of the phenomena understudy, this aggregation of data around a median value gives a more robust interpretation of the phenology data for a selected unit of analysis. Furthermore, we excluded sub-districts if there were fewer than 30 pixels detected for a given season.

The boro maps confirm the well-known patterns of agronomic practices. Fields start to be flooded (MIN boro map) in the southeast, red (Jan 2011), and then the season progresses northwest, orange (Feb 2011). The SoS boro map highlights this trend more clearly. On the other hand, the aman season starts in the north and heads south due to the north-south progression of freshwater in the river systems as depicted by the light green (Jun 2011) and green (Jul-Aug 2011) colours in the MIN map, respectively.

During the boro season, no rice crop is detected in the southwest since that corresponds to the dry season, where there is insufficient freshwater for irrigation and the region is exposed to high water salinity. Freshwater availability increases in the aman season in the southwest during that time. The correct detection of these two seasons in this part of the country suggests that this approach is well suited to the detection of seasonality.

The analysis of the aus map is more difficult because of the smaller and more fragmented area of the rice crop in this season. Aus rice detection is concentrated mainly in the central part of Bangladesh, along the Ganges River, and on the eastern region of Chittagong. The aus crop relies on both early-season irrigation and late-season rainfall, and this partially explains the limited extent. Two cropping patterns could appear: the first one, much earlier in the middle of the country, and the second, a later pattern for the rest of the country.

5. Discussion

5.1. Observations on the SAR analysis and results

Temporal analysis of moderate-resolution SAR, especially using the backscattering coefficient ($\sigma°$) from C-band time-series, is ideally suited for irrigated and lowland rice detection. It is particularly advantageous in monsoon Asia where much of the world's rice is produced over huge areas and under cloudy conditions.

The use of the 10-year ENVISAT ASAR Wide Swath data archive, even if not optimal for the targeted application – since it is irregularly acquired in temporal terms – provides a valuable

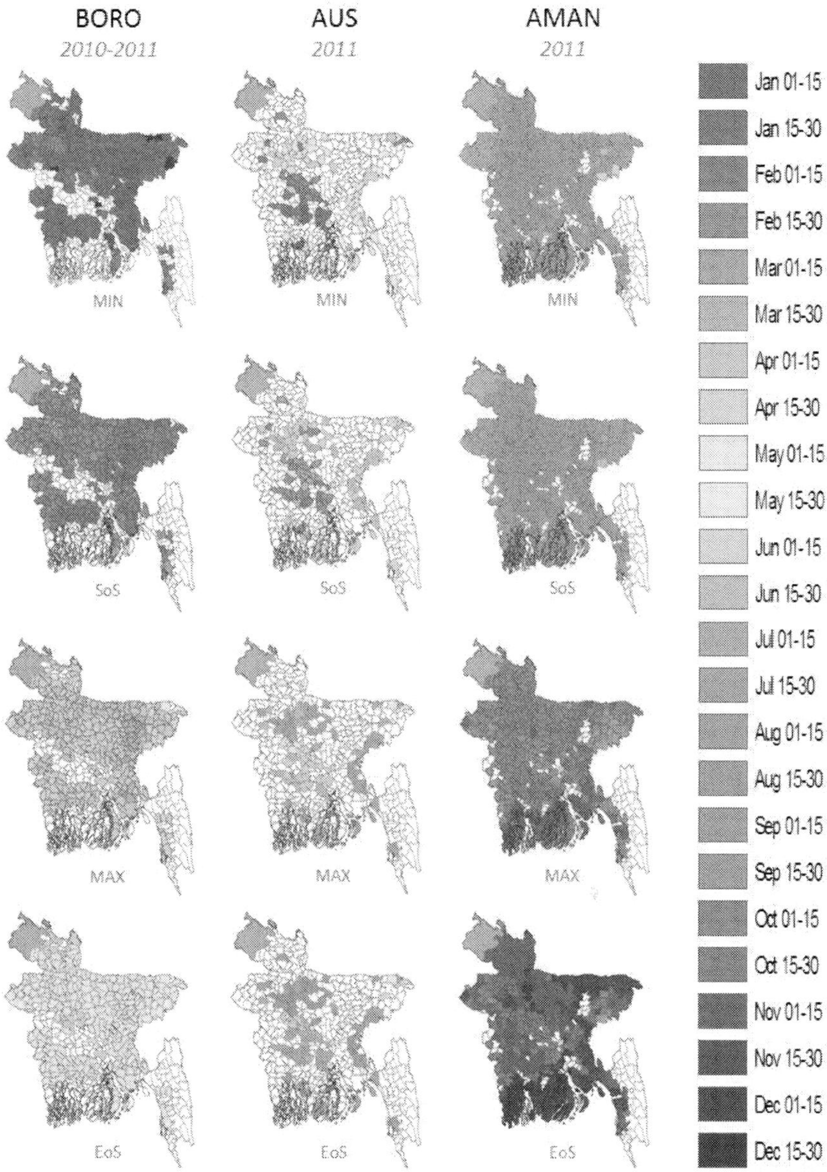

Figure 13. From top to bottom: Min, SoS, Max, and EoS phenological metrics for the boro, aus, and aman rice seasons in Bangladesh, in the analysed time window: June 2010-December 2011.

data source, enabling the generation of a consistent rice extent product, with 1-hectare resolution, nationally. In this respect, it is well-known that SAR data availability is currently problematic, mainly because of the recent failure of the ENVISAT ASAR and ALOS PALSAR-1 systems. Nevertheless, today, there are tens of thousands of archived unexploited SAR images that could be leveraged for baseline mapping to generate rice extent and area. This multi-year approach is applicable to Asia because most rice systems are generally not subject to seasonal rotations (upland rice is one exception, but it accounts for a small portion of the total rice area) and there is little scope for rice area expansion in Asia. This is obviously not feasible for all crops, where rotation mandatory and significant land-use change can occur.

The availability of systematic SAR acquisitions at an appropriate time interval — for instance, bi-monthly or weekly, as planned for in the Sentinel-1A/B mission — would allow the generation of annual and seasonal rice area data nationally. Moreover, high-resolution SAR time-series acquired along the whole season could complement the phenological monitoring provided by optical moderate resolution, particularly with respect to the detection of key phenological stages such as the start and peak of season. This would be facilitated if space agencies, with mandates for future SAR missions, could incorporate systematic background missions according to the geographical areas and applications, instead of building data archives based on sporadic/irregular acquisitions.

Concerning the SAR data processing aspect, it is essential that a multi-temporal data processing approach is performed, in primis, to enhance the data quality — hence, the level of detail — by significantly reducing the speckle. Future hyper-temporal data stacks acquired from Sentinel-1A/B will have an important role for the provision of high-quality data (i.e., high Equivalent Number of Looks) at the highest level of detail. This will permit a pixel-based approach, which is simpler and less time-consuming than a regional based one.

Due to the nature of the multi-year ENVISAT ASAR Wide Swath data (irregularly acquired in temporal terms), the resulting rice temporal signature is occasionally not optimal or not representative enough for rigorous date-by-date analysis. For this reason, we used a pixel-wise temporal signature analysis using selected representative temporal features (relative minimum and maximum, and the corresponding difference, minimum and maximum gradient between two subsequent acquisitions). Although this approach discriminated between rice and non-rice areas, the features are not sufficient to properly determine the country-wide extent of various rice seasons (single, double, etc.) and, even less, the rice phenology. There are some exceptional cases with sufficient coverage, where various rice seasons can be identified by detecting the amount of minimum and maximum gradients and related dates. In summary, assuming the availability of dense multi-year, regularly acquired SAR images, an average rice crop calendar could be generated together with the rice extent.

5.2. Observations on the MODIS analysis and results

Temporal analysis of moderate-resolution optical data time-series for detection of rice phenology is possible where there is a good rice extent base map from SAR or other sources, and good prior knowledge of the general cropping systems. Phenological occurrence maps show an agreement with local knowledge on rice cultivation, highlighting different gradients

http://dx.doi.org/10.5772/57443

of transplanting dates in the country. Analysis and interpretation of Vegetation Index time-series without prior knowledge or understanding of the crops, cropping patterns, and crop management is extremely challenging and difficult to interpret in terms of an operational crop monitoring system.

The proposed method is able to identify particular crop management systems such as single-, double-, and triple-cropping patterns. Moreover, a range of seasonalities and of crop durations from 90-to 150-day varieties could be detected. Further experiments in rice/wheat, rice/pulse, rice/shrimp and drought-prone areas should be conducted to assess the capability of detection in other commonly occurring rice cropping systems and environments.

The optical time-series data described the temporal pattern of rice systems across Bangladesh. The SAR data describes where rice is cultivated and the optical data describes when rice is cultivated. This is encouraging considering the aforementioned triple challenges of using moderate-resolution data in a region characterised by pervasive cloud cover, small fields, and complex cropping systems. The conservative nature of the phenological detection with moderate-resolution data suggests that pixel-level interpretation and pixel counting for area estimation should be avoided as they are likely to be unreliable estimates of crop system characteristics. Such estimates can always be improved through fine-tuning, expert classification, and other manual interventions, but these do not lend themselves to operational methods that require rapid and robust estimates that are free of user-bias.

Summarising the crop phenology at high levels of aggregation that match existing data or management units could provide one solution to this underestimation effect. PhenoRice is conservative in its detection criteria and, if we can assume that the detected pixels are representative of the cropping systems, then some statistical representation of the seasonality by sub-district, irrigation scheme, or other management unit would provide useful information. Most cropping calendars (e.g. Figure 2) provide ranges for the start and end of season, and remotely sensed phenolology metrics could be used in the same way to provide robust, year- and season-specific information to assess the onset of a late or early season. The same information is vital for yield modelling, since the phenology data can be used to drive/force crop growth simulation models in order to produce more reliable yield estimates.

5.3. Final comments

The current structure of the algorithm does not allow a near–real-time analysis since it needs to obtain all remote-sensing data through to the end of the year to provide complete phenological information. This can be overcome by interpreting the vegetation indexes or σ° trends at the end of the time-series, which would be refined as new information is added in near–real-time. Thus, for each phenological parameter, there would be a 'possible detection' counterpart that could be provided rapidly and which would be confirmed as a 'real detection' after several more images are acquired. This approach to crop and crop phenology detection is an analogy to what can be achieved with the advent of the Landsat continuity mission and GMES-ESA Sentinel mission, which will, for the first time, provide the possibility of high-resolution SAR and multi-spectral time-series analysis with weekly revisit periods. Moderate-resolution analysis will still play a role with Proba V, representing an alternative or a complement, to MODIS data.

In this processing chain, we have demonstrated the joint use of active and passive data as a smart method, based on the strengths of both sensors, to produce rice extent information (from multi-year ASAR analysis) and seasonal monitoring (from MODIS analysis). Crop monitoring systems should rely on a range of remote-sensing information sources in order to obtain the best possible information and smart combinations. The ever-increasing range of information from remote sensing is essential if such systems are to become operational, reliable, and accurate.

Acknowledgements

This work has been undertaken within the framework of the Remote Sensing-Based Information and Insurance for Crops in emerging Economies (RIICE) project financially supported by the Swiss Development Cooperation. It was also funded by the CGIAR Global Rice Science Partnership (GRiSP) programme. We are grateful to the European Space Agency for access to the archive of ENVISAT ASAR data.

Author details

Andrew Nelson[1*], Mirco Boschetti[2], Giacinto Manfron[2], Franceco Holecz[3], Francesco Collivignarelli[3], Luca Gatti[3], Massimo Barbieri[3], Lorena Villano[1], Parvesh Chandna[4] and Tri Setiyono[1]

*Address all correspondence to: a.nelson@irri.org

1 IRRI, Los Baños, Philippines

2 CNR, Milano, Italy

3 Sarmap, Purasca, Switzerland

4 IRRI, Dhaka, Bangladesh

References

[1] National Food Supply data from the Crops Primary Equivalent table. FAOSTAT. http://faostat.fao.org/site/609/defalt.aspx#ancor. (accessed 01-Nov-2013).

[2] Yearbook of Agricultural Statistics of Bangladesh. Bangladesh Bureau of Statistics. http://bbs.gov.bd/PageWebMenuContent.aspx?MenuKey=314. (accessed 01-Nov-2013).

[3] Estimates of Boro, Aus and Aman rice. Bangladesh Bureau of Statistics. http://bbs.gov.bd/PageWebMenuContent.aspx?MenuKey=144. (accessed 01-Nov-2013).

[4] Hussain SG., Iqbal A. editors. Research Priorities in Bangladesh Agriculture. Bangladesh Agricultural Research Council. Dhaka, Bangladesh. 2011.

[5] Kam SP., Holecz F., van Valkengoed E., Barbieri M., Casiwan CB., Asilo SL., Santos LA., Manalili RG., Collado WB., Adriano SA., Maunahan A. The makings of an internet-based rice information system: Piloting in the Philippines, First Symposium on Geoinformatics, Philippines. 2004.

[6] Xiao X., Boles S., Frolking S., Li C., Babu JY., Salas W., Moore III, B. Mapping paddy rice agriculture in South and Southeast Asia using multi-temporal MODIS images. Remote Sensing of Environment, 2006;100, 95–113.

[7] Manfron G., Crema A., Boschetti M., Confalonieri R., 2012. Testing automatic procedures to map rice area and detect phenological crop information exploiting time series analysis of remote sensed MODIS data. Proc. SPIE 8531, Remote Sensing for Agriculture, Ecosystems, and Hydrology XIV, 85311E (October 19, 2012); doi: 10.1117/12.974662. 8531, 85311E–85311E–11.

[8] Holecz F., Collivignarelli F., Barbieri M., 2013. Estimation of cultivated area in small plot agriculture in Africa for food security purposes, ESA Living Planet Symposium, Edinburgh. 09-13 September, 2013.

[9] Holecz F., Barbieri M., Collivignarelli F., Gatti L., Nelson A., Setiyono TD., Boschetti M., Manfron G., Brivio PA., Quilang E., Obico M., Minh VQ., Kieu DP., Huu QN., Veasna T., Intrman A., Wahyunto P., and Pazhanivelan S. An operational remote sensing based service for rice production estimation at national scale. ESA Living Planet Symposium, Edinburgh. 09-13 September, 2013.

[10] Ulaby FT., Allen CT., Eger G., Kanemasu ET. Relating the microwave backscattering coefficient to leaf area index. Remote Sensing of Environment, 1984;14 113-133.

[11] Le Toan T. Laur H. Multitemporal and dual polarisation observations of agricultural crops by X-band SAR images, Geoscience and Remote Sensing Symposium, IGARSS'88, pp. 1291-1294. 12-16 September 1988.

[12] Inoue Y., Kurosu T., Maeno M., Uratsuka V., Kozu T., Dabrowka-Zielinska K., Qi J. Season-long daily measurements of multifrequency (Ka, Ku, X, C, and L) andfull-polarization backscatter signatures over paddy rice field and their relationship with biological variables. Remote Sensing of Environment 2002;81(3) 194–204.

[13] Suga Y., Konishi T. Rice crop monitoring using X-, C-and L-band SAR data. Proc. SPIE 7104, Remote Sensing for Agriculture, Ecosystems, and Hydrology X, 710410 (October 02, 2008); doi:10.1117/12.800051.

[14] Bouvet A., Le Toan T., Lam-Dao N. Monitoring of the rice cropping system in the Mekong Delta using ENVISAT/ASAR dual polarization data. IEEE Transactions of Geoscience and Remote Sensing, 2009;47(2) 517-526.

[15] Oh Y., Kim Y., Hong JY., Kim YH. Polarimetric Backscattering Coefficients of FloodedRice Fields at L-and C-Bands: Measurements,Modeling, and Data Analysis.IEEE Transactions on Geoscience and Remote Sensing, 2009; 47(8) 2714–2720.

[16] Kim YH., Hong SY., Lee YH. Estimation of paddy rice growth parameters using L, C, X-bands polarimetric scatterometer. Korean Journal of Remote Sensing, 2009;25(1) 31-44.

[17] Kim S., Kim B., Kong Y., Kim YS. Radar backscattering measurements of rice crop using X-band scatterometer. IEEE Transactions on Geoscience and Remote Sensing, 2000;38(3) 1467–1471.

[18] de Grandi F., Leysen M., Lee J., Schuler, D. Radar reflectivity estimation using multiplicative SAR scenes of the same target: technique and applications. Geoscience and Remote Sensing Symposium, IGARSS'97. 3-8 August 1997

[19] Aspert F., Bach-Cuadra M., Cantone A., Holecz F., Thiran J-P. Time-varying segmentation for mapping of land cover changes. ESA ENVISAT Symposium, Montreux, Switzerland. April 23-27 2007.

[20] Boschetti M., Stroppiana D., Brivio PA., Bocchi S. Multi-year monitoring of rice crop phenology through time series analysis of MODIS images. International Journal of Remote Sensing. 2009;30, 4643–4662.

[21] Boschetti M., Nelson A., Manfron G., Brivio PA. An automatic approach for rice mapping in temperate region using time series of MODIS imagery®: first results for Mediterranean environment, in: European Geophysical Union (Ed.), EGU Geophysical Research Abstracts. 14068–1. 2012.

[22] Xiao X., Boles S., Liu J., Zhuang D., Frolking S., Li C., Salas W., Moore B. Mapping paddy rice agriculture in southern China using multi-temporal MODIS images. Remote Sensing of Environment 2005;95 480–492.

[23] Peng D., Huete AR., Huang J., Wang F., Sun H. Detection and estimation of mixed paddy rice cropping patterns with MODIS data. International Journal of Applied Earth Observation and Geoinformation. 2001;13 13–23.

[24] Rogers AS., Kearney MS. Reducing signature variability in unmixing coastal marsh Thematic Mapper scenes using spectral indices. International Journal of Remote Sensing 2004;25 2317–2335.

[25] Huete A., Didan K., Miura T., Rodriguez EP., Gao X., Ferreira LG. Overview of the radiometric and biophysical performance of the MODIS vegetation indices. Remote Sensing of Environment 2002;83, 195–213.

[26] Chen J., Jönsson P., Tamura M., Gu Z., Matsushita B., Eklundh L. A simple method for reconstructing a high-quality NDVI time-series data set based on the Savitzky–Golay filter. Remote Sensing of Environment 2004;91, 332–344.

[27] Boschetti L., Flasse SP., Brivio PA. Analysis of the conflict between omission and commission in low spatial resolution dichotomic thematic products: The Pareto Boundary. Remote Sensing of Environment 2004;91, 280-292.

Chapter 5

Change Detection and Classification Using High Resolution SAR Interferometry

Azzedine Bouaraba, Nada Milisavljević,
Marc Acheroy and Damien Closson

Additional information is available at the end of the chapter

http://dx.doi.org/10.5772/57246

1. Introduction

Change detection with Synthetic Aperture Radar (SAR) images involves often a pair of co-registered images acquired over the same area at different times [1]. To identify changes, different methods are commonly applied. These methods differ with respect to the parameters that are used to indicate changes. Since SAR data contain amplitude and phase information, both parameters can be used as change indicators [2, 3].

Incoherent change detection identifies changes by comparing sample estimates of the mean backscatter power taken from the repeat pass image pair [4, 5]. The Coherent Change Detection (CCD) technique uses the coherence of a SAR image pair to quantify changes in the observed amplitude and phase of the image pixels. As the Interferometric SAR (InSAR) phase is sensitive to changes in the spatial distribution of scatterers within a resolution cell, the CCD technique that generally detects the low-coherence areas as ground changes [6, 7] has the potential to detect centimeter changes that may remain undetected using only SAR intensity images (incoherent detection) [4, 8].

With the recent incoming of satellite constellations, delivering high-resolution SAR images, it becomes possible to detect surface changes with fine spatial details and with a short revisiting time. This aspect makes the CCD technique ideal for use in military and scientific applications such as border security and environmental monitoring. All these reasons lead to the development of new methods that enhance the change detection performance [4, 8, 9]. However, two main difficulties must be overcome in order to improve the analysis of CCD results.

The first difficulty concerns the InSAR coherence misestimation. Indeed, the sample coherence estimator is biased, especially for low-coherence values [7, 10]. In addition to the presence of speckle in SAR data [11], the consequence of this bias is the appearance of highly coherent pixels inside the changed areas. Within this context, the change detection performance degrades considerably which complicates more the CCD map interpretation particularly when using high resolution SAR data. Medium resolution SAR images obtained from satellites such as ERS, Envisat and Alos involve the complex multilooking operation to square the pixels. This operation leads to the improvement of the coherence estimation but at the expense of the spatial resolution. Multilooking is not necessary when using high resolution SAR data such as Cosmo-SkyMed (CSK) and TerraSAR-X since the image pixels are already nearly square. Nevertheless, the speckle is more noticeable in high resolution data and the coherence bias is more important.

In [7] a method is proposed to reduce the bias, based on space-averaging of coherence samples over a local area. This method allows for the enhancement of the probability of detecting changes, but a large window size is still needed to detect the entire changes. The Local Fringe Frequencies (LFF), which is a measure of the interferometric phase variability, is successfully applied in [8] as an additional change indicator to clean the aberrant highly coherent pixels inside changed areas. The use of the LFF to enhance the coherence map permits an important improvement in change detection performance and offers the advantage of preserving the spatial resolution.

The second difficulty of the CCD technique concerns the change identification. As the coherence is affected by several factors (such as baseline decorrelation and volume decorrelation), the coherence map can reveal changes that are not only due to man-made activities [1]. An area of low-backscatter strength (e.g., water surface, smooth surface, shadows) leads to decorrelation in the coherence image which is not truly the change of interest. The identification of the changes remains difficult and further investigation is needed.

In this chapter, a classification scheme is proposed with eight classes and the significance of these classes. The change classification is based on the improved coherence map combined with the pair of SAR intensity images to identify the types of change (man-made activity, natural surface decorrelation,...etc.). The two SAR intensity images are not used to detect changes as in [5], but only to support the coherence map analysis and interpretation. A set of three high resolution CSK SAR images is used, which concerns Goma airport in the Democratic Republic of Congo (DRC). High-resolution visible images are also used in the visual qualitative validation process. The results show that the proposed classification scheme is simple and effective for the change detection and identification, and that it contributes significantly to the overall scene analysis and understanding.

2. Methods

2.1. SAR Interferometry

InSAR is a technique that exploits the phase differences of at least two SAR images acquired from different orbit positions and/or at different times [3]. In Repeat-pass interferometry the SAR system flies on (ideally) parallel tracks and views the terrain from slightly different directions at different times. Fig. 1 shows a typical spaceborne InSAR configuration. SAR 1 and SAR 2 denote the satellite positions when the first and second SAR images were taken. The distance between them is called the (geometric) baseline and is denoted as B in this figure. The perpendicular baseline, $B_{\perp}$, is the component of B in the direction perpendicular to the SAR 2 look direction. R_1 and R_2 denote the ranges to the target T from satellite positions SAR 1 and SAR 2 respectively. The principle of the SAR Interferometry technique

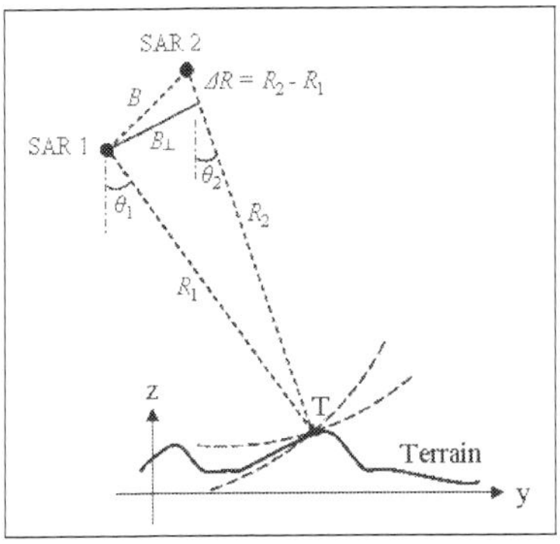

Figure 1. Interferometric configuration. Satellite positions SAR 1 and SAR 2 have ranges, to target T, R_1 and R_2 respectively. The separation B between the satellites is called the spatial baseline.

is the use of the phase information of every pixel which is directly related to the parallaxe $\Delta R = R_2 - R_1$. Let

$$f(R,\ x) = |f(R,\ x)|\ \exp\{j\phi_1(R,\ x)\} \text{ and } g(R,\ x) = |g(R,\ x)|\ \exp\{j\phi_2(R,\ x)\} \tag{1}$$

be the two co-registered SAR images forming the interferogram

$$I(\cdot) = f(\cdot)g^*(\cdot) \tag{2}$$

the phase of which

$$\phi(\cdot) = \phi_1(\cdot) - \phi_2(\cdot) \tag{3}$$

is the interferometric phase.

The phase of the SAR image response ϕ of a point scatterer is proportional to range (where k is the factor of proportionality) plus a possible shift ϕ_{scat} due to the scatterer itself [10], i.e.

$$\phi_1 = -2kR_1 + \phi_{\text{scat},\ 1} \tag{4}$$

$$\phi_2 = -2kR_2 + \phi_{\text{scat},\ 2} \tag{5}$$

Assuming that there is no possible ground displacement of the scatterers between the observations, the interferometric phase contains the following terms [3]:

$$\phi = \phi_{topo} + \Delta\phi_{scat} \tag{6}$$

where:

- $\phi_{topo} \cong \frac{4\pi}{\lambda} \frac{B_\perp}{R \sin\theta} z$ is the topographic term;

- $\Delta\phi_{scat}$ represents the influence of any change in scattering behavior. It may be a deterministic phase offset (i.e. change in dielectric constant) or a random phase (i.e. temporal decorrelation).

For Digital Terrain Model extraction, the first term ϕ_{topo} is improved when the other is minimized. In CCD applications, the term $\Delta\phi_{scat}$ is of interest and its variability is measured by the coherence defined in the next section.

2.2. InSAR coherence

2.2.1. *Coherence estimation*

In order to provide some measure of discrimination in the SAR image pair and to accommodate the random noise fluctuations, the InSAR coherence is commonly used. The sample coherence, which is defined as the magnitude of the estimated sample complex cross-correlation coefficient between the SAR image pair, encodes the degree of scene similarity as a value in the range [0, 1][6, 7]:

$$\gamma_N e^{j\phi'} = \frac{\sum_{i=1}^{N} f_i g_i^*}{\sqrt{\sum_{i=1}^{N} |f_i|^2 \sum_{i=1}^{N} |g_i|^2}} \tag{7}$$

where γ_N is the sample coherence obtained by N measurements, and ϕ' is the filtered interferometric phase.

As shown in [1], the coherence is affected by different contributions which are mainly related to:

1. the relative backscatter signal to radar receiver noise ratio (SNR) in the interferometric image pair,
2. the baseline decorrelation that is related to the satellite tracks separation, and
3. the temporal decorrelation caused by changes in the land surface, e.g., man-made objects, vegetation change, ploughing or wind action in desert areas, ... etc. It is the dominant factor in the repeat pass SAR interferometry [4].

In order to assess the suitability of using the InSAR coherence to detect man-made changes, some important results of interferometric SAR processing are presented in the following. Fig. 2 shows CSK intensity images and their coherence image in two different environment types. The white pixels of the intensity images indicate objects that reflect strong energy toward the satellite antenna, whereas the dark pixels indicate surfaces that do not reflect energy toward the satellite. In the coherence image, white pixels represent values of coherence near 1, while dark pixels represent values near 0.

In the port area, one can clearly distinguish the sea that is indicated by Fig. 2(a)-1, the dock (quay) in Fig. 2(a)-2, and the vehicles terminal in Fig. 2(a)-3. In the coherence image of Fig. 2(c), the sea area is characterized by low coherence values as it concerns incoherent medium. The coherence of the quay is preserved, as indicated in Fig. 2(c)-2, because of the absence of disturbance in the surface roughness. For the vehicles terminal, the coherence is

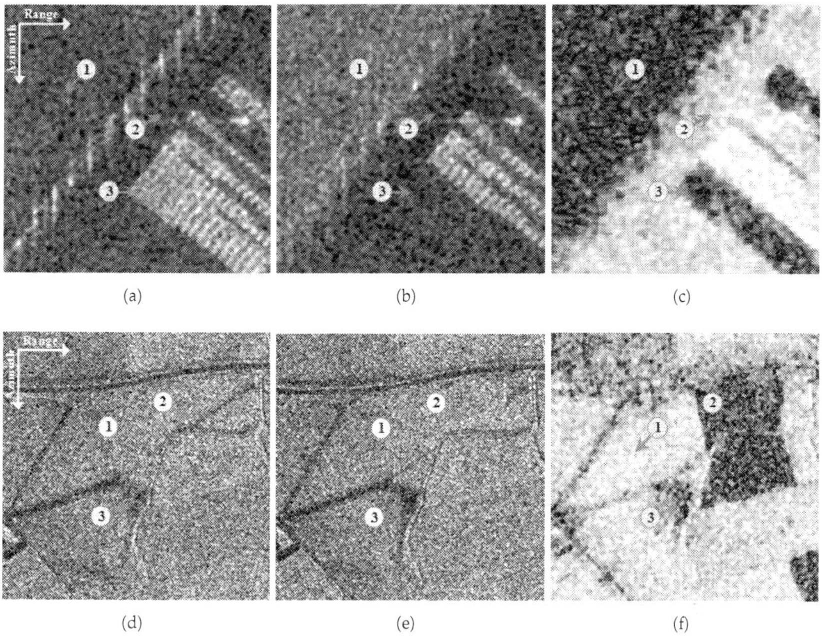

Figure 2. SAR intensity image pairs and the corresponding coherences. (a) CSK intensity image of June 15, 2011 acquired over a port area. (b) CSK intensity image of June 19, 2011. (c) Coherence image of June 15 and 19, 2011. (d) CSK intensity image of January 1, 2010 acquired over an agricultural area. (e) CSK intensity image of January 9, 2010. (f) Coherence image of January 1 and 9, 2010. Imaged scene size of 300×300 m^2.

low in Fig. 2(c)-3, as a result of moved vehicles between the two acquisition dates. The changes related to the man-made objects, such as vehicles and containers, may also be detected via incoherent techniques, which only use SAR intensity images [12]. However, in this situation the coherence can offer the advantage of being able to detect changes even if a vehicle are moved and replaced by another one, which is very similar. This is not possible with incoherent methods.

The use of coherence to measure surface changes is even more interesting for the analysis of examples in agricultural environment. The two intensity images of Fig. 2(d) and Fig. 2(e) are nearly identical showing the limited capability of incoherent methods. With an interferometric SAR processing, the cultivated parcels are clearly indicated by low coherence values (Fig. 2(f)-2), while the uncultivated parcels are characterized by high coherence values (Fig. 2(f)-1). In terms of the interferometric phase, which mainly measures the terrain topography, the obtained values are quite steady in the undisturbed areas, while the values are randomly distributed in the changed areas causing a loss of coherence. In military applications, most of the interesting man-made changes, such as vehicle tracks in a no-man's land, are not indicated by the intensity images but rather by the interferometric phase. This is true in particular for X-band SAR data that are highly sensitive to centimeter changes in the scatterers distribution within resolution cell [4].

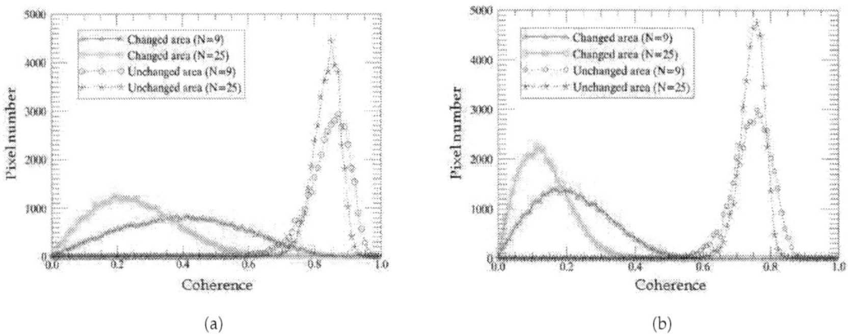

Figure 3. InSAR coherence histograms of the cultivated and uncultivated areas using different number N of samples. (a) Coherence evaluated without complex multilooking. (b) Coherence evaluated with complex multilooking of 4 pixels.

The results in Fig. 2 demonstrate the appropriateness of the SAR coherence, which uses both intensity images and interferometric phase to detect all man-made changes. However, there are two difficulties to overcome. On the one hand, the coherence can indicate some changes as in Fig. 2(f)-3, which are not of interest but only due to low backscattering (SNR) in the two SAR images (Fig. 2(d)-3 and Fig. 2(e)-3). This situation, that mainly occurs in presence of smooth surfaces (roofs, roads, shadows,...etc.) constitutes the main raison for what we propose in the present work: to develop a classification scheme that permits identification of man-made changes. On the other hand, there is a problem of coherence misestimation, causing highly coherent pixels inside changed areas which complicates the coherence image interpretation.

2.2.2. Coherence bias

As demonstrated in [7], the InSAR coherence estimator is biased especially for low coherence values. The bias is due to the limited sample size (N measurements) in the numerical computation of the Estimate (7). Fig. 3 depicts histograms of coherence images evaluated for different number of samples N. It can be seen that the coherence bias induces the presence of highly coherent pixels inside changed areas (light-colored pixels in Fig. 2(f)-2).

The coherence bias decreases when the number of samples used to estimate the coherence increase. For example, the coherence evaluated using $N = 5 \times 5$ represents more faithfully the coherence of the changed area than the coherence evaluated by $N = 3 \times 3$, as there are less samples of high coherence inside the changed area. According to Fig. 3-(a), the coherence mean value of the cultivated parcel is 0.24 with $N = 5 \times 5$ against 0.43 for $N = 3 \times 3$. Note that the coherence histograms correspond to the same area of Fig. 2(f)-2, and that the discrepancy in the results is only due to the increasing number N of samples. On the contrary, in case of an undisturbed area as in Fig. 2(f)-1, the coherence mean value is practically not affected by the increase of N. According to Fig. 3-(a), the coherence mean value of the unchanged area is 0.82 for both $N = 3 \times 3$ and 5×5, which may be explained by the fact that the coherence estimate is particularly biased for the low coherence values. Fig. 3-(b) shows that the use of the complex multilooking of the SAR images leads to a better

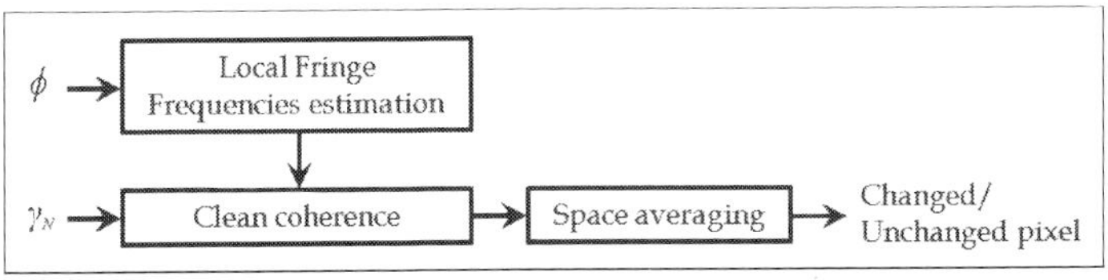

Figure 4. Scheme of the LFF-based CCD method.

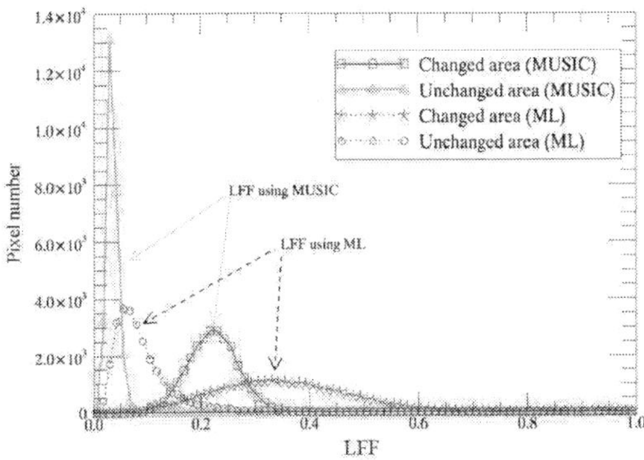

Figure 5. LFF histograms of the cultivated and uncultivated parcels of Figure 2(f).

separation between the changed and unchanged areas, but at the expense of the spatial resolution.

The results of Fig. 3, similar to those in [7], show the necessity to increase the number of samples to obtain a good separation between coherence values of the changed and unchanged areas. In CCD applications, the challenge consists in separating as much as possible the changed and the unchanged pixels, using small window-size N and without multilooking, to preserve subtle changes in the coherence image as much as possible.

In this work, the SAR interferometric processing is done in full resolution with a small window-size of $N = 3 \times 3$ pixels, corresponding approximately to an area of 2.4×2.4 m^2. According to Fig. 3, this corresponds to the most unfavorable situation of bad separation between the changed and unchanged classes, thus an additional processing step of coherence bias removal is necessary to improve the coherence map.

2.2.3. Coherence map improvement using LFF

The basic space averaging operation over M-pixel local area is commonly performed for further coherence bias removal [7]:

$$z = \frac{1}{M}\sum_{i=1}^{M} \gamma_{Ni} \underset{H_1}{\overset{H_0}{\gtrless}} T_c \tag{8}$$

where H_0 is a realization of the null hypothesis (scene changes of interest absent) and H_1 is the alternative hypothesis (scene changes of interest present). To make the decision, the statistic z is compared to a detection threshold T_c [8].

As shown in [8], a large number of samples N and a large window-size M are needed to detect the entire set of changes, which is obviously at the expense of the preservation of small changes. For that reason, another information source must be used to better analyze the coherence image. In the following paragraphs, we describe a method for the coherence map improvement, which is based on the exploitation of the LFF information.

Using a 2-D notation, the complex InSAR phase can be modeled by a first-order approximation (for simplicity, noise is neglected) [13]:

$$e^{j\phi(k,l)} = e^{j2\pi(kf_x + lf_y)} \tag{9}$$

where f_x and f_y are the LFF components in range and azimuth directions respectively.

According to Equation (6) and by assuming that the neighboring pixels have the almost identical terrain height, which is valid only using high resolution images, the change components may also be obtained by directly differentiating the interferometric phase,

$$\begin{cases} \Delta\phi_{scat-x} = \phi(k+1,l) - \phi(k,l) \\ \Delta\phi_{scat-y} = \phi(k,l+1) - \phi(k,l) \end{cases} \tag{10}$$

According to (9) and (10), the LFF estimates may correspond to the change components that can be obtained by differentiating the interferometric phase directly, a process which is not restricted to the first-order model that is due to the fact that $\phi(k,l)$ in (7) includes all the frequencies components (f_{xi}, f_{yi}) that exist inside the N-pixel local area [13]. Therefore, the measured interferometric phase is highly influenced by noise. In this context, the LFF estimation can be obtained via either the Maximum Likelihood (ML) method [14] or the MUltiple SIgnal Classification (MUSIC) method [15].

An analysis of Fig. 5 indicates that the two LFF histograms corresponding to the cultivated and uncultivated parcels of Fig. 2(f) are far apart. This confirms that the LFF components represent an additional indicator of changes. Here, the MUSIC method is used since it provides better LFF histograms separation as shown in Fig. 5. Various SAR image pairs have been used to evaluate LFF histograms for both changed (cultivated parcels, sea area) and unchanged areas. It was observed that the location of the histograms' intersection does not change significantly. As a results, it is possible to set the LFF threshold to $T_l = 0.1$ as shown in Fig. 5, an important aspect for window-size adaptation.

Here, the coherence improvement is achieved using the LFF information as an additional change indicator, to clean the highly coherent pixels inside the changed areas that are also characterized by high LFF values [8].

Fig. 6 depicts InSAR coherence maps obtained with and without the LFF information. The area of interest concerns fields with a dense agricultural activity. The result of a simple thresholding method (without space-averaging, i.e. $M = 1$) of the coherence, see Fig. 6(a), is not usable due to the presence of a large number of highly coherent pixels inside the changed

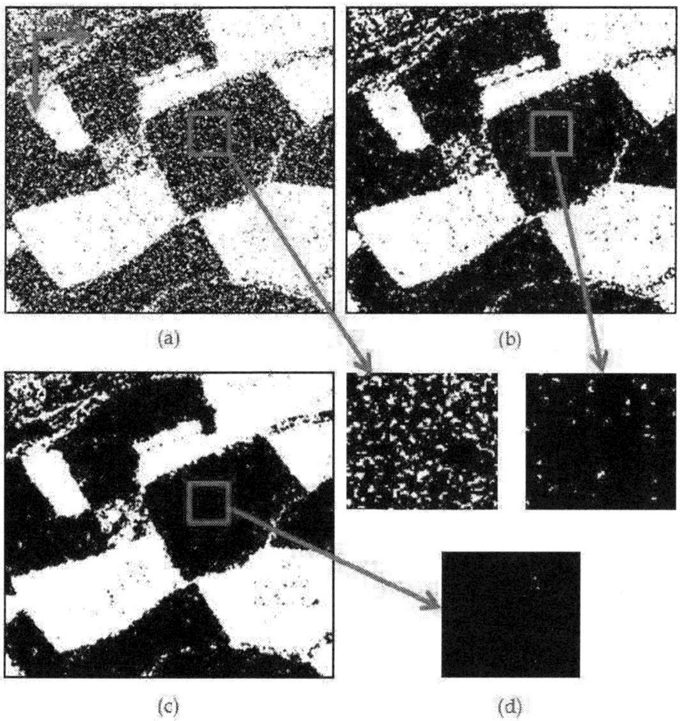

Figure 6. InSAR coherence result using $N = 3 \times 3$ samples. (a) Coherence map without space-averaging ($M = 1$). (b) Coherence map with space-averaging ($M = 3 \times 3$). (c) Change map using the LFF information. (d) Zoom on a cultivated parcel. White pixels indicate unchanged areas, while dark pixels indicate changed areas.

area. In this case, a probability of detection $P_d = 0.64$ is only achieved, leading to an unusable coherence map. An enhancement of the result is obtained by space-averaging coherence, see Fig. 6(b), using a 3×3 window. In this case, the detection probability improves to $P_d = 0.94$ but remains insufficient. The best result is obtained by using the LFF information, as presented in Fig. 6(c), with a detection probability $P_d = 0.99$. In this case, the highly coherent pixels inside changed areas are mostly reduced (detection of almost all changes). The zoom of the changed area, as presented in Fig. 6(d), clearly indicates that the method using the LFF information outperforms the existing coherence space-averaging method.

2.3. Proposed classification scheme

After the InSAR coherence improvement step, the detected changes must be identified as the coherence is affected by several decorrelation sources. To identify the man-made activities in a CCD map, the improved coherence map was combined with the two corresponding SAR intensity images. Fig. 7 presents a schematic overview of the processing chain used for the change classification.

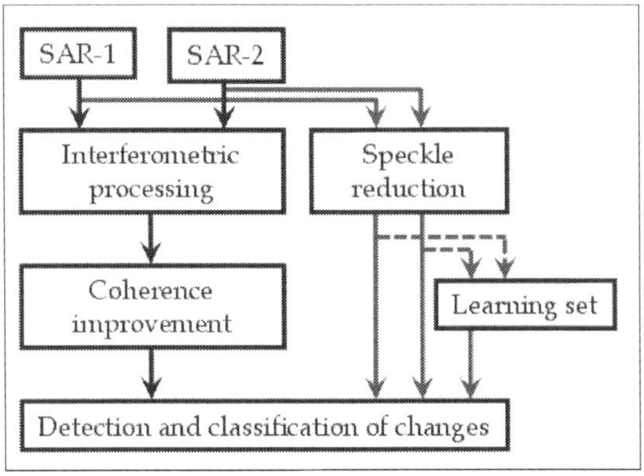

Figure 7. Schematic overview of the change detection and classification.

Class	Features			Interpretation
	Coherence	SAR-1	SAR-2	
C1	L	L	L	Specular surfaces: water, roads, roofs, shadows
C2	L	L	H	Man-made objects present in SAR-2, not in SAR-1
C3	L	H	L	Man-made objects present in SAR-1, not in SAR-2
C4	L	H	H	Man-made object present in both images but it changed from SAR-1 to SAR-2
C5	H	L	L	Bare soil or low vegetation
C6	H	L	H	Invalid class (problem of intensity thresholding caused by speckle)
C7	H	H	L	Invalid class (problem of intensity thresholding caused by speckle)
C8	H	H	H	Scatterers present in both scenes: fixed structures (e.g. parts of buildings, railways), undisturbed areas

Table 1. Overview of the eight classes resulting from the change detection and classification.

Human activity is thus characterized by a low coherence and a high intensity in at least one of the two SAR images. However, any decision based on SAR intensity is hampered by the presence of speckle which leads to an increase in the number of false alarms. Therefore, a speckle reduction is performed prior to the change classification: a 3×3 Lee filter [16] is applied to SAR intensity images in this work. The developed classification scheme is quite simple and is based on a combination of thresholds on the three features, i.e. the enhanced coherence and the two corresponding Lee-filtered intensity images. The coherence map threshold corresponds to the intersection of coherence histograms of the changed (cultivated field) and unchanged areas of Fig. 3 [8].

The intensity threshold is determined from the histograms intersection of the learning sets to sub-divide each of the two SAR images into a low (L) and a high (H) value area. The learning set consists in two classes: low backscattered area (smooth surface or water surface) and

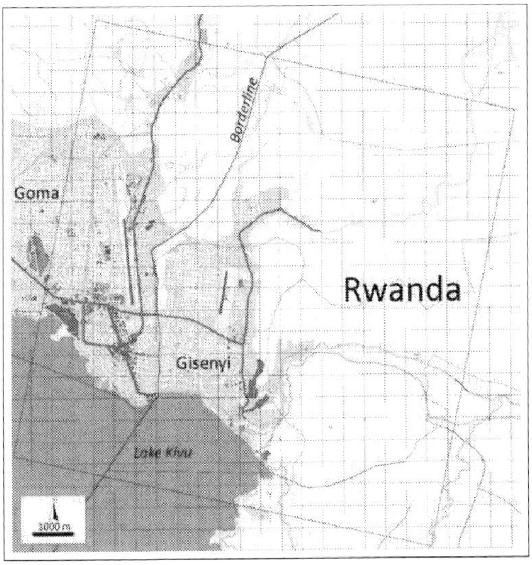

Figure 8. Geographical location of the test site in Goma (DRC) that is characterized by a flat topography. The grey rectangle indicates the borders of the imaged scene corresponding to the used CSK SAR images of 14568 × 14376 pixels.

high backscattered area (rough surface). For each of the two classes an area of approximately 200 × 200 pixels is identified in the scene. The actual change detection and characterization is a rule-based set of decisions applied to the thresholded SAR images and to the coherence map.

Dividing each of the three feature sets into two value regions leads to eight possible combinations, thus eight possible classes. Table 1 presents an overview of the properties of these classes. Classes of interest for change detection and activity monitoring are C2, C3 and C4. Classes C1, C5 and C8 contribute to the overall scene understanding. C6 and C7 represent classes in which the coherence is high but the intensity changes between the two images. If this situation occurs, it is most probably due to the fact that the intensity value exceeds the threshold in a region where it should not (the tails of the histograms).

It is also possible to divide the coherence into 3 regions; low (L), medium (M) and high (H). This situation would lead to 12 classes that complicate further the CCD map analysis and interpretation. For this reason a binary thresholding of the coherence had been chosen first, which proves to be sufficient for identification of all man-made changes. In future work, the influence of the intermediate (medium) level on the quality of the final result will be analyzed.

3. Experimental results

3.1. Data and study areas

The results presented in this work are obtained using the X-band CSK SAR images, horizontally polarized in spot-light mode. The CSK images are acquired on 24 March, 28

March and 1 April 2011, with an incident angle of 26°. The pixel resolution is 0.73 m in the ground range direction and 0.7 m in the azimuth direction. The test site, shown in Fig. 8, concerns the Goma airport in DRC. For the validation of the classification scheme, high-resolution visible images are used with the SAR intensity images to serve as ground truth. The co-registration of the SAR images was performed with Sarscape software (Sarmap 2012). Programs and algorithms were developed in Interactive Data Language (IDL).

The area of interest, shown in Fig. 9(a), concerns a part of the Goma airport that is a busy and flat test site. The airport was under extension, and an important earth moving activity was recorded during the acquisition period of 24 March to 1 April 2011. The scene concerns a wide open field surrounded by buildings and an urban area. Four zones are identified in the test site of Fig. 9(a). Firstly, the area occupied by the company in charge of the airport extension, as indicated by Fig. 9(a)-1, is surrounded by roads where excavators are clearly visible. The anthropic activities are mainly concentrated in the dike that is indicated by Fig. 9(a)-2 and in an area pointed by Fig. 9(a)-3. The dike causes a distortion (shadow) in the SAR image as indicated in Fig. 9(b)-2. The fourth zone concerns the airport runway that is indicated by Fig. 9(a)-4. It is characterized by a low backscattering power as shown in Fig. 9(b)-4. In the urban area, left side of Fig. 9(b), most of the roofs are characterized by a low backscattering power.

3.2. Results and analysis

Fig. 9(c) shows the InSAR coherence map of the imaged area obtained using the CSK acquisitions of March 24 and 28, 2011. Light-colored pixels represent values of coherence near 1, while dark pixels represent values near 0. Areas with high coherence indicate the absence of surface perturbation. Places with low coherence, corresponding to the areas disturbed between the two acquisitions dates, are located e.g. in Fig. 9(c)-1, 2 and 3. All roads leading to the holding area of Fig. 9(c)-1 have lost the coherence, due to the movements of the excavators.

The analysis and interpretation of the InSAR coherence map become complicated in the presence of man-made structures. Indeed, for example, the big building in Fig. 9(c)-1 is characterized by a low coherence in Fig. 9(c)-1, which is not truly the change that is of interest. The change classification results (or CCD map) corresponding to the period of 24 and 28 March 2011 are depicted in Fig. 9(d). Only the changes C2, C3 and C4 are of interest, and the other classes help to analyze the scene. Note that the incoherent change detection methods can detect the C2 and C3 changes, but fail to detect the C4 changes (type of change that is revealed by the interferometric phase). Besides the advantages of the CCD technique, the classification contributes to the scene analysis. The large building in Fig. 9(d)-1, and most of the urban area on the left side of Fig.9(d), is now classified as C1 instead of change in the coherence map.

The analysis of the results of Fig. 9(c) and Fig. 9(d) shows that in a simple environment without obstacles (i.e., open field in the middle of the scene shown in Fig. 9(a)), the classification method identifies the man-made changes as belonging to classes C2, C3 or C4 while the coherence map confuses the changes in a single category. In a complex environment, e. g. of the urban area in the left part of the scene presented in Fig. 9(a) and the airport runway in Fig. 9(c)-4, the InSAR coherence map becomes hard to interpret and the proposed change classification scheme identifies well the man-made changes and contributes

http://dx.doi.org/10.5772/57246

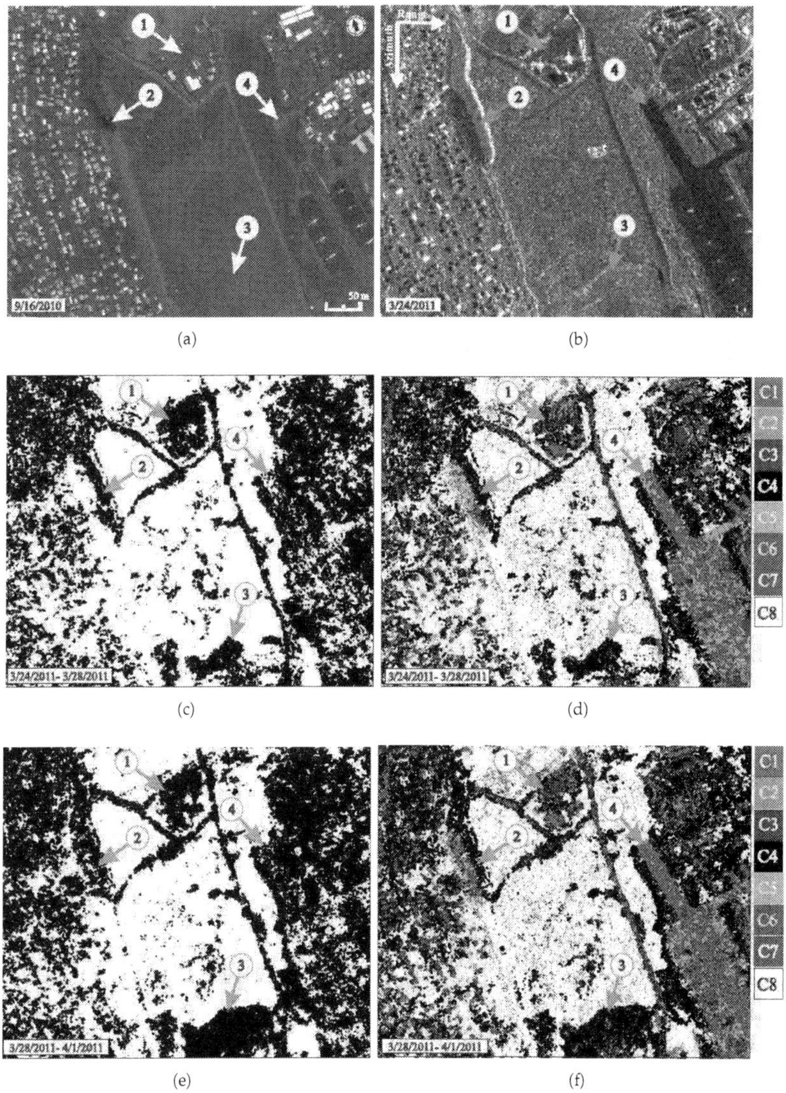

Figure 9. Change classification results. (a) Visible image of the test site. (b) CSK SAR image intensity of March 24, 2011. (c) Coherence map of March 24 and 28, 2011, normal baseline = 307 m. (d) Change classification results (CCD map) between March 24 and 28, 2011. (e) Coherence map of March 28 and April 1, 2011, normal baseline = 65 m. (f) Change classification results (CCD map) between March 28 and April 1, 2011. COSMO-SkyMed TM Product - ASI [2011] processed under license from ASI - Agenzia Spaziale Italiana. All rights reserved.

significantly to the overall scene understanding. The results of Fig. 9(e) and Fig. 9(f), obtained using an other image pair (28 March and 1 April 2011) confirm the validity of the classes proposed in Table 1. All areas of the scene are classified in the same way in Fig. 9(d) and Fig. 9(f), except in the presence of man-made changes, as in Fig. 9(d)-3. In addition, the results show that class C5 is encountered only in the specular surface and shadowed areas; it can be also assimilated to class C1. Despite the complexity of the environment, the invalid classes C6 and C7 are rarely present in the analyzed scene. In general, if classes C6 and C7 are significantly present in the classification result, it is an indication that the value of the intensity threshold should be revised.

4. Conclusion

This work deals with the development and the validation of a new method of coherent change detection and classification. The InSAR coherence map is enhanced and combined with the two corresponding SAR intensity images to build a CCD map using a simple change classification scheme. The proposed method is tested successfully using high resolution COSMO-SkyMed images. The test area of interest concerns the Goma airport, which is a flat and busy test site.

In a complex environment, the obtained results show that the InSAR coherence map reveals changes but remains hard to interpret. The proposed change classification scheme identifies well the man-made changes, and contributes significantly to the overall scene analysis and understanding. The results obtained using other image pairs confirm the validity of the proposed changed-unchanged classes. The proposed method is an improvement for the analysts in charge of the exploitation of information derived from radar imagery.

In future work, we will analyze the influence of the intermediate thresholding of the coherence on the quality of the final CCD map.

5. Acknowledgements

The Authors would like to thank the private companies Sarmap and EXELisvis for their support in providing a full SarscapeTM license (v.4.8, 2011) embedded in the latest version of ENVI software. The COSMO-SkyMed SAR images were obtained through collaboration within the SPARE-SAFE project, launched by the Belgian Ministry of Defense, and processed at the CISS Department of the Royal Military Academy of Belgium.

Author details

Azzedine Bouaraba[1], Nada Milisavljević[2], Marc Acheroy[2], Damien Closson[2]

1 Radar and Microwave Laboratory, École Militaire Polytechnique, Algiers, Algeria

2 Department of CISS, Royal Military Academy, Brussels, Belgium

References

[1] H. A. Zebker and J. Villasensor. Decorrelation in interferometric radar echoes. *IEEE Trans. and Geosci. Remote Sens.*, Vol. 30(5):950–959, 1992.

[2] E. J. M. Rignot and J. J. V. Zyl. Change detection techniques for ers-1 sar data. *IEEE Trans. and Geosci. Remote Sens.*, Vol. 31(4):896–906, 1993.

[3] D. Massonnet and K. L. Feigl. Radar interferometry and its applications to changes in the earth's surface. *Review of Geophysics*, Vol. 36(4):441–500, 1998.

[4] M. D. Preiss, A. Gray, and N. J. S. Stacy. Detecting scene changes using synthetic aperture radar interferometry. *IEEE Trans. Geosci. and Remote Sens.*, Vol. 44(8):2041–2054, 2006.

[5] X. Boli, M. C. Jing, and K. Gangyao. A change detection measure based on a likelihood ratio and statistical properties of sar intensity images. *Remote Sensing Letters*, Vol. 3(3):267–275, 2012.

[6] D. G. Corr and A. Rodrigues. Coherent change detection of vehicle movements. In *Proceedings of IGARSS*, pages 2451–2453, May 1998.

[7] R. Touzi, A. Lopes, J. Bruniquel, and P. W. Vachon. Coherence estimation for sar imagery. *IEEE Trans. Geosci.and Remote Sens.*, Vol. 37(1):135–149, 1999.

[8] A. Bouaraba, D. Borghys, A. Belhadj-Aissa, M. Acheroy, and D. Closson. Improving ccd performance by the use of local fringe frequencies. *Progress In Electromagnetics Research C*, Vol. 32:123–137, 2012.

[9] N. Oishi. A change detection measure based on a likelihood ratio and statistical properties of sar intensity images. In *Proceedings of the 6th European Radar Conference*, pages 278–281, Rome, Italy, October 2009.

[10] D. Just and R. Bamler. Phase statistics of interferograms with applications to synthetic aperture radar. *Applied Optics*, Vol. 33(20):4361–4368, 1994.

[11] C. L. Martinez. Coherence estimation in synthetic aperture radar data based on speckle noise modeling. *Applied Optics*, Vol. 46(4):544–558, 2007.

[12] J. Im, J. R. Jensen, and J. A. Tullis. Object based change detection using correlation image analysis and image segmentation. *Remote Sensing Letters*, Vol. 29(2):399–423, 2008.

[13] Z. Suo, Z. Li, and Z. Bao. A new strategy to estimate local fringe frequencies for insar phase noise reduction. *IEEE Geosci. Remote Sens. Lett.*, Vol. 40(7):771–775, 2010.

[14] U. Spagnolini. 2-d phase unwrapping and instantaneous frequency estimation. *IEEE Trans. Geosci. Remote Sens.*, Vol. 33(5):579–589, 1995.

[15] E. Trouvé, M. Caramma, and H. Maitre. Fringe detection in noisy complex interferograms. *Applied Optics*, Vol. 35(20):3799–3806, 1996.

[16] J. S. Lee. Digital image enhancement and noise filtering by use of local statistics. *IEEE Trans. on Pattern Analysis and Machine Intelligence*, Vol. 2:165–168, 1980.

Section 3

Topographic Applications

Chapter 6

High Resolution Radargrammetry – 3D Terrain Modeling

Paola Capaldo, Francesca Fratarcangeli,
Andrea Nascetti, Francesca Pieralice,
Martina Porfiri and Mattia Crespi

Additional information is available at the end of the chapter

http://dx.doi.org/10.5772/57483

1. Introduction

Radargrammetry is a methodology to extract 3D geometric information from Synthetic Aperture Radar (SAR) imagery. Similarly to photogrammetry, radargrammetry forms a stereo model. The continuous observations of the Earth's surface by satellite SAR sensors with short revisit times enable a near real-time 3D Earth surface mapping through Digital Surface Models (DSMs). Such products have a large relevance in territorial applications, such as topographic mapping, spatial and temporal change detection, morphological feature extraction, geographic data management, and visualization. SAR satellite systems provide information independently from logistic constraints on the ground (as for airborne data collection), illumination (daylight) and weather (clouds) conditions. Starting from the SAR data, two different approaches can be considered: the phase-based interferometric techniques and the intensity-based radargrammetric ones. Being aware that radar interferometry may suffer for lack of coherence, the two methods are thus complementary to achieve the most accurate and complete results. Radargrammetry was first used in the 1950s and then less and less exploited. The low resolution in amplitude supplied by the first spaceborne radar sensors (around 20 m) didn't raise more attractiveness. From 2007, the availability of very high resolution SAR data (up to 1 m Ground Sample Distance (GSD)) from COSMO-SkyMed, TerraSAR-X and RADARSAT-2 data allowed new developments. For instance Raggam et al. [1] studied the potentialities of TerraSAR-X, while Toutin [2] studied the RADARSAT-2 ones.

In this chapter, we focus on the radargrammetric approach and propose a complete procedure for generating DSMs starting from zero Doppler focused high resolution SAR imagery. A tool for radargrammetric processing of high resolution satellite SAR stereo pairs was implemented in the scientific software SISAR (Software per le Immagini Satellitari ad Alta Risoluzione), developed at the Geodesy and Geomatic Division of the University of Rome "La Sapienza" [3].

Two steps are necessary for radargrammetric DSM extraction: the image orientation and the image matching (automatic detection of homologous points).

As regards the orientation model (discussed in paragraph 2), it was established, starting from the rigorous model proposed in Leberl [4], in paying attention to the orbital model represented with Lagrange polynomials to exploit the potential of the novel high resolution (both in azimuth and in range). The Rational Polynomial Functions (RPFs) model based on Rational Polynomial Coefficients (RPCs) was also considered. The RPCs can be an useful tool in place of the rigorous model in the processes of image orthorectification/geocoding or as the DSMs generation. This generalized method is standard and unique for all sensors. The performances of the RPFs model using the RPCs can reach the level of the rigorous model. A terrain independent procedure to generate RPCs starting from radargrammetric model was defined. The possibility to generate RPCs sounds of particular interest since, at present, most part of SAR imagery is not supplied with RPCs (only in RADARSAT-2 metadata the RPCs are available), although the RPFs model is available in several commercial software.

The matching process (discussed in paragraph 3) is the automatic identification of pixels representing the same object in two or more images. If corresponding pixels are recognized, then a simple geometric intersection is needed to compute the position of their corresponding object in space. The development of a fully automatic, precise and reliable image matching method that adapts to different images and scene contents is a challenging problem. Dissimilarities between SAR images due to occlusion, illumination differences, radiometric differences and speckle noise must be taken into account. Many different image matching approaches have been developed in recent years, both within the photogrammetry and computer vision research fields. In all matching algorithms, there are two fundamental aspects that must be taken into account: the definition of a primitive model, and consequently of an identification criterion, and the choice of a strategy for the search of corresponding pixel (also named homologous points in photogrammetry) on a couple (or more) images.

An additional problem for SAR imagery is the speckle, which significantly impedes the image matching. Speckle filtering strategies were investigated, to be applied before matching. The images geometric configuration also impacts image matching. The optimum stereo imagery configuration for the radargrammetric application is when the target is observed in opposite-side view. However it causes large geometric and radiometric disparities, hindering the image matching. A good compromise is in the use of a same-side configuration stereo pair with a convenient base to heigh ratio, to increase the efficiency in the correlation image process [5].

As regards image matching, an original strategy was defined and implemented. The algorithm is based on an hierarchical solution with a geometrical constrain and the correspondences are looked using an area based matching criterion and analysing the Signal to Noise Ratio (SNR).

To demonstrate the radargrammetric mapping potential of high resolution SAR satellite imagery, several tests were carried out using data acquired in SpotLight mode and coming from COSMO-SkyMed, TerraSAR-X. They are presented in paragraph 4. Finally, conclusion and ideas for future work are outlined in paragraph 5.

2. Stereo SAR geometry

Radargrammetry is based on stereogrammetry, a classic method for relief reconstruction using optical remote sensing images. Stereo viewing reproduces the natural process of stereovision. Figure 1 represents the radargrammetric SAR geometry, S_1, S_2 are the satellites, B_x, B_z the horizontal and vertical baseline respectively; R_1, R_2 the distances between the sensors and the ground target P. Target P is seen as P_1 and P_2 in both SAR images from S_1 and S_2.

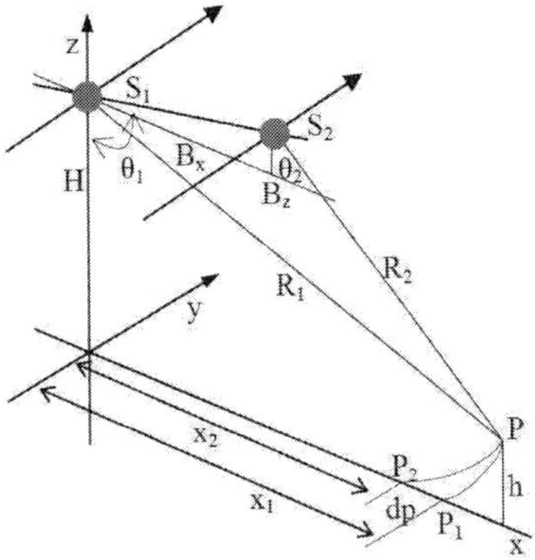

Figure 1. The different observation positions and geometry for radargrammetry

The differences between images is measured to establish a disparity map. It is used to compute the terrain elevation from the measured parallaxes between the two images [6].

In the 1960s, stereoscopic methods were applied to radar images to derive ground elevation leading to the development of radargrammetry [7]. It was shown that some specific SAR stereo configurations would produce the same elevation parallaxes as those produced by aerial photos.

To orientate the SAR imagery, several models were developed, based on classical radar [4] equations and physical constrains [8].

2.1. SAR imagery rigorous orientation model

2.1.1. Observation equation

The radargrammetric rigorous model implemented in SISAR is based on the equation of radar target acquisition and zero Doppler focalization. Radargrammetry performs a 3D reconstruction based on the determination of the sensor-object stereo model, in which the position of each point on the object is computed as the intersection of two radar rays coming from different positions and therefore with two different look angles (Figure 2).

These radar rays are modeled as two segments centered along two different satellite orbits. The intersection generating each object point is one of the two possible intersections between two circumferences centered in the two different positions and laying into two planes

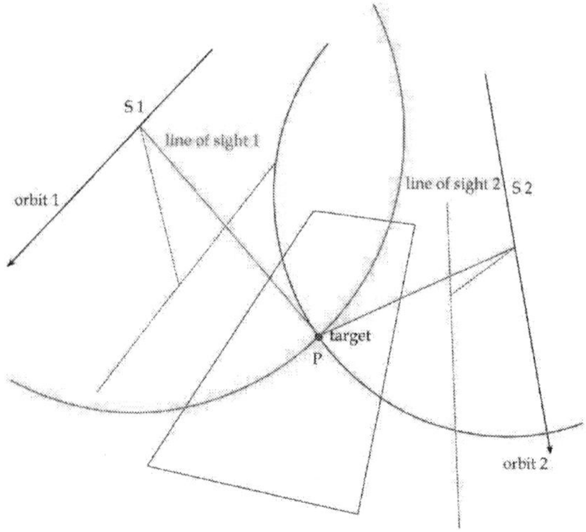

Figure 2. SAR acquisition system in zero Doppler geometry

orthogonal to the two satellite orbits whose radii are equal to the segment measured lengths [9]. The first equation of (1) is the slant range constrain. The second equation of (1) represents the orthogonality condition between each radar ray heading and the flying direction of the satellite. The couple of equations in a ECEF (Earth Centered Earth Fixed) system (for example WGS84) reads:

$$\begin{cases} \sqrt{(X_P - X_S)^2 + (Y_P - Y_S)^2 + (Z_P - Z_S)^2} - (D_s + \Delta r \cdot I) = 0 \\ u_{S_X} \cdot (X_P - X_S) + u_{S_Y} \cdot (Y_P - Y_S) + u_{S_Z} \cdot (Z_P - Z_S) = 0 \end{cases} \quad (1)$$

where:

- X_P, Y_P, Z_P are the coordinates of the generic ground point P in the ECEF coordinate system
- X_S, Y_S, Z_S are the coordinates of the satellite in the ECEF coordinate system
- u_{X_S}, u_{Y_S}, u_{Z_S} are the Cartesian components of the satellite velocity in the ECEF coordinate system
- D_s is the so-called "*near range*"
- Δr is the slant range resolution or column spacing
- I is the column position of point P on the image

The acquisition time of each line t can be related to line number J through a linear function (2), since the satellite angular velocity can be considered constant along the short orbital arc related to the image acquisition:

$$t = start\ time + \frac{1}{PRF} \cdot J \tag{2}$$

Start time is the time of start of acquisition *PRF*-Pulse Repetition Frequency, *J* the line number. *Start time*, *PRF* and *near range* are available in the metadata file of COSMO-SkyMed, TerraSAR-X and RADARSAT-2 products, for TerraSAR-X products, *start time* and *near range* corrections are supplied [10].

2.1.2. Orbit computation

The first step for the image orientation is the orbit computation. The goal is to estimate the satellite position for each line according to zero Doppler geometry. In the metadata file, with SAR imagery, the ECEF position and velocity of the satellite related to the time are supplied through state vectors at regular intervals (orbital segment), whose number N depends on SAR sensor.

The orbit is then interpolated using Lagrange polynomials (3):

$$\begin{aligned} &L_k(x) = \prod_{i=0, i \neq k}^{N} \frac{x - x_i}{x_k - x_i} \quad k = 0, ..., N \\ &p_n(x) = y_0 \cdot L_0(x) + y_1 \cdot L_1(x) + ... + y_N \cdot L_N(x) = \sum_{k=1}^{N} y_k \cdot L_k(x) \end{aligned} \tag{3}$$

Polynomials parameters are computed by Newton formula (4) to reduce the computational cost using all orbital state vectors available in the metadata:

$$\begin{aligned} &p_n(x) = f[x_0] + f[x_0, x_1] \cdot (x - x_0) + f[x_0, x_1, x_2] \cdot (x - x_0) \cdot (x - x_1) + \\ &... + f[x_0, x_1, ..., x_n] \cdot (x - x_0) \cdot (x - x_1) \cdot ... \cdot (x - x_{n-1}) \end{aligned} \tag{4}$$

where $f[x_0, ..., x_k]$ is, for $k = 0, ..., n$, the element (k, k) of matrix A diagonal (5):

$$A = \begin{vmatrix} f(x_0) & & & & \\ f(x_1) & \frac{f(x_1) - f(x_0)}{(x_1 - x_0)} & & & \\ f(x_2) & \frac{f(x_2) - f(x_1)}{(x_2 - x_1)} & \frac{A[2,1] - f[x_0, x_1]}{(x_2 - x_0)} & & \\ ... & ... & ... & ... & \\ f(x_n) & ... & ... & \frac{(A[n-1, n-2] - A[n, n-2]}{(x_n - x_{n-1})} & \frac{A[n, n-1] - f[x_0, ..., x_n]}{(x_n - x_0)} \end{vmatrix} \tag{5}$$

The interpolation polynomial on the ties x_0, x_1,...,x_n can be written in recursive form (6):

$$p_{k+1}(x) = p_k(x) + f[x_0, x_1, ..., x_{k+1}] \cdot (x - x_0) \cdot (x - x_1) \cdot (x - x_{k+1}) \tag{6}$$

where $p_k(x)$ is the interpolation polynomial of degree k on ties x_0, x_1,...,x_k. The ties x_0, x_1,...,x_k are the state vector time and the $f(x_0)$, $f(x_1)$,..$f(x_k)$ represent either the position state vector to define satellite flight path or the velocity state vector to calculate the satellite velocity [10]. Lagrange polynomial interpolation is enough accurate to model the short orbital segment and its well-known problems at the edges do not affect the modeling since the images are acquired in the central part of the orbital segment. Additionally, using a standard divide and conquer algorithm, it is possible to find in a rapid and accurate way the epoch when satellite orbit is perpendicular to the line of sight between the sensor and the generic ground point.

2.1.3. Metadata Tie Points: a tool for orientation model check

TerraSAR-X and RADARSAT-2 products have in the metadata a number of Tie Points(TPs) distributed on a regular grid. These points were calculated directly by the image providers, using the intrinsic radar geometry acquisition model. These points have image and ground coordinates, so the object coordinates of each point were used to determinate the image coordinates with the model implemented in SISAR. The image coordinates residuals shown that the defined and implemented model is quite compliant with the intrinsic SAR acquisition geometry at sub-pixel level (Table 1).

SAR image	BIAS TPs		ST. DEV. TPs		RMSE TPs	
	I	J	I	J	I	J
Hannover 05/12/2007	0.25	0.32	0.28	0.26	0.38	0.41
Hannover 10/12/2007	0.31	0.39	0.29	0.29	0.43	0.49
Hannover 29/12/2007	0.47	0.40	0.30	0.28	0.55	0.48
Trento 19/01/2011	0.30	0.41	0.30	0.31	0.42	0.52
Trento 14/01/2011	-0.50	0.27	0.30	0.58	0.58	0.39

Table 1. Accuracy result on metadata Tie Points (example for TerraSAR-X) [pix]

2.1.4. Impact of GCPs errors on stereo orientation accuracy

The Ground Control Points (GCPs) selection on radar imagery is much more difficult than on the optical one, and it is possible to misregister their positions (Figure 3). To evaluate the impact of GCPs collimation errors on the accuracy of stereo orientation, a Monte Carlo simulation was used. Starting from a no errors configuration, a Gaussian error was applied, with 1 to 6 pixel standard deviation (common values are around 1 to 4 pixels) to GCPs image coordinates, to simulate collimation errors of different magnitude. The stereo pair was oriented using a different number of GCPs (3, 6, 9); for each orientation (3, 6, 9 GCPs) 100 orientations affected by random errors were computed, at last the orientation accuracy was evaluated computing the RMSE calculated on Check Points (CPs) ground coordinates residuals. The results shown that the RMSE error due to the collimation on radar image is at level of 1 - 2 m in horizontal coordinates and 1 m in vertical ones even if the number of GCPs increases, the accuracy does not significantly.

TerraSAR-X Spotlight, SAR image   Google Earth, aerial image

Figure 3. Visual comparison between radar and optical collimation of the same detail

Therefore, the collimation errors have a great impact on the finally accuracy of stereo orientation, so that the possibility to orientate the images directly, on the bases of the refined orbital model and the metadata parameters represent an opportunity to remove the effect of collimation errors and increase the orientation accuracy.

The accuracy results of orientation model performed without GCPs are closer to the results computed with a model based on GCPs refinement [11].

2.2. SAR imagery orientation with RPCs re-parametrization

The Rational Polynomial Functions model is a method to orientate optical satellite imagery. Currently all optical images supply, together imagery, the RPCs file generated starting from the own rigorous model. On the contrary, regarding SAR images only RADARSAT-2 have a RPCs file whereas COSMO-SkyMed and TerraSAR-X images are devoid. On the other hand the use of the RPFs model is common in several commercial software for two reasons: the implementation of the RPFs model is standard, unique for all sensors, and much simpler that the one of a rigorous model, which has to be customized for each sensor, and the performance of the RPFs model can be at the level of the ones from rigorous models. The RPCs generation on the basis of rigorous orientation sensor models is an important tool.

RPFs model relates the object point coordinates (latitude ϕ , longitude λ and height h) to the pixel coordinates (I, J) in the form of ratios of polynomial expressions (7) whose coefficients (RPCs) are often supplied together with imagery:

$$I = \frac{P_1(\phi, \lambda, h)}{P_2(\phi, \lambda, h)} \qquad J = \frac{P_3(\phi, \lambda, h)}{P_4(\phi, \lambda, h)} \tag{7}$$

The number of the RPCs depends on the polynomial order (usually limited to the third one), so that each of them takes the generic form (8):

$$P_n = \sum_{i=0}^{m_1} \sum_{j=0}^{m_2} \sum_{k=0}^{m_3} t_{ijk} \cdot \phi^i \cdot \lambda^j \cdot h^k \tag{8}$$

with $0 \leq m_1 \leq 3$; $0 \leq m_2 \leq 3$; $0 \leq m_3 \leq 3$ and $m_1 + m_2 + m_3 \leq 3$ where t_{ijk} are the RPCs. In case of third order polynomials, the maximum number of coefficients is 80 (20 for each polynomial); actually, it is reduced to 78, since the two equations (7) can be divided for the zero order term of their denominators. The image and ground coordinates in equation are usually normalized to (-1, +1) range to improve the numerical accuracy, using the formula (9):

$$T_n = \frac{T - T_{offset}}{T_{scale}} \tag{9}$$

where T_n are the normalized coordinates; T_{offset}, T_{scale} are the normalization parameters available in the metadata file; and T is the generic original ground or image coordinate ($T = I, J; \phi, \lambda, h$).

In principle, RPCs can be generated by a terrain-depended scenario without using the physical sensor model or by a terrain independent scenario.

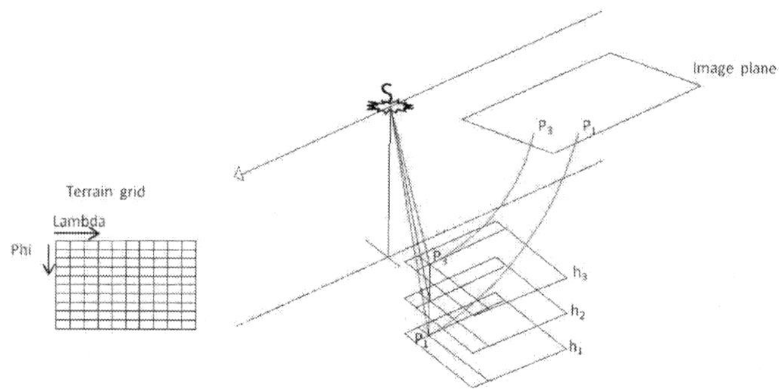

Figure 4. RPCs-terrain independent approach generation

Nevertheless, the first approach is not recommended for two relevant reasons. At first, it is likely to cause large deformations in areas far from the GCPs and it is very weak and vulnerable in presence of outliers. Further, it is not convenient, since the number of required GCPs could be very high. For example, at least 39 GCPs are necessary if RPCs up to the third order are looked for. On the contrary, following the second approach, RPCs can be generated using a known physical sensor model. This is the standard for some sensor managing companies, which supply through imagery metadata a re-parametrized form of the radargrammetric sensor model in term of RPCs, generated from their own secret physical sensor models [12, 13]. The developed and implemented procedure to generate

RPCs within SISAR includes three main steps: 1) at first the image is orientated through the already established radargrammetric rigorous orientation model; 2) further, 3D object grid with several layers slicing the entire terrain elevation range is generated; the horizontal coordinates of a point of the 3D object grid are calculated from the image corner coordinates with a regular posting, the corresponding (I, J) of the image grid are calculated using the computed orientation model; 3) finally, the RPCs are estimated in a least squares solution, having as input the coordinates of the 2D image grid points and of the 3D object grid points (Figure 4).

Investigations with optical imagery clearly underlined that many RPCs are highly correlated, so that the least squares problem is basically overparametrized. To avoid instability due to high correlations, leading to a pseudo-singular design matrix, usually a Tickhonov regularization is adopted, adding a damping factor to the diagonal of the normal matrix. On the contrary, in SISAR procedure, just the actually estimable RPCs are selected to avoid overparametrization (parsimony principle) [14, 15]). The Singular Value Decomposition (SVD) and QR decomposition are employed to evaluate the actual rank of the design matrix. To perform this selection, the remaining RPCs need to be constrained to zero [16]. Moreover, the statistical significance of each estimable RPC is checked by a Student T-test and the estimation process is repeated until only RPCs are selected.

3. SAR image matching

In general, the term image matching means automatic correspondence establishment, between primitives (homologous points) extracted from two or more digital images, depicting at least partly the same physical objects in space (Figure 5).

Figure 5. Example of homologous points in optical imagery

The fact that humans solve the correspondence problem, especially with optical imagery, so effortlessly should not lead us to think that this problem is trivial. On the contrary, humans exploit a remarkable amount of information to arrive at successfully declaring correspondence, including analysing context and neighbouring structures in the image and a prior information of the content of the scene. An image matching algorithm is a procedure able to solve the *correspondence problem* to obtain automatically a great number of homologous points [17].

3.1. An original matching strategy

The development of a fully automatic, accurate, and reliable image matching approach that adapts to different imagery (both optical and SAR) and scene contents is a challenging problem. Many different approaches to image matching have been developed within the photogrammetry and computer vision research fields [17–19]. Dissimilarities between SAR images due to occlusion, geometric distortions, radiometric differences and speckle noise must be taken into account leading to various approaches. Hereafter the basic features of an original matching strategy, presently under patenting by the University of Rome "La Sapienza", are outlined [20].

3.1.1. Area selection and filtering

At the beginning of the image matching procedure, it is mandatory to select an area of interest and a coarse height range (approximate maximum and minimum terrain ellipsoidal heights), to reduce the object space and to remarkably decrease the processing time.

SAR imagery are affected by speckle hindering target recognition and correct matching (Figure 6). To reduce speckle, three different adaptive spatial filters (Lee, Kuan, GammaMap) have been considered for a preprocessing enhancement. Thanks to a number of tests, it was highlighted that these spatial filters significantly increase the number of points at the expense of vertical accuracy, since they mitigate the speckle but smooth the image features.

Starting from this experimental awareness, an original filtering procedure *dynamic filtering* has been developed, to maximize not only the number of points, but also their quality. Unlike the traditional preprocessing techniques, the image filtering is done directly during the matching procedure; the leading idea is to find out all possible matched point using the raw imagery and, only after, to apply filters to search points in areas where the previous search failed. This allows to operate at several pyramidal levels (with different resolution) independently and in different ways, e.g. making one or more filtering cycles.

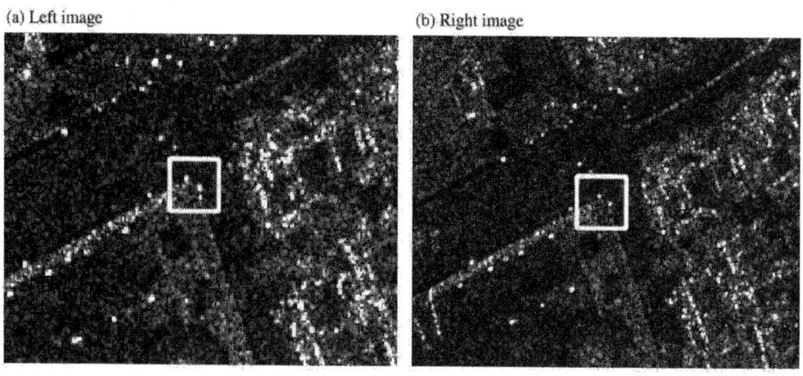

Figure 6. Example of homologous points in SAR imagery

http://dx.doi.org/10.5772/57483

3.1.2. Image matching strategy

The image matching strategy is based on an hierarchical solution with a geometrical constrain, and the corresponding points (actually, so-called primitives) are searched using an area based matching criterion and analysing the signal-to-noise ratio (SNR) [17]. In this sense, the peculiarity of the proposed algorithm is to use the image orientation model to limit the search area of the corresponding primitives, allowing a fast and robust matching. Primitives are searched directly in the object space re-projecting and re-sampling the stereo images on a regular grid in the ground reference system. Starting from a ground point with a selected height, the orientation model provides point image coordinates. It is thus possible to back-transfer the SAR radiometric information from slant-range to ground geometry.

From the practical point of view, after images preprocessing and area selection, a 3D grid is generated in ground geometry, with several layers slicing the entire height range. Starting from this 3D grid, by means of the orientation model, the two images are re-projected on each layer creating two voxel sets (one for left and one for right image). Through this process (Figure 7), the two generated voxel sets contain the geometrically corrected radiometric information in the same ground reference system.

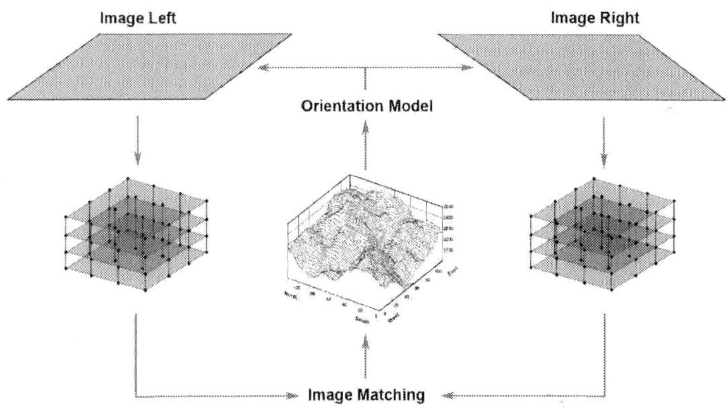

Figure 7. Geometrical constrain and voxel generation

At this point, for each horizontal position (X,Y) of the 3D grid, the main objective is to identify the correct height comparing the two voxel sets. This correct height corresponds to the best matching of the two voxels (for left and right image) at the same height; therefore, to this aim, the search can be conveniently carried out along vertical paths. During the algorithm development, different primitive models have been considered (i.e. Area Based Matching or Feature Based Matching [21, 22]), and the experimental results have highlighted that a normalized cross-correlation (NCC) linked with a signal-to-noise ratio analysis is the more efficient and accurate method. Overall, for each horizontal position (X,Y), the search of the corresponding primitives consists of the following steps:

- compute the NCC values along the vertical search path
- find out the maximum NCC value along the vertical search path (Figure 8)

- analyse the NCC profile and compute the vertical SNR according to the formula (10):

$$SNR_v = \frac{1 + \rho_{max}}{1 + \overline{\rho}} \tag{10}$$

where ρ_{max} and $\overline{\rho}$ are respectively the maximum and mean value of NCC along the vertical search path; note that this search mainly examines the correspondence of primitives in West-East direction for each horizontal layer, that is orthogonally to the direction of the orbits of the considered SAR satellite

- to strength the matching, a second search is performed moving the correlation windows in North-South direction in the selected horizontal layer (Figure 8), starting from the height corresponding to the found NCC maximum value; accordingly to the same formulation (10) a second value SNR_p is computed
- if ρ_{max} and both SNR_v and SNR_p are higher than the respectively chosen thresholds, the primitives are considered matched and the height value for the horizontal position (X,Y) is finally determined

At the end of this process, after investigating and finding all the corresponding primitives for each (X,Y) position, an irregular DSM (point cloud) in (X,Y,Z) coordinates is obtained.

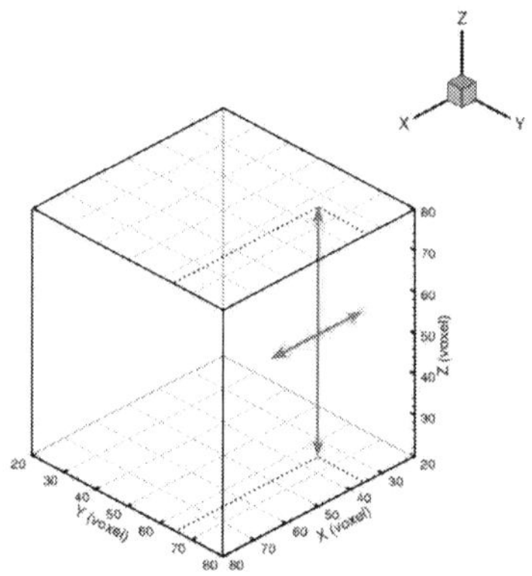

Figure 8. Search paths

3.1.3. Pyramidal approach

The described matching strategy is used in a coarse-to-fine hierarchical solution, following a standard pyramidal scheme based on a multi-resolution imagery approach. The advantage of this technique is that, at lower resolution, it is possible to detect larger structures whereas, at higher resolutions, small details are progressively added to the already obtained coarser DSM. The procedure starts choosing a suitable image multi-looking considering the original image resolution.

In this way, at each pyramid step, an intermediate DSM is extracted and it is modeled by the triangular irregular network (TIN) using a 2D Delauney triangulation method. Further,

http://dx.doi.org/10.5772/57483

DSM is interpolated on a regular grid in the ground reference system, becoming the input for the next pyramid level. Correspondingly, for each horizontal position (X,Y), the height coming from the DSM obtained in the previous pyramid step is selected as starting point for the vertical search path, whereas at the first iteration just a plane with a mean elevation is set as reference DSM. In this respect, it is worth to underline that, differently from other approaches [1], no external DSMs (for example SRTM DEM or ASTER DEM) are needed to guide matching.

As long as the resolution and pyramid level increase, and the DSM approaches the final solution, the mentioned discretization of the entire height range is correspondingly refined, so that the height step between the layers of the 3D grid, and also the number of the considered layers, decreases (Figure 9).

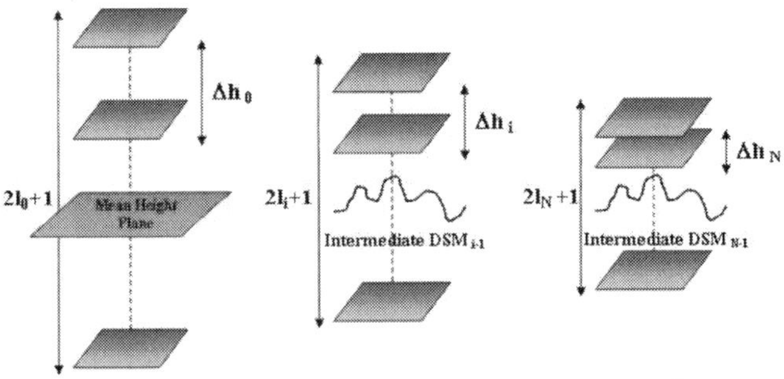

Figure 9. Coarse-to-Fine approach

Finally, in the algorithm flowchart (Figure 10), the complete radargrammetric approach is summarized and schematically illustrated.

4. Applications

Several tests were carried out to evaluated the effectiveness of the proposed radargrammetric tool. At first, the accuracy of the radargrammetric rigorous model and RPCs model were evaluated. Then, the accuracy of the extracted DSMs was computed comparing the DSMs extracted with a more accurate reference DSM (ground truth) obtained with the airborne LiDAR technology, to assess the potential and the reliability of the overall radargrammetric DSMs generation strategy. Three test sites, Merano, Trento, Como (Northern Italy) have been selected considering their main features, and different analyses were developed as outlined hereafter:

- Merano test site: analysis of foreshortening and layover effect on DSM accuracy and comparison between Spotlight imagery;
- Trento and Como test sites: analysis of radargrammetric potentiality over forested areas and study of advantages using ascending and descending stereo pairs.

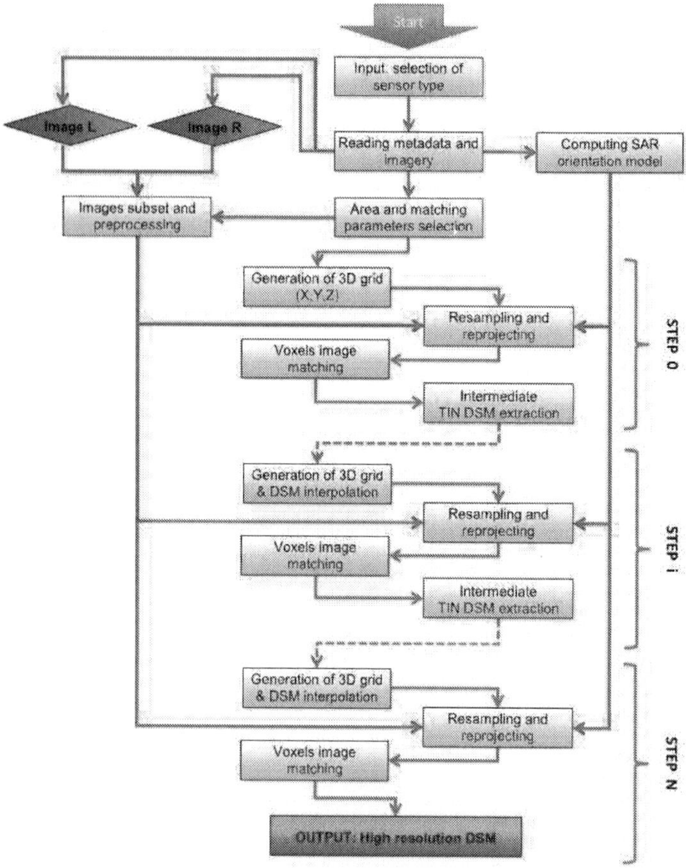

Figure 10. Radargrammetric workflow

4.1. Orientation and DSM assessment strategy

As regards orientation, both the rigorous radargrammetric and the RPFs model with RPCs obtained through the implemented generation tool were considered. The accuracy was evaluated as RMSE of the residual between estimated and known coordinates of the Check Points (CPs).

The RPCs were generated on the basis of the rigorous orientation model, without the use of GCPs. In all these cases, the RPCs generation tool estimated about 20 coefficients only, instead of the 78 coefficients generally employed in the third order RPFs, avoiding the overparametrization, and selecting only the estimable and significant parameters as mentioned before.

A homogeneous DSM assessment procedure has been considered in the different tests. It is based on a comparison with a reference (ground truth minus assessed DSM) using the scientific software DEMANAL (developed by K. Jacobsen - Leibniz University of Hannover, Germany) and the accuracy statistics are computed at the 95% probability level.

4.2. Merano test site

Merano is situated in the Autonomous Province of Bolzano, in Trentino Alto Adige Italian region. The area is characterized by a great cultivated plain area (the Adige river valley), where the city of Merano is located, at a mean altitude of 300 m, surrounded by mountains up to 2000 m. The available data for the experiment is a COSMO-SkyMed Spotlight stereo-pair; the imagery belongs to the Level 1A (SCS) category products, that is focused data in complex format, in slant range and zero-Doppler projection. The imagery main features are listed in Table 2.

Area	Acquisition date	Coverage [km^2]	Incidence Angle [deg]	Orbit	B/H
Merano	30/11/2009	10 x 10	25.90	Desc	0.60
	02/11/2009	10 x 10	42.30	Desc	

Table 2. CSK Merano: features of test site imagery

As concerned the ground truth, a LiDAR DTM (mean elevation accuracy of 0.25 m and horizontal posting 2.50 m), has been used as reference for the DSMs assessment. These data are available on the Provincia Autonoma di Bolzano website. Unfortunately, it has not been possible to get a DSM, and not a DTM (vegetation and buildings filtered out), for performing the comparison. Two different tiles with extension of 2-3 Km^2 were considered, which were selected to test the potentialities of the radargrammetric approach with different morphologies and in some difficult cases, where SAR imagery distortions (i.e. foreshortening, layover) inherit the image matching.

Twenty Ground Control Points were used to evaluated the stereo pair orientation. The horizontal coordinates of which were derived from cartography (scale 1:5000), whereas the heights come from the LiDAR DTM.

BIAS CPs			ST. DEV. CPs			RMSE CPs		
East	North	Up	East	North	Up	East	North	Up
0.69	-1.96	-0.15	3.43	2.31	1.94	3.50	3.03	1.95

Table 3. CSK Merano: radargrammetric model accuracy [m]

The horizontal accuracy is at level of 3.0 - 4.0 m, and the vertical one is around 2.0 m (Table 3).

BIAS CPs			ST. DEV. CPs			RMSE CPs		
East	North	Up	East	North	Up	East	North	Up
0.03	-1.64	1.35	3.42	2.31	2.22	3.42	2.84	2.60

Table 4. CSK Merano: RPFs model accuracy [m]

The generated RPCs were used to orientate the stereo pairs. The accuracy level was close to the one achieved by the rigorous orientation model. Proving the effectiveness of the RPCs generation tool implemented in SISAR (Table 4).

Starting from the point clouds, two DSMs were generated and assessed on a 2 m posting. As regards the DSMs accuracy, results of Tile 1 (Figure 11 (a)) highlight that over a flat area the RMSE is better than 3 m (Table 5). Thanks to a quite dense point cloud generated even over forested areas, the DSM standard deviation raises to 4 m. A large negative bias is present, due to to the forest (mean height canopy about 15 m); on the contrary, the details of urban areas were not correctly reconstructed (Figure 11 (b)).

		Tile 1		
Land Cover	BIAS	ST.DEV.	RMSE	LE95
Flat	-2.03	1.94	2.80	4.54
Forested	-14.40	4.28	15.02	9.89
		Tile 2		
Land Cover	BIAS	ST.DEV.	RMSE	LE95
Urban 1	-4.34	3.59	5.63	8.63
Urban 2	-4.92	3.40	5.98	9.11

Table 5. CSK Merano: DSMs accuracy [m]

In this case, it should be taken into account that the generated DSMs were compared with the reference DTM, which does not include vegetation and buildings. This is the reason of a part of the differences between the compared surfaces and this turns of particular interest over forested areas, where the radargrammetric approach was able to generate a dense points cloud despite the quite low coherence between images, in comparison to the InSAR technique. In these areas the differences are mainly due to the forest and the bias values are representative of the canopy height (apart from the already known effect of radar penetration into the canopy, causing a height underestimation around 25-30%) [23].

The urban area represented in Tile 2 (Figure 12 (a)) has been chosen because it also contains some of the most common geometric distortions that characterize SAR imagery. The relief is affected by foreshortening and layover. Foreshortening compresses features which are tilted toward the radar. Two urban areas not affected by distortions have been selected to evaluate the accuracy of the extracted DSM. The accuracy was about 6 m for both areas. In these cases, unlike the forest canopy in Tile 1, the buildings are not correctly reconstructed. The bias is only 5 m and it is not representative of the average building heights.

To see the effect of radar distortions, an image of Tile 2 has been orthorectified using the extracted DSM. During orthorectification process, layover and foreshortening situations are stretched back to their correct positions and pixels are stretched or smeared, creating areas where the matching algorithm cannot find homologous points due to of lack of radiometric information. These areas are easily recognizable in the error map (Figure 12 (b) below, red zone) because they are characterized by the highest values of height discrepancies (about 30 m) between the extracted DSM and the reference.

http://dx.doi.org/10.5772/57483

Figure 11. CSK Merano Tile 01: image screeshot (a) and extracted DSM (b)

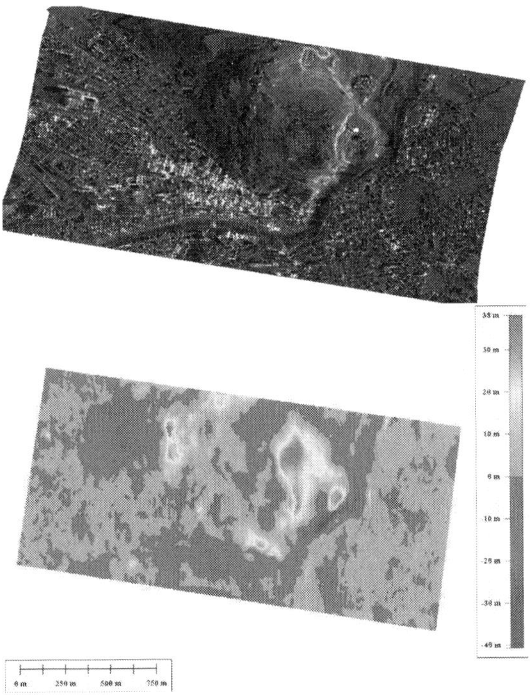

Figure 12. CSK Merano Tile 02: orthorectified image (a) and error map (b)

4.3. Trento test site

Trento city area is characterized by a dense urban morphology located inside typical hills of mountainous alpine territory.

Twelve TerraSAR-X SpotLight images have been acquired, six in descending and six in ascending mode (Table 6). A LiDAR airborne DSM supplied for free by the Provincia Autonoma di Trento was available as reference.

Area	Acquisition date	Coverage km^2	Incidence Angle (deg)	Orbit	B/H
Trento	**19/01/2011**	5 x 10	24.10	Desc	0.35
	01/01/2012	5 x 10	24.13	Desc	
	27/03/2013	5 x 10	24.15	Desc	
	14/01/2011	5 x 10	38.95	Desc	
	07/01/2012	5 x 10	38.89	Desc	
	10/03/2013	5 x 10	38.91	Desc	
	22/01/2011	5 x 10	31.10	Asc	0.35
	09/01/2012	5 x 10	31.14	Asc	
	04/04/2013	5 x 10	31.15	Asc	
	16/01/2011	5 x 10	44.19	Asc	
	31/03/2012	5 x 10	44.22	Asc	
	24/02/2013	5 x 10	44.19	Asc	

Table 6. TSX Trento: features of SpotLight imagery

13 GPs were used to evaluate the orientation accuracy coming from GPS survey. The accuracy is similar to COSMO-SkyMed, around 3 m (Table 7).

BIAS CPs			ST. DEV. CPs			RMSE CPs		
East	North	Up	East	North	Up	East	North	Up
-1.17	-1.07	0.06	5.78	3.12	3.24	5.89	3.29	3.24

Table 7. TSX Trento: radargrammetric model accuracy [m]

The generated RPCs were used to orientate the stereo pairs and the results of RPCs applications are presented in Table 8 for TerraSAR-X data. The accuracy level is close to the one achieved by the rigorous orientation model, showing the effectiveness of the RPCs generation tool implemented in SISAR.

As regards the pre-processing step for the DSM extraction, a multi-temporal filter has been used to reduce speckle and enhance images features. The three images with the same incidence angle have been co-registered (stack generation).

BIAS CPs			ST. DEV. CPs			RMSE CPs		
East	North	Up	East	North	Up	East	North	Up
-1.35	-0.93	0.90	6.13	3.23	3.55	6.28	3.36	3.66

Table 8. TSX Trento: RPFs model accuracy [m]

http://dx.doi.org/10.5772/57483

In Table 6, the master images of each stack are highlighted in bold. Multi-temporal averaging filtering technique has been performed using a kernel (11x11 pixels) available in the ESA Next SAR toolbox v. 5.05. Starting from twelve speckled images, two same-side filtered stereo-pairs (one ascending, one descending) were formed and used for the DSMs generation after separate processing.

The two stereo pairs were processed separately and the corresponding point clouds were assessed. The height differences were computed by interpolating with a bilinear method the analysed DSM over the reference LiDAR DSM. A tile featured by a mixed morphology was selected for the analysis. The results of the accuracy assessment are presented in Table 9. The point clouds derived from the ascending and descending stereo pairs, directly produced by matching procedure without any further post-processing, have been analyzed.

Point cloud assessment			
Ascending			
BIAS	ST.DEV.	RMSE	Points
1.90	6.59	6.86	603420
Descending			
BIAS	ST.DEV.	RMSE	Points
0.60	6.79	6.82	702990
SRTM filtered point cloud assessment			
Ascending (about 6 % points removed)			
BIAS	ST.DEV.	RMSE	Points
1.86	6.36	6.63	563598
Descending (about 7 % points removed)			
BIAS	ST.DEV.	RMSE	Points
0.61	6.29	6.32	650019

Table 9. TSX Trento: point cloud assessment results [m]

The accuracy was around 7 m. Some outliers were detected in the point clouds, probably due to mismatching, causing incorrect morphological reconstruction in small areas.

Ascending				
Filter	BIAS	ST.DEV.	RMSE	Points
No	1.54	6.55	6.73	259351
Yes	1.57	6.43	6.62	252628
Descending				
Filter	BIAS	ST.DEV.	RMSE	Points
No	0.27	7.93	7.93	268552
Yes	0.33	6.67	6.68	254468
Final DSM				
	BIAS	ST.DEV.	RMSE	Points
	1.05	6.37	6.45	275827

Table 10. TSX Trento: DSMs assessment results (4x4 meters posting) [m]

To remove these outliers, a free available low resolution DSM (SRTM DEM, 3′ grid posting) was used as reference. The point clouds were compared with SRTM, and the height differences computed. When the difference was greater than a fixed threshold (25 meters), the corresponding point was rejected. As shown in Table 9, no significant improvement in term of RMSE were detected; on the other hand, about 6-7% of points were removed. Three DSMs were generated and assessed on a 4 m posting. In Table 10, the results of the interpolated DSMs are shown. The ascending and descending DSMs were generated using the points clouds (both filtered and no filtered). A merged DSM was generated using a combination of the filtered point clouds to achieve the best result. The accuracy was around 7 m and 8 m for the ascending and the descending no filtered DSMs respectively. An improvement of 1 m in term of RMSE only for the descending filtered DSM was detected. Overall, the merging slightly improved the results in term of RMSE (about 0.5 meters), though more details were observed through a visual inspection.

4.4. Como test site

Como city area is characterized by a dense urban morphology.

Area	Acquisition date	Coverage km^2	Incidence Angle (deg)	Orbit	B/H
Como	07/08/2011	10 x 10	28.90	Asc	0.60
	17/06/2011	10 x 10	50.80	Asc	
	24/06/2011	10 x 10	27.80	Desc	0.80
	28/06/2011	10 x 10	55.40	Desc	

Table 11. CSK Como: features of SpotLight test site imagery

Imagery suitable for radargrammetric application, were acquired on Como test fields. Two same-side stereo pair were available, acquired on ascending and descending orbits respectively. Images features are listed in Table 11.

As a ground truth, a LiDAR DSM provided by the "Regione Lombardia" was used. The reference DSM has an horizontal spacing of 1 m and its vertical accuracy is about 0.25 m.

Considering that a descending and an ascending stereo pairs were available, the DSM could be reconstructed using two different points of acquisition. Starting from the ascending and descending points clouds, three DSMs were generated and assessed, estimating the heights on a 5 m x 5 m grid by a linear interpolation, after a Delaunay triangulation. The merged DSM was generated using a combination of the point clouds that were previously filtered, removing the matched points with lower correlation. Table 12 shows the assessment results. The accuracy ranges from 8 to 10 meters, for the ascending and the descending DSMs, and decrease to 7 m in the merged product. This test highlights that, the use of at least two stereo pairs acquired under different look side seems to be an effective strategy to overcome the limitations arising from SAR imaging system such as, layover, foreshortening and shadow.

5. Conclusions and future prospects

5.1. Conclusions

This chapter discussed on high resolution satellite SAR imagery for DSMs generation with a radargrammetric stereo-mapping approach. The main features of the radargrammetric

Total				
DSM	BIAS	ST.DEV.	RMSE	LE95
Ascending	-1.07	7.79	7.86	21.87
Descending	1.53	10.24	10.35	33.14
Merged	-1.10	6.94	7.02	18.14
Wooded				
DSM	BIAS	ST.DEV.	RMSE	LE95
Ascending	-0.69	7.10	7.14	18.10
Descending	1.32	8.53	8.63	27.06
Merged	-0.88	6.07	6.14	15.55
Urban				
DSM	BIAS	ST.DEV.	RMSE	LE95
Ascending	-1.45	8.32	8.45	22.01
Descending	1.73	11.86	11.98	38.10
Merged	-1.40	7.59	7.72	20.18

Table 12. CSK Como: DSM Absolute Error [m]

procedure implemented in SISAR package were defined. It outlined the orientation model and it focused on the original matching strategy, presently patent pending by the University of Rome "La Sapienza". It is established on area based primitive model and on an hierarchical solution with geometrical constrain. The leading idea was to search the corresponding primitives directly in the object space, re-projecting and re-sampling the stereo images into a 3D ground grid. The correspondences are looked analysing the signal-to-noise ratio (SNR) along two perpendicular search paths. A specific speckle dynamic filtering technique was designed and embedded into the radargrammetric procedure, based on three standard speckle filters (Lee, Kuan, GammaMap).

The complete radargrammetric processing chain was developed and implemented using the IDL development environment. To demonstrate its mapping capabilities, several tests were carried out using high resolution SAR satellite imagery acquired in Spotlight mode and coming from different platforms (COSMO-SkyMed, TerraSAR-X). A homogeneous DSM assessment procedure was considered in different tests, based on the comparison with a reference ground truth using the scientific software DEMANAL. Summarizing the main results of other tests, the DSMs vertical accuracy was strictly related to the terrain morphology and land cover. In case of limited SAR distortions (layover and foreshortening) the observed RMSE values ranged from 3-4 meters over bare soil and forest to 6-7 meters in more complex urban areas. Regarding the area of Como, the accuracy became worse.The terrain morphology might be conveniently reconstructed using at least two same-side stereo pairs in ascending and descending modes.

Finally, radargrammetric stereo-mapping approach appears a valuable tool to supply topographic information. It is likely to become an effective complement/alternative to InSAR technique, since it may work using a couple of images with a good performance even over areas (forested or vegetated areas) characterized by low coherence values.

5.2. Some suggestions for the future

Although the experimental results have demonstrated that StereoSAR approach has the capability to give good and encouraging results, there are still a lot of challenging issues which need to be considered for further improvements. Hereafter are listed some ideas for the future:

- orientation model customization and possible refinement: additional tests should be performed on data acquired by the expected Sentinel-1 SAR satellite sensor;
- self-tuning matching parameters: the automatic determination of the matching parameters is necessary to improve success rate and decrease mismatches. These parameters: size of the correlation window and search distance and the threshold values, should be evaluated in analysing the radiometric information of the higher-level image pyramid matching and in using them at the current pyramid level;
- efficient quality control measures: a quality control procedure of matched points should be developed by well-defined precision measures that are derived from the primitive model and from the parameter estimation;
- efficiency improvement in urban areas: to model the complicated urban morphology, specific algorithms must be developed, accounting for remarkable features as double bounces or building shadows. Some preliminary investigations applying semiglobal matching [24] to quasi-epipolar image previously generated gave promising results;
- algorithm optimization: a speed-up of the matching process could be achieved exploiting the high computational performance of Graphic Processing Units (GPUs). Reliability and accuracy could be improved, allowing concurrent processing of multiple stereo-pairs (i.e. ascending and descending ones);
- accounting for polarimetric information: studying algorithms and techniques for optimizing DSMs generation from full SAR polarimetric data [25] through radargrammetry. In particular, the potential of polarimetric imagery and their derived products (i.e. span, entropy, H-A-α classification maps) should be investigated to enhance the image matching;
- interferometry and radargrammetry tight integration: the two techniques should be considered to exploit the 3D mapping potential of high resolution satellite SAR data: radargrammetric DSMs can be used within the InSAR processing chain to simplify the unwrapping process to avoid areas affected by phase jumps.

6. Acknowledgements

The Authors are indebted with:

- Prof. K. Jacobsen for the DEMANAL software
- Dr. R. Lanari, PI of the Italian Space Agency Announcement of Opportunity for COSMO-SkyMed project "*Exploitation and Validation of COSMO- SKyMed Interferometric SAR data for Digital Terrain Modelling and Surface Deformation Analysis in Extensive Urban Areas*", within which the Como imagery were made available
- Regione Lombardia, for making available the LiDAR DSM
- Prof. Uwe Soergel, PI of the international DLR project "*Evaluation of DEM derived from TerraSAR-X data*" organized by the ISPRS (International Society for Photogrammetry and Remote Sensing) Working Group VII/2 "SAR Interferometry", within which the TerraSAR-X imagery were made available

http://dx.doi.org/10.5772/57483

Author details

Paola Capaldo*, Francesca Fratarcangeli, Andrea Nascetti, Francesca Pieralice, Martina Porfiri and Mattia Crespi

Geodesy and Geomatic Division - DICEA, University of Rome "La Sapienza", Italy

*Address all correspondence to: paola.capaldo@uniroma1.it

References

[1] H. Raggam, K. Gutjahr, R. Perko, and M. Schardt. Assessment of the Stereo-Radargrammetric Mapping Potential of TerraSAR-X Multibeam Spotlight Data. *IEEE Transactions on Geoscience and Remote Sensing*, 48(2), 2010.

[2] T. Toutin. Stereo RADARSAT for Mapping Applications. *2nd International ADRO Symposium, Montreal, Canada*, 1998.

[3] P. Capaldo, M. Crespi, F. Fratarcangeli, A. Nascetti, and F. Pieralice. High resolution SAR imagery. Radargrammetric application with COSMO-SkyMed. *IEEE Geoscience and Remote Sensing Letters*, 8.

[4] Franz W. Leberl. *Radargrammetric image processing*. Norwood, MA : Artech House, 1990.

[5] S. Meric, F. Fayard, and E. Pottier. Radargrammetric SAR image processing. *In: Pei-Gee Peter Ho (Eds.), Geoscience and Remote Sensing, Intech*, pages 421–454, 2009.

[6] T. Toutin and L. Gray. State-of-the-art of elevation extraction from satellite SAR data. *ISPRS Journal of Photogrammtery & Remote Sensing*, 55:13–33, 2000.

[7] G. La Prade. An Analytical and Experimental Study of Stereo for Radar. *Photogrammetric Engineering*, 29(2):294–300, 1963.

[8] T. Toutin. Geometric processing of remote sensing images: models, algorithms and methods (review paper). *International Journal of Remote Sensing*, 10:1893–1924, 2004.

[9] P. Capaldo, M. Crespi, F. Fratarcangeli, A. Nascetti, and F. Pieralice. A radargrammetric orientation model and a RPCs generation tool for COSMO-SkyMed and TerraSAR-X high resolution SAR. *Italian Journal of Remote Sensing / Rivista Italiana di Telerilevamento*, 44(1):55–67, 2012.

[10] P. Capaldo. High resolution radargrammetry with COSMO-SkyMed, TerraSAR-X and RADARSAT-2 imagery: development and implementation of an image orientation model for Digital Surface Model generation, 2013. PhD Thesis - Supervisor: Crespi M.

[11] P. Capaldo, M. Crespi, F. Fratarcangeli, A. Nascetti, F. Pieralice, M. Porfiri, and T. Toutin. DSMs generation from COSMO-SkyMed, RADARSAT-2 and TerraSAR-X imagery on Beauport (Canada) test site: evaluation and comparison of different radargrammetric approaches . *ISPRS Hannover Workshop 2013*, 2013.

[12] C.V. Tao and Y. Hu. The rational function model. A tool for processing high resolution imagery. *Earth Observation Magazine*, 10(1):13–16, 2001.

[13] H.B. Hanley and C. S. Fraser. Sensor orientation for high-resolution satellite imagery: further insights into bias-compensated RPC. *Proceeding of XX ISPRS Congress, Istanbul, Turkey*, 2004.

[14] F. Sansò, A. Dermanis, and A. Gruen. An overview of data analysis methods in geomatics. . *Geomatic methods for the analysis of data in the earth sciences, Springer*, 2000.

[15] M.A. Box, M. Jenkins, F. Reinsel, and F. Time Series Analysis-Forecasting and Control. *Prentice Hall.*

[16] G. Strang and K. Borre. *Linear Algebra, Geodesy and GPS*. 1997.

[17] Yi Ma, Stefano Soatto, Jana Kosecka, and S. Shankar Sastry. *An invitation to 3D Vision: from images to geometric models*, volume 1 of *1*. Springer, 2004.

[18] Christian Heipke. Overview of image matching techniques. *OEEPE workshop on Application of digital photogrammetric workstations*, 1996.

[19] T. Heuchel. Experience in applying matching techniques using images from digital cameras. *Photogrammetric Week Wichmann, Verlag*, 05:181–188, 2005.

[20] A. Nascetti. High resolution radargrammetry: development and implementation of an innovative image matching strategy, 2013. PhD Thesis - Supervisor: Crespi M.

[21] W. Forstner. A feature based correspondence algorithm for image matching. *ISP Comm III - Int Arch. of photogrammetry*, pages 150–166, 1986.

[22] C. Harris and M. Stephens. A combined corner and edge detector. *Fourth Alvey Vision Conference*, pages 147–151, 1988.

[23] R. Perko, H. Raggam, J. Deutscher, K. Gutjahr, and M. Schardt. Forest assessment using high resolution SAR data in X-Band. *Remote Sensing*, 3(4):792–815, 2011.

[24] Hirschmuller H. Stereo Processing by Semiglobal Matching and Mutual Information. *IEEE Trans on Pattern Analysis and Machine Intelligence*, pages 328–341, 2008.

[25] Lee J-S. Pottier E. *Polarimetric Radar Imaging: From Basics to Applications*. CRC Press, Taylor and Francis Group, 2009.

Chapter 7

Fusion of Interferometric SAR and Photogrammetric Elevation Data

Loris Copa, Daniela Poli and Fabio Remondino

Additional information is available at the end of the chapter

http://dx.doi.org/10.5772/57584

1. Introduction

Digital Elevation Models (DEM) are digital representations of the topographic information of a geographic area and are based on a high number of points defined by the X-, Y- and Z coordinates describing the Earth's three dimensional shape. The term DEM is generic and includes two distinct products: Digital Terrain Models (DTM), describing the ground of the imaged area, and Digital Surface Models (DSM) describing the surface including the above ground objects elevation. DEMs may either be arranged regularly in a raster or in a random point cloud. These products have become a major component for an extensive number of remote sensing and environmental applications, such as mapping, orthoimage generation, virtual terrain reconstruction and visualization, simulations, civil engineering, land monitoring, forestry, flood planning, erosion control, agriculture, environmental planning, archaeology and others. Different techniques based on Earth imaging products belonging to different families are commonly used in order to obtain the elevation information. The most common techniques are based on Optical imagery, LiDAR (Light Radar) and Synthetic Aperture Radar (SAR). In this chapter the main focus will be on techniques exploiting Optical and SAR sensors: InSAR (Interferometric SAR) and Photogrammetry. The generic term DEM will be used, even if usually the data recorded with optical sensors refers to above ground objects (DSM), while the one obtained with SAR usually is a DSM/DTM composite. An overview and some results will be shown, outlining the main strengths and weaknesses of the approaches, thus leading to the fusion hypothesis in order to exploit the inter-platform complementarity and the intrinsic advantages of both techniques. Two different product levels (raster and point cloud) – on which the fusion approaches have to be built – are also proposed, as well as examples of such approaches.

2. DEM generation from Synthetic Aperture Radar sensors

The generation of digital elevation models has been among the first techniques exploiting the phase information of SAR data. Being an active sensing technology, SAR offers day and night operability. However, one of the major drawbacks of this technology is posed by the temporal decorrelation between passes of SAR data. This drawback has been addressed mainly in two different ways. The first one is the minimization or suppression of the temporal separation between two or more acquisitions exploited in the DEM generation process. The second way relies on the selection of different frequencies/polarizations.

The different approaches available for the spaceborne platform as well as the basic interferometric framework will be introduced in the following Sections.

2.1. SAR interferometry

Interferometry is among the main techniques applied to derive elevation data from SAR images exploiting their phase information, while radargrammetry [1], which has been developed in the eighties, exploits their intensity information (see the radargrammetry chapter for further information). In this section, a brief description of the InSAR technique, as well as an example of processing chain to generate InSAR DEMs, are given. A more in-depth overview of SAR interferometry can be found in [2].

SAR sensors belong to the so-called "active" family of instruments, in that they emit a wave which rebounds on the target and returns to the satellite where it is recorded. The wavelength of the outgoing signal is known as it is consequently its corresponding phase. This implies the possibility to compare the return phase with the reference one, thus allowing their coherence estimate. Ideally, the return phase should be a function of the satellite-target and back distance plus a phase difference which can be measured. In reality, however, several other factors affect the phase value, i.e. systematic effects which have to be considered in order to obtain useful information.

In interferometry, two images covering the same area acquired from the same or slightly different position are used in order to obtain the topographic information. The phase difference between the two acquisitions is used to produce the interferogram. The difference is measured in radians and is recorded in fringes each one representing a 2 π cycle. The signal path length to the ground and back consists of a number of whole wavelengths plus some fraction of a wavelength, this means that the return wave phase is function of the distance between the target and the emitter/receiver. Said fraction can be seen as a phase shift in the returning wave. The total number of whole wavelengths is unknown, as it is the total distance to the satellite, but the phase shift defined by the extra fraction of a wavelength can be measured by the receiver with a high accuracy.

The measured phase is influenced by different factors throughout the sensor-target-sensor path. One of the most important among them coming into play is the interaction with the

http://dx.doi.org/10.5772/57584

ground surface. Reflection, depending on the physical properties of the target material, may cause phase changes. As a consequence, the return signal for every pixel is the result of the summation of the sub-pixel interactions between the signal and a multitude of smaller targets included in the imaged ground area. These targets may show different physical and topographic properties between one another, e.g. different dielectric properties, a different satellite-target distance or varying surface roughness. This leads to a signal which can be uncorrelated with the ones from the surrounding pixels and which is defined as "speckle". Orbital effects, which are other important signal noise contributors, should be considered as well. The importance of these factors derives from geometrical constraints, since images from satellite platforms with different orbits cannot be compared. The spatial position from which the images are acquired should be as close as possible and the data comparable, hence they must come from the same orbital track of a single satellite platform. Their baseline difference, i.e. their perpendicular distance difference, is known within a few centimeters and causes a regular phase difference in the interferogram, which is modelled and removed. The position offset also implies a difference in the topographic effects on phase between the images. The topographic height needed to produce a phase change fringe, known as altitude of ambiguity, becomes smaller when the baseline becomes longer. The latter aspect can be exploited in order to obtain the topographic height of the target, allowing the generation of a DEM. The topographic phase contribution can be quantified and subtracted from the interferogram by means of a previously available coarser DEM in conjunction with the orbital information.

Once these contributions have been identified and removed, the final interferogram contains the residual signal, despite the presence of remaining phase noise. The residual signal represents the phase change caused by the varying distance of the ground pixel from the satellite. In the interferogram, one phase difference fringe is recorded when the ground motion corresponds to half the radar wavelength. In the two-way travel distance this tallies with an increase of a whole wavelength. The absolute movement of a point can either be obtained by means of Ground Control Points (GCPs) recorded from GPS or similar instruments or by assuming that an area in the interferogram did not experience any deformation.

In the following paragraphs, an example of processing chain used to produce interferograms and subsequently DEMs is outlined. First, the two-pass SAR images have to be focused, exploiting a phase preserving SAR focusing. Subsequently, the images have to be coregistered, to quantify their offset and geometric difference, thus ensuring their inter-comparability. A coarse and fine coregistration procedure exploiting a coarser DEM is used to define the parameters for the inter-image point alignment. The parameters are iteratively refined, resampling the slave image to match the ground area of the master image. A common Doppler bandwidth filtering is an additional step to apply before the interferogram generation in order to minimize the decorrelation which can be caused by the presence of an offset in the Doppler centroids.

The Interferogram is then generated, a spectral shift filtering is performed and the Hermitian product is computed. The interferometric phase due to the curvature of the Earth is also removed by DEM flattening. In this process synthetic fringes are generated either from the ellipsoidal height or from a coarser DEM exploiting a backward geocoding approach. These fringes are cross-multiplied by the SAR interferogram, allowing the removal of the wrapped phase low frequency components. After the phase flattening, the complex interferogram is put through a filtering process to improve its signal to noise ratio, which in turn will result in a higher height accuracy of the final product. The interferogram must be unwrapped, interpolating it over the whole 0 to 2π phase values to obtain a continuous deformation field. This has to be carried out in order to solve the 2π ambiguity induced by the cyclicity of the phase values in the interferogram. A common approach is to apply a region growing based approach where a coherence threshold is applied in the growing process. If the images are characterized by large areas of low coherence, limiting the growing ability, a Minimum Cost Flow approach can be applied. This approach considers a regular square grid over the whole image. All the pixels whose coherence is lower than a user defined threshold are masked out. A third approach is derived from the latter, here the Delaunay polygons are used only on those areas exceeding the aforementioned coherence threshold. As a result only the points showing good coherence are unwrapped, avoiding any influence from the less coherent ones. The use of the Delaunay triangular polygons also allows the exclusion of eventual irregular areas of small size showing poor coherence, thus avoiding the representation of phase jumps in the interferogram. Eventual unwrapping errors can then be corrected in a following phase editing step, which can be both a semi-automatic or a fully manual procedure.

The following step aims at optimizing the geometric characteristics of the orbits, in order to improve the phase to height conversion and to ensure a precise geo-location. With this purpose, ground control points (GCPs) have to be provided, introducing absolute location information. Several ways to collect GCPs are possible. The most precise one being the "in-situ" collection through GPS, although it is the most time consuming as well. Another option is to manually collect the GCPs on the image itself, obtaining the height from a reference DEM. This process can also be accomplished in an automatic fashion, with the software providing the user with a series of candidate pixels – ideally the best suited for the process. The synthetic phase, which has been subtracted in the flattening process, is then added back to the unwrapped phase. This ensures that the final DEM is obtained only from the SAR data. Ultimately, the DEM is obtained in the phase-to-map conversion, in which the relation between the phase and the topographic distance is exploited in order to obtain the elevation data composing the DEM. The final product is projected onto the desired cartographic reference system, taking into account all the geodetic and cartographic parameters needed. During this process the DEM precision can also be extrapolated keeping in mind that phase and topography are proportional, the theoretical elevation standard deviation can be derived from the interferometric phase dispersion. The latter can be obtained from a relation based on the signal wavelength, and represents the dispersion of the estimates around an expected value. This last aspect is of main importance to quantify the quality of a DEM, as well as providing useful information for further processing steps such as mosaicing or, as it will be explained in the following sections, in DEM fusion. An example of a DEM along with its respective precision map is shown in Figure 1.

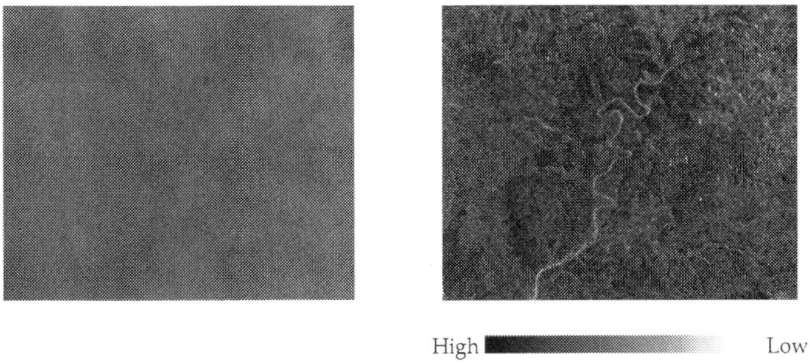

Figure 1. Example of a TanDEM-X DEM (left) and the respective Precision map (right) over a region of Australia.

2.2. Spaceborne InSAR DEM generation

The ERS-1/2 (European Remote Sensing satellite 1 and 2) Tandem mission (1995, [3][2]) has been the first to give the ability to cover the same area with two instruments over 1 day. The two satellites shared the same orbital plane, satisfying the orbital requirements imposed by an interferometric process. Moreover, the 1 day coverage interval is a very valuable characteristic to generate DEMs, since it lessens the temporal decorrelation issue. A nation-wide operational example of a DEM produced using SARscape® is given in Figure 2, see [4] for a technical description. The DEM covers the whole surface of Switzerland, approximately 42'000 km^2, with a 25 m horizontal resolution and height accuracies ranging from 7 to 15 metres, going from moderate to steep topography respectively.

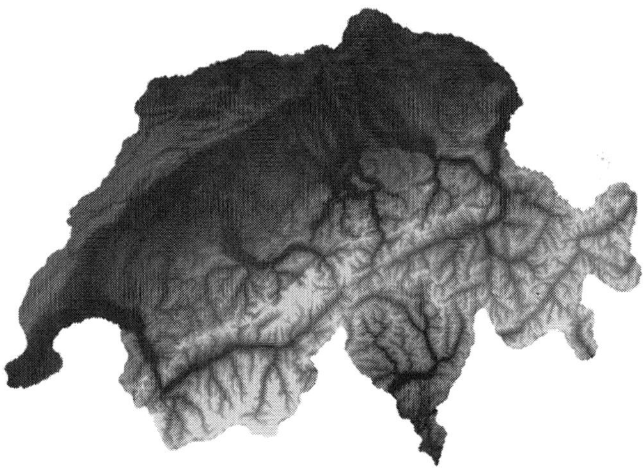

Figure 2. Example of a nation-wide ERS-1/2 DEM over Switzerland.

The decorellation issue of spaceborne SAR data has then been the main object of the Shuttle Radar Topography Mission [5]-[7]. The design of the instrument allowed the acquisition of space-based single-pass interferometric data, which is the best suited for InSAR DEM generation. The mission was flown on board of the Space Shuttle in February 2000 and was designed to obtain elevation radar data on a near-global scale. The main objective was to generate the most complete high-resolution digital topographic database of planet Earth. The instrument's C-Band and X-Band radar sensors were used to cover most of Earth's (land) surface lying between latitude 60 degrees North and 56 degrees South, providing a coverage of some 80 per cent of Earth's land.

The second version (V2) of the SRTM-C digital topographic data, also known as the finished version, has been released in recent years. This version resulted from an extensive editing effort led by the National Geospatial Intelligence Agency (NGA). Some of the most valuable features of this version are represented by the well-defined water bodies and coastlines, as well as the absence of single pixel errors such as spikes and wells. Note however that some areas of missing data (voids) can still be found. Moreover, in Version 2 directory one can also find a vector mask defining the coastline. The vector layer has been defined by NGA and is commonly known as the SRTM Water Body Data. The data is available at reproduction cost or as free download, and offers a 3 arc-seconds (about 90 m) spatial resolution.

After the availability to the public of the first SRTM data, an important effort has been invested in its accuracy validation. The accuracy assessment methodologies and tests have at first been applied to limited areas, some examples can be found in [8]-[11]. Finally, the accuracy assessment of the whole dataset has been accomplished as well [12]-[15]. The complementarity between C and X has also been studied [16][17].

An important and peculiar issue related to the generation of the SRTM-C final DEM has been the filling of the voids plaguing the processed SAR data. The main areas in which this issue was present were characterized by severe topography or very low backscatter, characteristics renowned for their effect on SAR signal acquisitions. The filling subject has been covered in a number of publications, for example in [18]-[23], in which the authors proposed approaches either based on pure spatial interpolation or also incorporating the height information derived from other datasets. A first example of a completely filled SRTM dataset is the SRTM DEM Version 3, generated by CGIAR. A later version (Version 4), has also been released. In the latter, the voids have been further filled by including more auxiliary DEMs and SRTM30 data where available. Moreover, improved interpolation techniques [24] have been applied to enhance the product. These datasets are freely available online [25].

The SRTM dataset is definitely suitable as input layer in a vast number of applications for different topics and disciplines. The major contribution offered by SRTM can be resumed in the availability of a dataset with very good height accuracy, even considering its spatial resolution. Moreover, a good consistency of its characteristics over a very large area is available, allowing studies focussing on very large regions, where DEMs at higher resolution / height accuracy might also be obtained afterwards (in a preliminary study) or at the same time.

http://dx.doi.org/10.5772/57584

The application fields in which SRTM data can be used are disparate. Applications to geomorphology [26][27], volcanology [28], hydrological studies [22],[29]-[31], forestry research [32]-[34], archaeology [35], glaciers volume monitoring [36], and many others have been proposed.

As discussed, SRTM data offers a large coverage, but at the expense of spatial resolution. In order to improve the results obtained from the SRTM mission and to exploit the existing high- or very-high resolution SAR missions, the research has currently focused on the identification of satellite SAR missions to complement the existing ones with the main purpose of very high resolution DSMs. The main concept, introduced among others by [37][38], is the definition and assembly of satellite constellations composed by either passive or active SAR satellites orbiting around a master sensor. Single-pass capabilities are another main requirement that should be fulfilled. The concept has been studied in [39]-[43], finally resulting in the TanDEM-X (TerraSAR-X add-on for Digital Elevation Measurement, [44]) mission, launched in 2010. The latter will work in conjunction with the TerraSAR-X mission, providing a second satellite orbiting around the first one. This type of pairing provides a continuous combination of along-and across-track baseline, suitable respectively for currents monitoring and for DEM generation. This satellite constellation will be used to produce the WorldDEM™, a DTED (Digital Terrain Elevation Data) level 3 DSM [46], which will be available in 2014. Some initial examples of such very-high resolution DEMs, have been shown by [47] and [48]. One example of TanDEM-X product is shown in Figure 3.

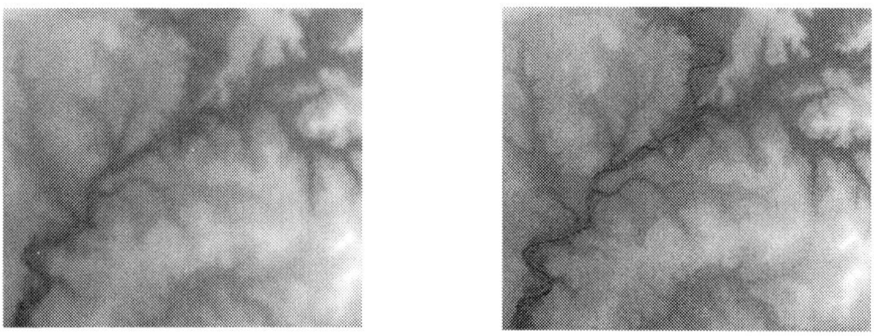

Figure 3. Example of a TanDEM-X DEM (right) compared to an SRTM-V4 DEM (left) over Australia.

A similar concept has been developed for the COSMO-SkyMed mission (see [45]), also completed in 2010 with the launch of the fourth satellite of the constellation. The missions considered in these studies take into account high frequency (X-Band) sensors. A more in-depth example of a COSMO-SkyMed DEM compared to the SRTM-V4 equivalent is given in Figure 4.

In Figure 4, the substraction of the two DEMs is given, showing the residuals in meters (mid-left image). This map and the corresponding normal distribution (mid-right image), show that on average the relative positioning between the two products is quite similar, with the majority of the observations being between the -12/12 meters range. However, the COSMO-SkyMed

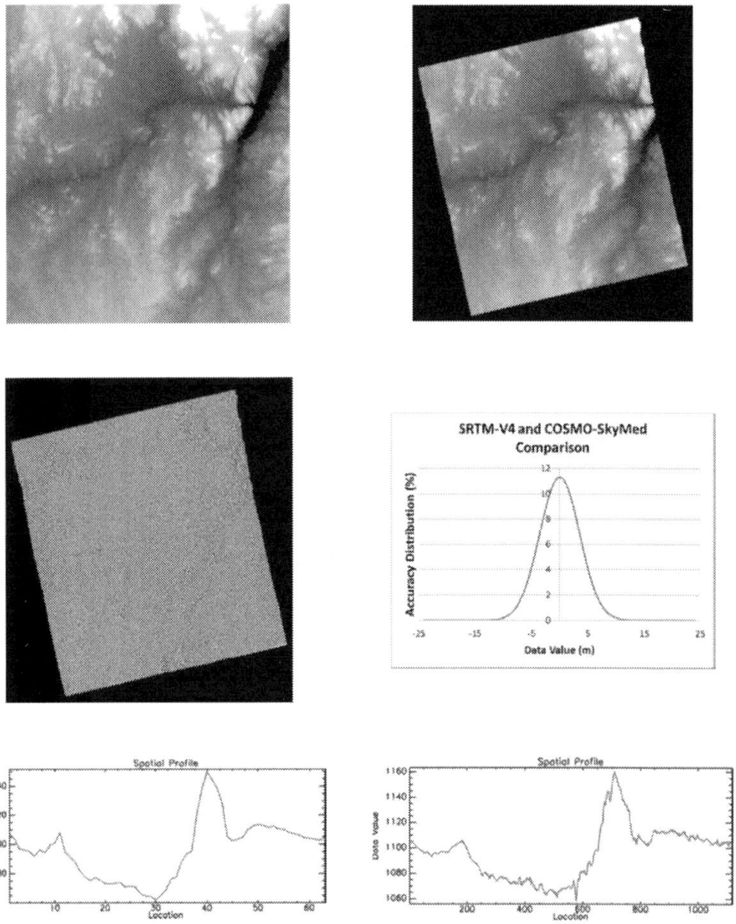

Figure 4. Comparison between an SRTM-V4 DEM (top-left) and a COSMO-SkyMed DEM (top-right). Height residual in meters between the DEMs (mid-left) and corresponding residual distribution (mid-right). Elevation profile over a test zone of SRTM-V4 (bottom-left) versus COSMO-SkyMed (bottom-right), the test profiles were collected over the input DEMs as shown by the red line in the top images. Image collected over the Malawi region. Image courtesy of A.S.I. Agenzia Spaziale Italiana.

DEM has a much higher spatial resolution and altimetric precision than the SRTM data, allowing for a better slope definition as shown by the profiles at the bottom of Figure 4. In the SRTM data one can clearly see that the altimetric and geometric detail is lost. The better slope representation of COSMO-SkyMed also allows a better absolute radiometric correction of intensity data, while, having such a higher three dimensional spatial definition, the data is particularly suited in geo-information processing.

An alternative to these solutions is the exploitation of ALOS PALSAR L-Band data, if one looks at stable scattering mechanisms at lower frequencies. The lower frequency of the PALSAR system offers higher penetration in vegetated areas, resulting in a higher temporal correlation. This results in a lower phase noise compared to C-Band and X-Band systems [50]. In Figure 5, an example of PALSAR DEM is given in comparison with the respective SRTM counterpart. A better slope definition is appreciable with ALOS-PALSAR as well, as shown by the profiles at the bottom of Figure 5.

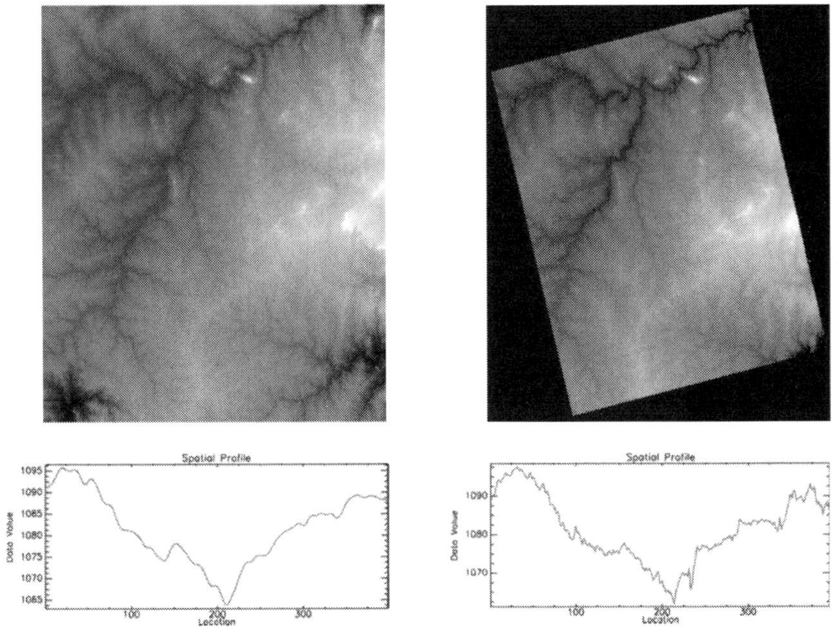

Figure 5. Comparison between SRTM-V4 DEM (top-left) and ALOS-PALSAR DEM (top-right) over the Malawi region. Elevation profile over a test zone of SRTM-V4 (bottom-left) versus ALOS-PALSAR (bottom-right), the test profiles were collected over the input DEMs as shown by the red line in the top images

3. DEM generation from spaceborne optical sensors

3.1. VHR optical sensors

Today an increasing number of Earth-observation platforms are equipped with Very High-Resolution (VHR) optical sensors characterized by a ground resolution of less than 1m [51], enabling the discrimination of fine details, like buildings and individual trees. The radiometric and geometric quality of the satellite images can be compared with original digital aerial images. The image orientation has been simplified by using rational polynomial functions [52] and the direct sensor orientation has been improved, allowing, in some cases, the image processing and DEM generation without ground control information [53]-[54]. Complementing the high spatial resolution, VHR sensors are mounted on highly agile platforms, which enable rapid targeting, a revisit time up to 1 day and the stereoscopic coverage within the orbit for 3D information recovery and DEM generation. The cost of the stereo acquisition (two stereo images) is in general equal to the double of a single acquisition. The availability of stereo acquisitions in the archives of image distributors is much lower than the single acquisition mode, but stereo acquisition can be planned on demand.

3.2. Image acquisition

Earth observation optical sensors mounted on satellites acquire mainly in pushbroom mode. The imaging system of a pushbroom sensor consists of an optical instrument (usually a lens or a telescope) and Charge Coupled Devices (CCD) lines assembled in the focal plane. While the platform moves along its trajectory, successive lines are acquired perpendicular to the satellite track and stored in sequence to form a strip. In case of multispectral sensors, a strip is generated for each channel. In most cases the sensors are placed along a single line or in two or more segments staggered along the flight direction (i.e. EROS-B) or butted with some overlap (i.e. Quickbird) to increase the image ground resolution through a specific geometric and radiometric post-processing.

In order to produce DEMs, stereoscopic coverage is mandatory. Amongst satellites using a stereo acquisition mode, the following can be distinguished: (1) standard across-track systems, (2) standard simultaneous multi-view along-track systems, and (3) agile single-lens systems.

In the standard *across-track configuration*, the CCD lines and one optical system are generally combined with a mirror that rotates from one side of the sensor to the other across the flight direction, up to 30°, with the overlapping area across the flight direction. According to this configuration, stereo images are collected from different orbits at different dates, with temporal variation between the images. Examples are very popular high resolution (HR) satellite sensors, like SPOT-1-4 HRV and SPOT-5 HRG by CNES, IRS-1C/1D PAN by ISRO, Beijing-1 CMT by the People Republic of China and Kompsat-1 EOC by Korea Aerospace Research Institute (KARI). In the standard *along-track configuration*, two or more strips are taken simultaneously from the same orbit at different angles along the flight direction. For each viewing direction there are one lens and one set of CCD lines placed on the focal plane. The along-track angles are generally fixed and the base-over-height (B/H) ratio constant for each satellite. This same-date along-track configuration thus minimizes the temporal variation between the images. Examples of HR sensors with this configuration are SPOT-5 HRS by CNES, ALOS-PRISM by JAXA, Cartosat-1 by ISRO and ZY-3 by the People Republic of China.

The last generation VHR sensors use an *agile configuration* that was first introduced on IKONOS-2 in 1999. These sensors are carried on small and flexible satellites flying along sun-synchronous and quasi-polar orbits and use a single-lens optical system [59]. For stereo viewing or frequent temporal coverage, they have the ability to rotate on command around their cameras axes and view the target from different directions, in forward mode (from North to South) or reverse mode (from South to North). Therefore, along-track and across-track stereo pairs of a particular area of interest are planned in advance and acquired on almost any part of the Earth's surface. The general limit for off-nadir angles is 45º, but larger values are used in some specific situations. Some agile sensors have a synchronous acquisition mode, thus the satellite speed and the scanning speed are equal, and the viewing angle is constant during one image acquisition. Examples are IKONOS-2, Kompsat-2 and Formosat-2 RSI by National Space Program Office of China (NSPO). On the other hand more agile sensors, including Quickbird, WorldView 1 and WorldView-2 (DigitalGlobe), GeoEye-1 (GeoEye), EROS-A and –B (ImageSat International), Orbview-3 (Orbimage) and TopSat (QinetiQ), Pléiades 1A and 1B (Astrium) scan the Earth in asynchronous mode: the platform velocity is higher than the scanning one, therefore the satellite rotates continuously during the acquisition and a CCD line scans longer a line on the ground. The limitation of this scheme is that the geometry is less stable. The success of agile single-lens systems for the acquisition of VHR stereo imagery is confirmed by its planned use in future missions too. GeoEye-2, planned for launch in 2013, and Worldview-3, planned for 2014, will have the stereo capability of their processors GeoEye-1 and WorldView-2 respectively.

Images acquired by VHR sensors are distributed with a certain level of processing; unfortunately, the range of terminology used to denominate the same type of image data is different at each data provider. In general, three main processing levels can be distinguished: a) raw images with only normalization and calibration of the detectors without any geometric correction (this level is preferred by photogrammetrists working with 3D physical models); b) geo-referenced images corrected for systematic distortions due to the sensor, the platform and the Earth rotation and curvature; c) map-oriented images, also called geocoded images, corrected for the same distortions as geo-referenced images and North oriented. Generally, very few metadata related to sensor and satellites are provided; most of metadata are related to the geometric processing and the ellipsoid/map characteristics.

3.3. Image processing

The standard photogrammetric workflow for digital surface modeling and 3D information extraction from stereo optical images is summarized in Figure 6. Two major steps, the image orientation and image matching for DEM generation, are discussed in the following sections.

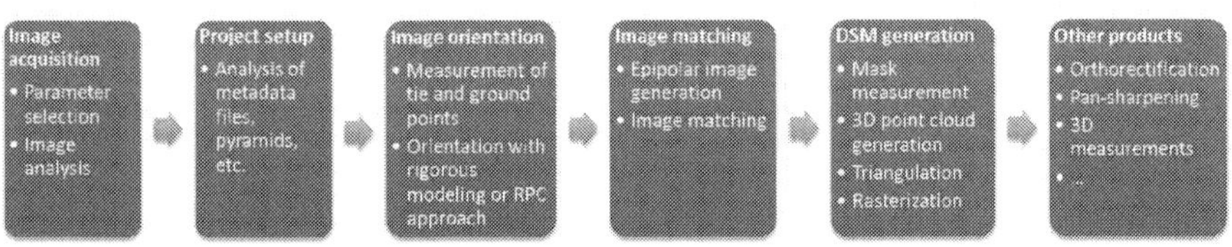

Figure 6. Photogrammetric workflow for DEM and 3D information extraction from stereo imagery.

4. Image orientation

For DEM generation, the orientation of the images is a fundamental step and its accuracy is a crucial issue during the evaluation of the entire production pipeline. Over the last three decades various models based on different approaches have been developed [57]. Rigorous models try to describe the physical properties of the image acquisition through the relationship between the image and the ground coordinate system [58]. In case of images acquired by pushbroom sensors, each image line is the result of a perspective projection. The mathematical model is based on the collinearity equations, which relate the camera system (right-hand 3D system centered on the instantaneous perspective center, with the y axis parallel to the image line and the x axis perpendicular to the y axis, closely aligned with the flight direction) to a 3D local coordinate system. The collinearity equations are extended to include exterior and interior orientation modeling of pushbroom sensors. Usually ephemeris information is not precise enough for accurate mapping, and the exterior orientation has to be further improved with suitable time-dependent functions and estimated in a bundle adjustment. Simple third-order Lagrange polynomials [60][61], quadratic functions [62] and piecewise quadratic polynomials [56] have been proposed for this purpose. In [63] a study of a number of trajectory models is given. For initial approximation of the modeling parameters, the physical properties of the satellite orbit and the ephemeris are used. With respect to the interior orientation, suitable parameters are added to model the optical design (number of lenses, viewing angles), the lens distortions and the CCD line distortions. A detailed description of the errors can be found in [64].

In recent years, 3D rational functions (RFs) have become the standard form to approximate rigorous physical models of VHR sensors. They describe the relation between normalized image and object coordinates through ratios of polynomials, usually of the third order. The corresponding polynomial coefficients, together with scale and offset coefficients for coordinate normalisation, form the so-called rational polynomial coefficients (RPCs) and are distributed by image vendors as metadata information. Anyway to obtain a sensor orientation with sub-pixel accuracy, the RPCs need to be refined with linear equations requesting more accurate GCPs or, more commonly, with 2D polynomial functions ([65], [66] and others). For the latter solution, one or two GCPs are used for zero-order 2D polynomial functions (bi-directional shift) and six to ten GCPs for first and second-order 2D polynomial functions to compute their parameters with a least squares adjustment process. In the case of stereo-images, a block adjustment with RPC is applied [66] for the image orientation. RPCs are widely adopted by image vendors and government agencies, and by almost all commercial photogrammetric workstation suppliers.

4.1. Image matching

Image matching is referred to the establishment of correspondences between primitives extracted from two (stereo) or more (multi-view) images. Once the image coordinates of the correspondent points are known in two or more images, the object 3D coordinates are estimated via collinearity or projective model. In image space this process produce a *depth map* (that assigns relative depths to each pixel of an image) while in object space we normally call it *point cloud*. For more than 50 years, image matching has been an issue of research, development and practical implementation in software systems. Gruen [67] reports the development of image matching techniques over the past 50 years while in [68] more critical analyses and examples are reported.

Today many approaches for image matching exist. One first classification is based on the used primitives, which are then transformed into 3D information. According to these primitives, the matching algorithms can be *area-based matching* (ABM) or *feature-based matching* (FBM) [69]. ABM, especially the LSM method with its sub-pixel capability, has a very high accuracy potential (up to 1/50 pixel) if well textured image patches are used. Disadvantages of ABM are the need for small searching range for successful matching, the large data volume which must be handled and, normally, the requirement of good initial values for the unknown. Problems occur in areas with occlusions, lack of or repetitive texture and if the surface does not correspond to the assumed model (e.g. planarity of the matched local surface patch). FBM is often used as alternative or combined with ABM. FBM techniques are more flexible with respect to surface discontinuities, less sensitive to image noise and require less approximate values. Because of the sparse and irregularly distributed nature of the extracted features, the matching results are in general sparse point clouds which are then used as seeds to grow additional matches.

Another possible way to distinguish image matching algorithms is based on the created point clouds, i.e. *sparse* or *dense* reconstructions. Sparse correspondences were the initial stages of the matching developments due to computational resource limitations but also for a desire to reconstruct scenes using only few sparse 3D points (e.g. corners). Nowadays all the algorithms focus on dense reconstructions-using stereo or multi-view approaches. A dense matching algorithm should be able to extract 3D points with a sufficient resolution to describe the object's surface and its discontinuities. Two critical issues should be considered for an optimal approach: (i) the point resolution must be adaptively tuned to preserve edges and to avoid too many points in flat areas; (ii) the reconstruction must be guaranteed also in regions with poor textures or illumination and scale changes.

A rough surface model of the terrain is often required in order to initialize the matching procedure. Such models can be derived in different ways, e.g. by using a point cloud interpolated from the tie points measured at the orientation stage, from already existing terrain models, or from a global surface model generated with 3-second DEM by Shuttle Radar Topography Mission (SRTM). The matching procedures for terrain modelling are generally pyramid-based, that is, they start from a high image pyramid level, and at each pyramid level matching is performed; the results at one level are used to update the terrain range at the next lower pyramid level. In this way, the ambiguity of terrain variation is reduced at each pyramid level and search range are reduced as well. The algorithm convergence is a function of terrain slope, accuracy, and pixel size.

As dense matching is a task involving a large computing effort, the use of advanced techniques like parallel computing and implementation at GPU / FPGA level can reduce this effort and allow real-time depth map production.

The accuracy and reliability of the derived 3D coordinates rely on the accuracy of the sensor calibration and image orientation, the accuracy and number of the image observations and the imaging geometry (i.e. the effects of camera optics, image overlap and the distance camera-object). A successful image matcher should also employ strategies to monitor the matching results and quality with statistic parameters. The weaknesses of the image matching technique for DEM generation is the requirement of good texture in the images, the absence of dark and shadowed areas and the relative small baseline over flying height (B/D) ratio in order to have very small perspective distortion in the images.

5. DEM Quality assessment

The following evaluation is the result of two studies conducted on DEM estimation from VHR optical sensors from spaceborne platforms, and deeply described in [55], [70] and [71]. In these studies stereo scenes from three latest VHR resolution sensors were taken into account: GeoEye-1 (GE1, launched in 2011, WorldView-2 (WV2, launched in 2012) and Pleiades-1A (PL1, launched in 2012).

In the first study DEMs were generated from stereo images acquired by GE1 on Dakar (Senegal) and Guatemala City (Guatemala) and by WV2 on Panama City, Constitucion (Chile), Kabul (Afghanistan), Teheran (Iran), Kathmandu (Nepal) and San Salvador (El Salvador). The aim of this work was to evaluate the potential of VHR satellite images for the modeling of very large urban areas for the extraction of value-added products for hazard, risk and damage assessment.

The main characteristics of the images and areas are briefly reported. Due to the large extent of the cities (up to 1'500 km^2), the datasets generally consist of multiple pairs (or couples) of stereo images acquired by the same sensor and cut in tiles. If the stereo pairs are acquired in the same day, the time difference between their acquisitions is less than 1 hour. In case of multiple dates, differences are up to 3 months (i.e. Dakar). The viewing angles and consequently the convergence angle and B/H ratio are not the same for all stereo pairs. Different situations occur: a) one quasi-nadir image (acquisition angle close to the vertical) and one off-nadir image (backward or forward viewing); b) one backward and one forward image with symmetric angles, and c) one backward and one forward image with asymmetric angles and large convergence angle. Large viewing angles determine the presence of occlusions, mainly in urban areas, larger shadows, and larger GSD, with respect to quasi-nadir acquisitions. Areas like San Salvador and Kabul, smaller than 15-17 km in width, were scanned in one path. In case of larger areas the images were acquired in the same day from two (Panama City, Guatemala City, Constitucion) or three (Teheran or Kathmandu) different paths, or in different days (Dakar). In each path the acquisition angles are almost constant, but different between paths. This might cause differences in the DEM on the overlapping areas between paths, as the sensor performance for DEM generation depends on the B/H ratio, and consequently on the incidence angles of the stereo images.

GE1 stereo images were provided at geo-referenced processing level, while the WV2 ones were provided in raw level. In all cases, the rational polynomial coefficients (RPC or RPB formats) available for each image or tile were used as geo-location information for the geometric processing.

In general the geo-location accuracy of the RPC depends on the image processing level, the terrain slope and the acquisition viewing angles ([57],[58]). In flat areas like in Panama City (level 2A,-16º and 16º viewing angles) the relative accuracy of the RPC between two images composing a stereo pair is approximately 30m, while in the mountain area around Kathmandu (level 1B,-31º and 15º viewing angles) the relative accuracy between the stereo images reaches 300m.

The datasets mainly cover inhomogeneous urban areas with different layout: dense areas with small buildings, downtown areas with skyscrapers, residential areas, industrial areas, forests, open areas and water (sea, lakes, and rivers). In addition images include rural hilly and mountain areas surrounding the cities, with important height ranges: 2'400m in case of Kathmandu, almost 2'300m in case of Teheran, 1'100m in case of Guatemala City, 1'000m in case of Kabul. Some regions were not visible in the images because occluded by clouds or very dark cloud shadows, as in Panama City and Kathmandu cities.

The processing for DEM generation was applied separately to each dataset. On the orientation point of view, the geometric model for spaceborne pushbroom sensors based on RPC was used.

Ground points were not available, so at least the relative orientation of the images was guaranteed. Common tie points in two or more images were measured homogeneously in the images in order to ensure the relative orientation between the two images of the same stereo pair and between different stereo pairs that overlaps along or across the flight direction, and get a stable block. A minimum of 5 tie points for each pair were measured manually by an operator on well-defined and fixed/stable features on the terrain (i.e. crossing lines, road signs). Taking into account the ground spatial resolution of the input images, the DEMs were generated with 2m grid spacing (about 4 times the pixel size of the panchromatic channels). In case of projects with overlapping stereo pairs (i.e. Teheran, Kathmandu, Dakar, etc.), the DEM was computed separately for each stereo pair and then the single DEMs were merged using linear interpolation.

The second work on DEM quality assessment was carried out on a testfield with accurate ground reference data [70]. The testfield is located in Trento, in the Northeast of Italy, and varies from urban areas with residential, industrial and commercial buildings at different sizes and heights, to agricultural or forested areas, and rocky steep surfaces, offering therefore a heterogeneous landscape in term of geometrical complexity, land use and cover. The height ranges from 200m to 2100m. The testfield includes several heterogeneous datasets at varying spatial resolution, ranging from satellite imagery to aerial imagery, LiDAR data, GNSS surveyed points, GIS data, etc. For the scope of this Chapter, the stereo images from VHR satellite sensors used for DEM quality assessment are:

a. A WorldView-2 (WV2) stereo-pair, acquired on August 22, 2010. The first image was recorded in forward scanning mode with an average in-track viewing angle of 15.9°, while the second one was acquired in reverse scanning mode with an average in-track viewing angle of-14.0°. The processing level is Stereo 1B, i.e. the images are radiometrically and sensor corrected, but not projected to a plane using a map projection or datum, thus keeping the original acquisition geometry. The available channels are the panchromatic one and eight multispectral ones. The images cover an area of 17.64×17.64 km. The images were provided with Rational Polynomial Coefficients (RPCs).

b. A GeoEye-1 (GE1) stereo-pair, acquired on September 28, 2011. Both images were recorded in reverse scanning position, with a nominal in-track viewing angle of about 15° in forward direction and-20° in backward direction. The stereo images are provided as GeoStereo product, that is, they are projected to a constant base elevation. The available bands are the panchromatic one and four multispectral bands (blue, green, red, and near infrared). The images cover an area of 10×10 km. For each image the RPCs were provided.

c. A triplet from Pléiades-1A (PL1) sensor, acquired on August 28, 2012. The average viewing angles of the three images are, respectively, 18°,-13° and 13° in along-track direction with respect to the nadir and close to zero in across-track direction, while their mean GSD varies between 0.72m and 0.78m, depending on the viewing direction. The Pleiades images were provided at raw processing level called "Primary", that is, with basic radiometric and geometric processing. This product is indicated for investigations and production of 3D value-added products [72]. The three images cover an area of about 392km^2.

For the geometric processing of the stereo scenes, the commercial software SAT-PP (SATellite image Precision Processing) by 4DiXplorer AG [73] was used. Information about the software functionalities and the approaches for image orientation and DEM generation is given in [74]. The images were oriented based on the RPC-based model, by estimating the parameters modelling an affine transformation to remove systematic errors in the given RPCs. For this operation, a selection of available ground points visible in both images was used. A sub-pixel accuracy was reached in the orientation both for WV2 and GE1 stereo-pairs and PL1 triplet. The DEMs were generated with a multi-image least-square matching, as described in [74], using a grid space equal to 2 times the GSD, which leads to 1 m geometric resolution surface models. Few seed points were manually measured in the stereo images in correspondence of height discontinuities to approximate the surface. The DEMs were neither manually edited nor filtered after their generation. Table 1 shows the DEM analysis in three test areas: 1. Trento city center, characterized by small and buildings closed to each other, 2. Trento train station, with large flat buildings and open areas and 3. Fersina residential area, with separated buildings and open areas. The reference LiDAR DEM are shown together with the error maps of GE1, WV2 and PL1 DSMs. In the error maps the height differences in the range [-1 m, 1 m] were plotted in white, as they are within the intrinsic precision of the sensors, while large positive and negative errors were highlighted in red (LiDAR above the DEM) and blue (DEM above LiDAR), respectively.

By visual analysis of the DEMs, it was observed that in all datasets the surface shape was well modelled with both sensors (GE1 with processing level 2A and WV2 with processing levels 1B and 2A). In mountain areas with large elevation difference, like Trento, Teheran, Kabul and Guatemala City, the shape of valleys and mountain sides and ridges is well modeled in the DEM (Figure 7). In comparison to SRTM, using VHR images it is possible to extract finer DSM and to filter the Digital Terrain Model (DTM) at higher grid space. This is confirmed by the comparison of the height profiles of the WV2 DTM and the SRTM in Teheran (Figure 8). In rural areas, cultivated parcels can be distinguished, together with paths and lines of bushes and trees along their sides. In case of forest, the DEM clearly shows a different height with respect to adjacent cultivated areas or grass. It is even possible to distinguish roads and rivers crossing forests (Figure 9). In general rural areas are well modelled both in mountain, hilly and flat terrain.

In urban areas building agglomerations, blocks with different heights, the road network, some infrastructures (i.e. stadium, bridges, etc.) and rivers are generally well outlined both in flat and hilly terrains (Figure 10, Table 1). In residential and industrial areas it is possible to distinguish single buildings and lines of trees (Figure 11, Table 1). In some cases the roof structures of large and complex buildings are modeled (Figure 12). Errors are encountered between buildings, as narrow streets are not visible in the stereo-pairs due to shadows or occlusions. In these cases the DEMs overestimates the height of the terrain, as they do not model the street level (Table 1). In case of Trento dataset (Table 1), the two large red spots (highlighted in red and blue in the LiDAR DEM) occur on churches, in all three DEMs. The matching failure is likely due to the homogeneous material used for the roof cover, the shadowing, and, eventually, the structure geometry. The manual measurements of seed points on the two buildings would certainly help the matching procedure. In the train station test area, some differences are due to the presence of trees (blue circle, LiDAR is below the DEMs) and to a change in the area (red circle, LiDAR is above the DEMs), that is, a building was demolished between LiDAR and WV2/GE1/PL1 acquisitions. Regarding the third test area (Trento Fersina), significant height differences occur in correspondence of trees (blue areas) and between tall buildings.

By comparing the statistics of the error maps, in general there is a good agreement between the three surface models generated by PL1, WV2 and GE1 and the values in the statistics confirm the above analysis. The three surface models give similar results in terms of minimum and maximum values, mean value and RMSE, being Pleiades slightly better than the other two.

From the analyses above reported, it can be concluded that failures in surface modelling from VHR optical sensors can be caused by a number of factors, which are summarized in Table 2.

The image absolute geolocation accuracy, which depends on the viewing angle, on the processing level and on terrain morphology, influence the estimation of the height in object space, and therefore the quality of the absolute geolocation of the final DEM and related products (DTM, orthophotos, 3D objects). The measurement of accurate and well distributed ground control points (GCPs) in the images would solve the problem, but this information is often not available. The accuracy of the relative orientation between two images forming a pair is crucial for the epipolar geometry and image matching, while the relative accuracy between overlapping stereo pairs is responsible of height steps in the final DEM (Figure 13). The size of the height steps depends on the accuracy of the geometric orientation of each stereopair, and generally shows a systematic behavior. In both cases the relative orientation can be improved by manually measuring a sufficient number of common tie points between the images.

Figure 7. Trento. DEM from the GE1 stereo-pair; visualization in colour-shaded mode in SAT-PP software.

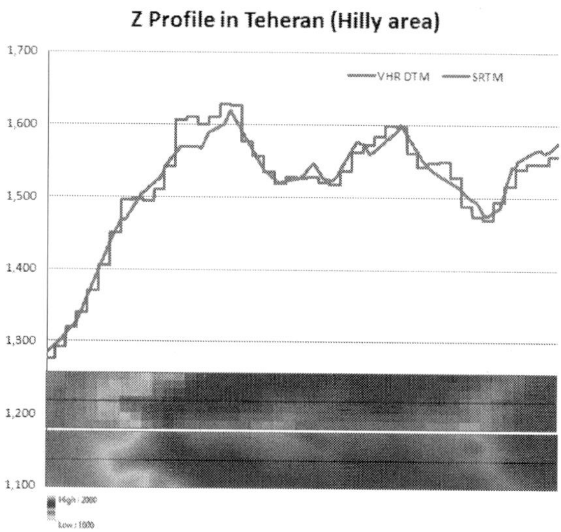

Figure 8. Teheran. Height profile in WV2 DTM (blue) and SRTM (red) and zoom in the corresponding surface models (above: SRTM, below: WV2 DTM).

http://dx.doi.org/10.5772/57584

Figure 9. Left: DEM of Rural area in Constitucion. Right: Original image (panchromatic)

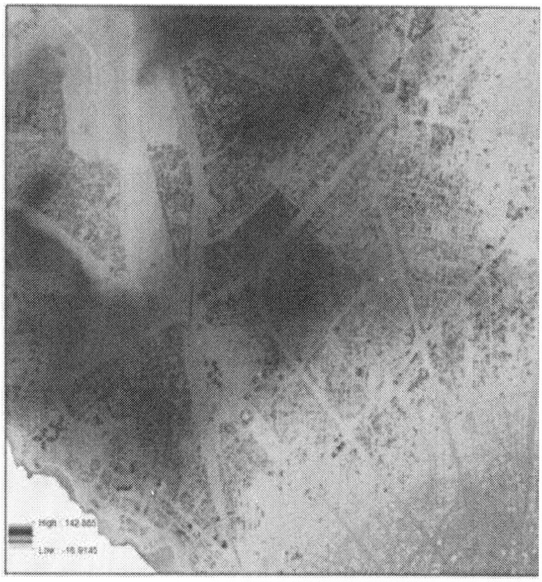

Figure 10. DEM of dense urban area on flat and hilly terrain in Dakar.

Figure 11. DEM of residential and rural area on flat terrain in Panama City. The black oval highlights a line of tree.

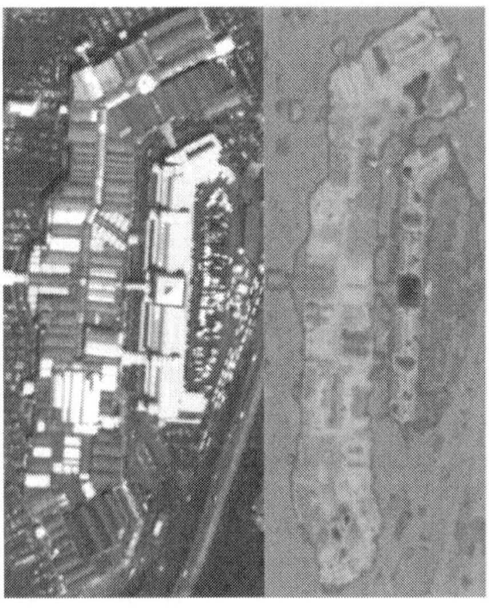

Figure 12. Left: DEM of Panama City Airport. Right: original image (pan sharpened).

http://dx.doi.org/10.5772/57584

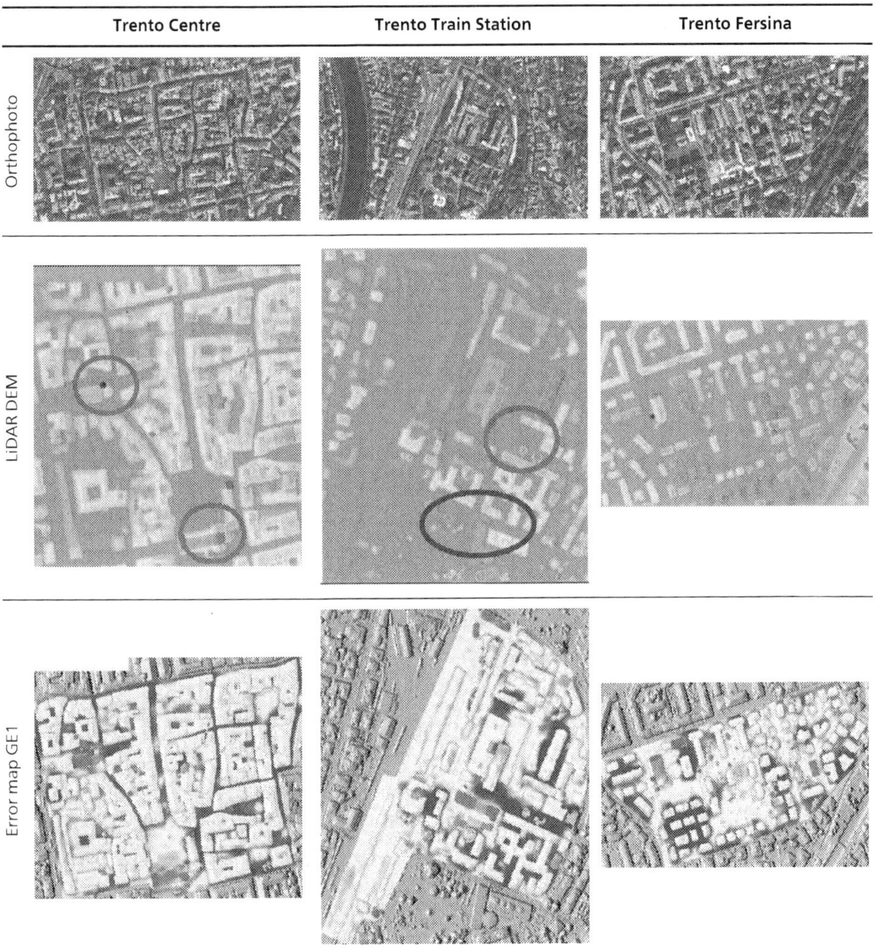

m: -44.4,M: 26.5	μ: 2.3	m: -28.1, M: 53.5	μ: 1.7	m: -46.3, M: 28.8	μ: 0.5
σ: 6.6	RMSE: 6.9	σ: 6.9	RMSE: 7.1	σ: 7.4	RMSE: 7.9

	Trento Centre		Trento Train Station		Trento Fersina	
Error map WV2	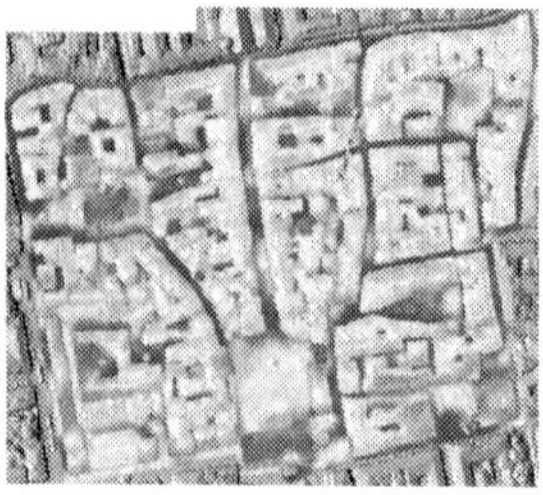					
Statistics	m: -53.9, M: 26.7	μ: 1.9	m: -39.0, M: 58.4	μ: 0.7	m: -43.7, M: 27.7	μ: 0.3
	σ: 7.1	RMSE: 7.4	σ: 7.1	RMSE: 7.4	σ: 8.5	RMSE: 8.5
Error map of PL1 triplet DEM						
	m:-42.39,M: 24.22	μ: 1.38	m:-33.10, M:54.69	μ: 1.05	m: -42.91, M: 24.22	μ: 0.47
	σ: 0.06	RMSE: 6.12	σ: 0.22	RMSE: 6.52	σ: 0.25	RMSE: 6.73

Table 1. DEM analysis on Trento testfield. Quality evaluation of WV2, GE1 and PL1 DEMs with respect to the LiDAR DEM: orthophoto with test areas contour (light blue) and profile transect (yellow), error planimetric distribution, statistics (minimum m, maximum M, mean μ, standard deviation σ, RMSE) and height profiles. Measures are in meters.

Low-textured and homogenous areas origin blunders in the DEM, as the automatic matching of the homologous points fails. This is typical in homogeneous land cover (i.e. bare soil, parking lots) and shadow areas and is caused by a combination of sun and satellite elevations and surface morphology (i.e. mountain faces). In Figure 14 building shadows, highlighted in the yellow ellipse, bring inaccuracies in the DSM. The use of a better initial DEM as initial approximation can help the matching procedures in these critical areas. If a DEM is not available, so-called seed points can be measured in stereo mode in the pairs and imported in the matching procedure as mass points. In addition, an ad-hoc radiometric processing can enhance details in low-textured regions and help the matching procedure.

Occlusions are generally present in urban areas and are due to tall buildings or trees, in combination with the acquisition viewing angles. In case of occlusions, corridors between buildings are not modeled correctly (Figure 15). To overcome this drawback, multi-angular acquisitions, like in Pleiades constellation, could reduce occlusions in the images.

Objects moving during the acquisitions of the stereo images, like vehicles, lead to small blunders, as highlighted in the blue circles in Figure 14. They can be removed with manual editing or filtering. Local blunders in correspondence of special radiometric effects, like spilling and saturation on roof faces due to the acquisition geometry and the surface type and inclination (i.e. roof faces in grass), may also occur.

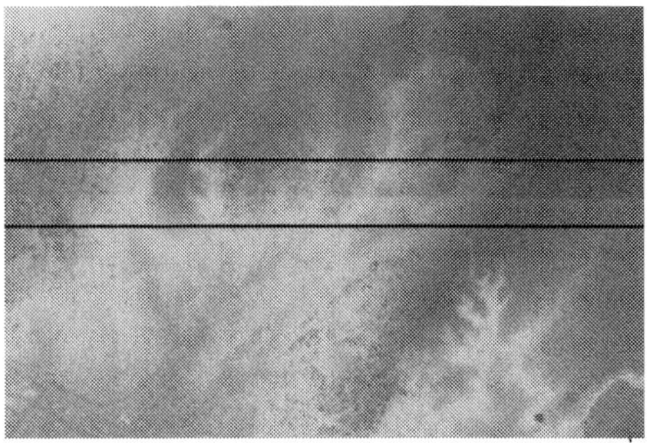

Figure 13. Height step (mean value: 3.5m, black rectangle) between the DEMs obtained from two different stereo pairs (Dakar).

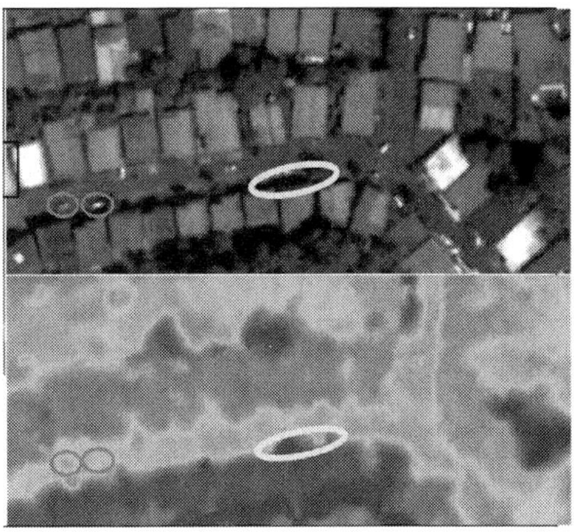

Figure 14. Example of effects of shadows and moving objects in the DEM (Panama City). Above: pansharpened nadir scene; below: DEM

Figure 15. Example of corridor occlusion due to tall buildings; (a) and (b): stereo images; (c) resulting DEM (Dakar).

Factor	Cause/Dependency	Effect in DEM	Possible solution
Poor initial absolute geo-location accuracy	Viewing angles Processing level Terrain shape	Height steps Poor final absolute geolocation quality	GCPs
Time interval between overlapping acquisitions	Large extent Swath width of VHR sensors	Height steps in overlapping areas	Tie points measurements
Differences in the images	Moving objects (vehicles)	Local blunders	DEM editing Filtering
Shadows	Large viewing angle Sun inclination Surface morphology	Mismatches	Radiometric preprocessing
Low texture areas	Land covers (parking lots, bare soil, etc.)	Mismatches	Radiometric preprocessing Seed points
Spilling/saturation of roof	Radiometry Sun and satellite elevation Surface inclination Surface material	Local blunders	Masking
Occlusions in urban areas	Convergence angle Tall buildings	Local blunders Wrong heights on streets	Seed points Multi-angular acquisitions
Height range	Steep mountain Vertical steps	Low details	Seed points
Cloud cover	Whether conditions	Mismatches	Masking
Water (lakes, sea)		Bad quality DEM	Masking

Table 2. Summary of factors influencing the DEM quality.

http://dx.doi.org/10.5772/57584

6. DEM Fusion

DEMs of large areas are usually generated either by SAR interferometry, laser scanning (mainly from airborne platforms) or photogrammetric processing of aerial and satellite images. As demonstrated in the previous sections, each sensing technology and processing algorithm (interferometry, image matching) shows its own strengths and weaknesses (see Table 3). Even within a single technology a highly varying DEM quality may be found.

DEMs can be generated at different coverage levels, ranging from very local areas to near-global coverage. Especially taking into account digital models offering a vast land coverage, one can see that the accuracy, error characteristics, completeness and spatial resolution offered by these products can vary wildly. These products and the possibility of their joint exploitation, with their respective issues, are the object of the techniques introduced in this Section.

6.1. The motivation: Complementarity

As mentioned, each different DEM generation approach and platform of origin have their advantages and drawbacks. Such characteristics seem to point into an interesting direction. Especially if one directs its attention on a cross-platform comparison. In Table 3 some characteristics between optical and SAR DEM generation are resumed. It can be observed that the two technologies almost perfectly complement each other.

Characteristic	SAR	Optical
Influenced by clouds	No	Yes
Influenced by atmospheric water vapour	Yes	No
Influenced by sun illumination	No	Yes
Performing on poorly textured areas	Yes	No
Performing on edge features	No	Yes
Influenced by layover effect	Yes	No
Surface height estimate	Yes	Yes

Table 3. Comparison between SAR and Optical sensing technologies: advantages and drawbacks.

The complementarity between these platforms is also visible if one takes into account the respective DEM quality maps shown in Figure 1 for the SAR example and in Figure 16 for the Optical one. Note that the quality scales are the opposite between the two, since the optical one describes matching cross-correlation combined with other factors (the higher the better), while the other defines the precision in meters (the lower the better). By analysing these examples, one can see that where the SAR result does not excel, i.e. on edges, the optical one shows better performance. And vice-versa where the optical result is not as strong, i.e. poorly textured areas, the SAR one shows consistently better results.

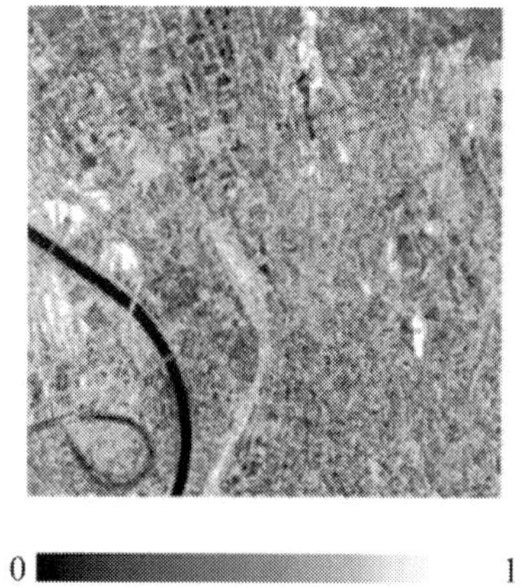

Figure 16. Example of a matching Pléiades-1A cross-correlation map over Trento, Italy.

As a consequence, a joint exploitation of the elevation products deriving from these platforms can potentially be highly profitable, since the information on which one is weak can be replaced/completed by the one supplied by the other.

The second typology of data that seems a legitimate candidate for the fusion can be defined as intra-platform, in the sense that two products coming from the same platform may as well be complementary to one another. In SAR, for instance, one may have both a DEM generated with ascending data and one with descending data, as shown in Figure 17, evidently complementing each other.

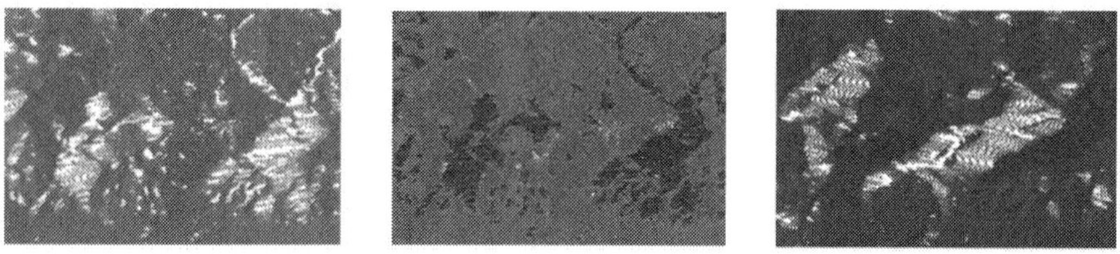

Figure 17. Example of intra-platform complementarity. Point cloud representation of an ascending ERS-1/2 DEM (left), a descending ERS-1/2 DEM (right) and the combination of the two (center).

In Figure 17 the digital terrain models are reproduced as point clouds (i.e. not interpolated). This is a less biased way to represent elevation data, even if at the expense of the ease of visualization and manipulation.

There are, however, constraints that directly arise in order to be able to compare and combine the fusion candidates. The first one, easier to solve, is dictated by the different pixel spacing. This issue is simply solved by oversampling the coarser DEM and its corresponding quality

http://dx.doi.org/10.5772/57584

map. The resampling does not imply that the data has acquired the same precision as the finer one, and, that the oversampling procedure implies a re-estimation of the data values. In the intra-platform case this can be avoided, as in the given example, since the products already possess comparable characteristics non requiring a re-estimate.

The second constraint is to have a comparable quality index. As shown, the quality maps proposed in Figure 1 and Figure 16 imply different types of measures, which must be re-interpreted in order to acquire a cross-significance. The way these indexes can be obtained is given in Table 4.

Quality Map	Factors
SAR DEM Precision Map	- baseline - wavelength - interferometric coherence - local incidence angle - spatial ground resolution
Optical DEM Accuracy Map	- matching parameters - slope, roughness and aspect (geomorphological criteria)

Table 4. Quality map estimation by technique.

In the following sections two different data representation levels, on which the fusion approaches are based, will be defined. Example approaches will be outlined as well.

6.2. Raster level fusion

The primary level which is directly available for fusion is the raster level. In fact, when represented in raster format, DEMs define a height value (when available) for each cell location on a regularly spaced grid covering the whole study area, and are a suitable input for raster-based fusion approaches.

By exploiting the quality estimates from the previous section, the easiest and sometimes more appropriate method to fuse two raster data is the weighted combination of the two observations, namely, a weighted average. This process is executed between DEMs covering the same surface and having the same pixel spacing, resulting in a very fast cell-to-cell calculation. The simplicity of the approach, however, may badly influence the results. The most important aspect to keep in mind when fusing two different datasets is their accuracy with respect to their spatial resolution. When the latter greatly differs between the two, their combination may result in an over-smoothing effect as well as single pixel aberrations and abrupt discontinuities (which in turn represent over-fitting). This results in a major degradation of the information contained in the datasets. When over-smoothing, the quality of the product with higher spatial resolution will be degraded, while when over-fitting the spatial characteristics will be extremely exaggerated/accentuated. As such, this kind of processing better suits products having the same pixel spacing from the start, rather than oversampled, and shall be avoided if the original pixel spacing difference is too large.

More advanced methods have been proposed to overcome the issues due to an over simplistic approach. These techniques keep into account the values and statistics not only in the single pixels but also in their neighbours, by exploiting statistics windows, extending the process to a less local estimate and thus ensuring a spatial consistency of the results. This type of approaches are conceptually less sensitive to the over-smoothing and over-fitting effects, since the spatial characteristics of both input data are taken into account. In general, during the fusion process, a higher priority is given to the input with finer spatial resolution. These approaches are more desirable than the first one based on weighted average, since they are less prone to produce outputs greatly inferior to the inputs and, moreover, they are less influenced in case of oversampling, since the window on the finer input will have a higher weight. The difference in pixel spacing should ideally be kept quite small even in this case.

One example of approach following these assumptions can be found in [75], where also additional information is used as input. The latter is constituted of a database (dictionary) of small DEM patches collected over existing DEMs showing similar characteristics to the ones to be fused. Sparse representations are used to solve the fusion problem. In this approach Orthogonal Matching Pursuit is used to identify the most suitable input patches to be used in weighted linear combination defining the output. An example of raster level fusion product is given in Figure 18.

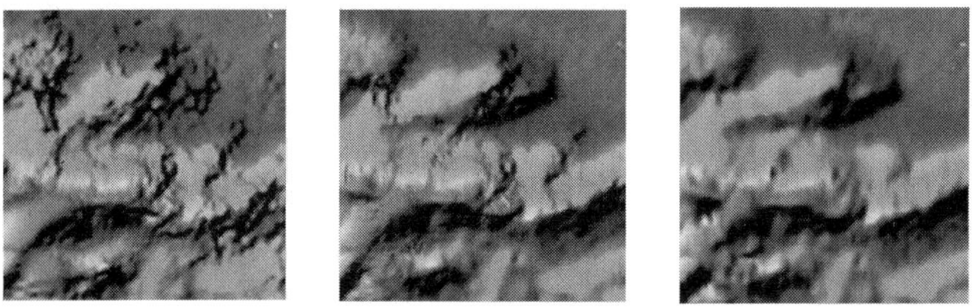

Figure 18. Example of raster based DEM fusion (center) between a SPOT-5 product (right) an ALOS-PALSAR-1 product (left).

6.3. Point cloud level fusion

The second product level available for data fusion is the Point Cloud level. Fusing point clouds, instead of rasterized surfaces, comes natural if one takes into account how Interferometry and Image Matching work, i.e. in a point-wise fashion. This allows to lessen an effect that occurs during the rasterization step of the outputs, i.e. error propagation. By avoiding rasterization, the error value computed in the height estimation process is preserved as is, and corresponds to the input data alone, thus preserving the quality of each observation.

http://dx.doi.org/10.5772/57584

The approaches based on point cloud level, however, show a greater complexity. Firstly, it is a kind of data which is conceptually more abstract to handle than the raster one. While the latter can easily be associated to matrices, the former are vectorial point objects in three dimensions (see Figure 19). This also implies a conceptual complexity of the approach to design for their fusion. The second aspect to take into account is the data size and point density. In case of SAR DEMs covering very large areas, for example, the study area may show very high coherence over the whole scene, implying a large amount of points with very large height variability over a restrained two dimensional (coordinates) space. Finally, with respect to raster-based approaches, more complex procedures must be designed for the inclusion of the information provided by the quality measures into the process.

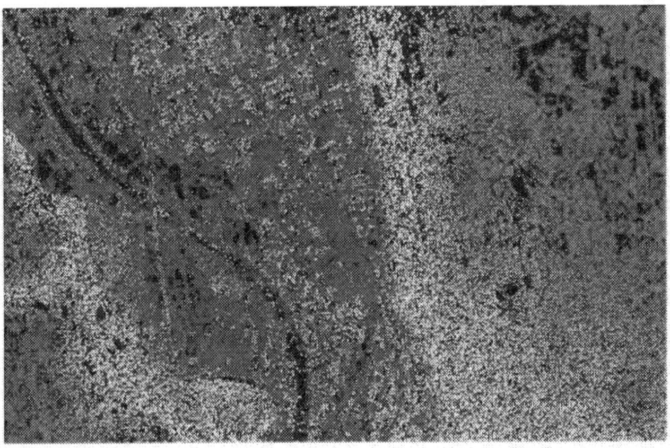

Figure 19. Example of image matching results, shown as point cloud, obtained from Pléiades-1A imagery over the city of Trento.

A possible solution to the overabundance of data is a preliminary sample selection/substitution phase, in which the information about the data can be exploited. In SAR-SAR fusion, precision can be used to drive the choice between redundant points. In SAR-Optical fusion, the instrument behaviour resumed in Table 3 can be exploited as well. An ideal approach would for instance prefer measurements coming from Optical edge matching rather than its interferometric counterpart, vice-versa points produced by matching a regular grid over poorly textured areas should be substituted by the interferometric ones. Note however that slope also remains a key factor to keep into account (see Table 3).

During the image matching process it is possible to identify which one is which, and this information should be taken into account and stored, as shown in Figure 20. The selection phase can also be used as an initial combination phase, computing local statistics such as the slope between points, in order to eliminate outliers and substitute the observations according to their quality measure and characteristics.

Figure 20. Breakdown of a subset of Figure 19 between: grid-based matching points (left), feature-based matching points (center), edge-and area-based matching points (right).

The second step to accomplish is the generation of the fused DEM itself, hence the definition of an algorithm capable to handle such data and produce the final model. Interpolators can be applied to estimate the final model from the newly defined point cloud. In this perspective, two main types of interpolators can be considered. The first one are the so-called exact interpolators [76], according to which the output estimating function must pass by fixed points (input observations). The second interpolators are the confidence-interval aware interpolators, for which the output function should pass between an "envelope" from the points at which the interpolant is fitted. This concept is shown in Figure 21. In the left image the output function is fitted to well selected samples. In the centre image, some samples showing a high error and outlying the correct estimate were not filtered, resulting in an estimate function (solid line) not reproducing the correct function (dashed line). In the right image, the outliers are still present, but thanks to the error aware interpolator, the function would ideally be better or correctly estimated. Note that ideally this error envelope should be user defined, exploiting the a priori quality knowledge offered by SAR Precision, Optical cross-correlation, slope and feature type, for each input point.

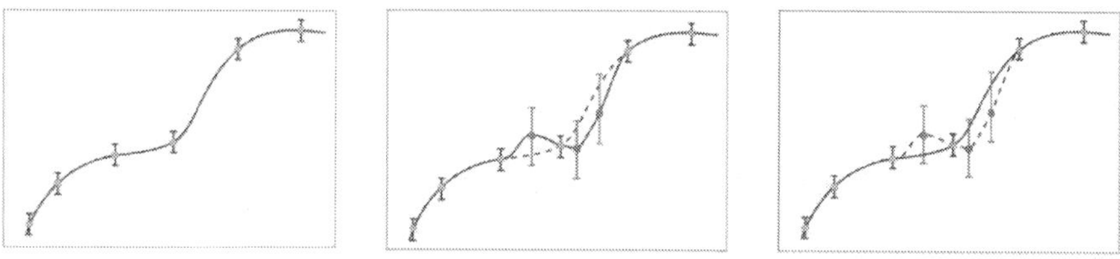

Figure 21. Example of a function estimated from an exact interpolator (left), from an exact interpolator with outliers (center) and from an error aware interpolator (right). The solid line is the estimate function, the dashed line the correct function. Green dots are good selected samples, red dots bad selected samples, vertical lines define the error envelope.

The first family of interpolators include, for instance, the Radial Basis Function [77], which is widely recognized as reliable estimator. This kind of interpolators, however, does not take into account the quality index of each point, but, after a well-executed sample selection step, the fusion can be performed. Ideally, interpolators taking into account the confidence interval are better suited, especially if they offer the possibility to specify the confidence range at each data point while computing the interpolant.

The size problem, however, still remains. By using advanced interpolators, the results improve greatly in accuracy, but increase in computational cost, memory cost and complexity. A way to avoid these problems is to implement the RBF following a structure similar to the one of the Shepard Interpolator, as proposed in [78]. The process is resumed in Figure 22.

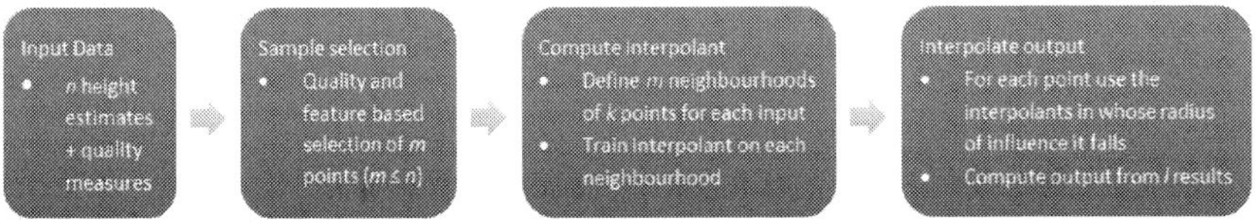

Figure 22. Diagram summarizing the steps of the proposed fusion approach.

This approach also leaves the possibility to decide the regionality and distance influence of the interpolator, allowing for a fine tuning and a good trade-off between over-smoothing and over-fitting. Being very dense datasets, the possibility to control how local the interpolant should be is very important. An example of DEMs fused with this approach is shown in Figure 23.

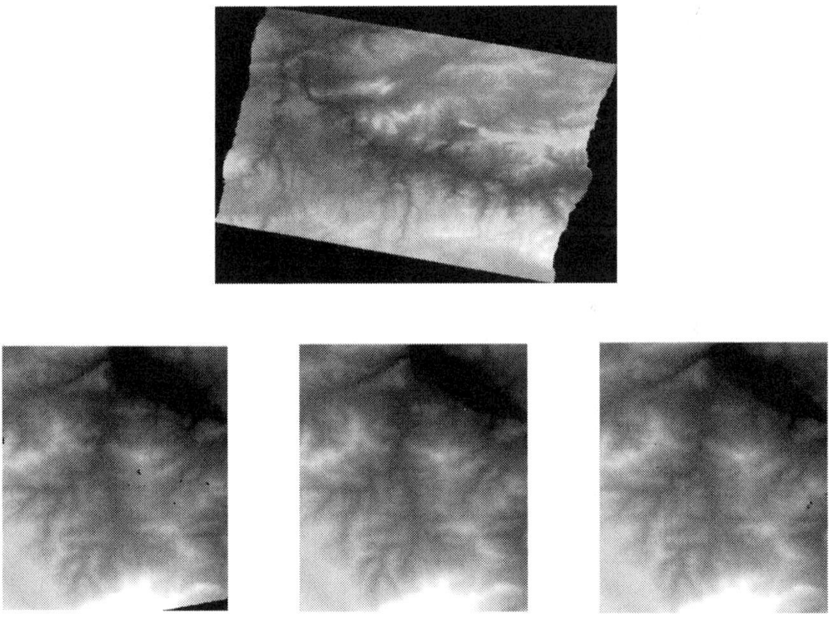

Figure 23. Examples of an ascending ERS-1/2 DEM (bottom-left), descending ERS-1/2 DEM (bottom-right), DEM fusion results (bottom-center) and the full DEM resulting from the fusion (top), over a region in Turkey.

The second family of interpolators is composed by more advanced techniques, which are currently used in a wide variety of study fields, as well as environmental studies. See [76] to have an extensive collection of such approaches. When taking into consideration these approaches, however, the very large size of the datasets becomes even more an issue. The computational and memory costs may well become prohibitive. A possible solution could be to apply a concept similar to the one introduced for Radial Basis Functions, reducing considerably the issues.

7. Conclusions

In this chapter, the SAR and Optical techniques applied to produce DEMs were introduced, leading to the topic of elevation data fusion. The main reasons strengthening the desirability of such an approach, especially at an inter-platform level, have been outlined. The fusion topic is still, however, an ongoing issue which has to be further investigated. The interest in developing this topic may even grow in the future. Especially considering the pace at which new instruments capable to extract elevation information are introduced, bringing with them new issues and shortcomings which may well be overcome by exploiting their complementarities. The production of worldwide "absolute" single-date single-sensor elevation models still remains utopic, even for new missions such as TanDEM-X, Sentinel or Pléiades. The shortcomings of the instruments will be lessened, their spatial resolution improved, but their intrinsic characteristics will remain, along with their complementarity and therefore the interest in their fusion. Temporal resolution is also an important aspect, since throughout the year the ground's physical characteristics can induce highly varying height estimates. The fusion of multi-date elevation models would allow to obtain products with a much more reliable terrain reproduction ability. Each fusion processing level has its advantages, the raster level surely is easier to handle than the point cloud one, while the latter offers a less biased data reproduction. For the sake of accuracy and reliability, the point cloud level is undoubtedly the better option on which to base future approaches. The dataset size-related issues opposing these approaches can be overcome by improving sample selection or by improving the interpolant computation step, greatly increasing the speed and manageability of the approaches. Finally, approaches giving the ability to finely exploit the data intrinsic information to guide the fusion should be investigated.

Acknowledgements

The authors would like to acknowledge Dr Giorgio Agugiaro (Bruno Kessler Foundation, Trento, Italy), Dr Emanuele Angiuli and Ivano Caravaggi (Joint Research Center, Ispra, Italy) for their support in satellite image processing, DEM extraction and quality assessment.

The work on Trento testfield was partly supported by the 3M project (co-founded Marie-Curie Actions FP7 – PCOFOUND – GA-2008-226070, acronym "Trentino Project"). The GeoEye-1

http://dx.doi.org/10.5772/57584

images acquired over Panama City (Panama) and San Salvador (El Salvador) belong to WorldBank. The authors would like to thank WorldBank for giving JRC the opportunity to use the data for research purposes, the Autonomous Province of Trento (PAT) and the Municipality of Trento, for providing spatial data for the Trento testfield, and Astrium GEO-Information Services for providing the Pleiades triplet for research and investigation purposes.

Author details

Loris Copa[1*], Daniela Poli[2] and Fabio Remondino[3]

*Address all correspondence to: lcopa@sarmap.ch

1 Sarmap, Cascine di Barico, Purasca, Switzerland

2 Terra Messflug GmbH, Imst, Austria

3 Bruno Kessler Foundation, Povo-Trento, Italy

References

[1] Leberl F.. Radargrammetric Image Processing. London, Artech House, 1990.

[2] Rocca F., Prati C., Ferretti A.. An Overview of SAR Interferometry. In: Proceedings of the 3rd ERS Symposium, 1997. September 01. Florence, Italy. ESTEC Publishing Division.

[3] European Space Agency: European Remote Sensing satellites: ERS. https://earth.esa.int/web/guest/missions/esa-operational-eo-missions/ers (accessed 12 August 2013).

[4] C. Poidomani, D. Costantini, and P. Pasquali, National-scale DEM generation using ERS-Tandem data in alpine regions, In Proceedings of the ERS-ENVISAT Symposium "Looking down to Earth in the New Millennium", 2000. October 16-20. Gothenburg, Sweden.

[5] National Aeronautics and Space Administration. NASA: Jet Propulsion Laboratory (JPL): SRTM. http://www2.jpl.nasa.gov/srtm/ (accessed 13 August 2013).

[6] Farr, T. 2001. The shuttle radar topography mission. *Proceedings of IGARSS 2001.*

[7] Rabus, B., Eineder, M., Roth, A. & Bamler, R.. The shuttle radar topography mission-a new class of digital elevation models acquired by spaceborne radar. ISPRS Journal of Photogrammetry and Remote Sensing 57, 2003, 4: 241–262

[8] Heipke, C., Koch, A. & Lohmann, P.. Analysis of SRTM DTM-Methodology and practical results. In: Armenakis C., Lee Y.C.. (eds.) Geospatial Theory, Processing and Applications: ISPRS Commission IV, Symposium 2002, July 9-12, 2002. Ottawa, Canada. ISPRS.

[9] Audenino, P., Rognant, L. & Chassery, J.. Qualification of SRTM DEM. A first approach toward an application dependant qualification framework. In: Geoscience and Remote Sensing Symposium, 2003. IGARSS '03. July 21-25 2003. Toulouse, France. Vol.5: 3082–3084. IEEE International.

[10] Bourgine, B. & Baghdadi, N.. Assessment of C-band SRTM DEM in a dense equatorial forest zone. Comptes Rendus Geosciences, Journal 2005. Vol.337, 14: 1225–1234.

[11] Carabajal, C.C. & Harding, D.J. 2006. SRTM C-band and ICESat Laser Altimetry Elevation Comparisons as a Function of Tree Cover and Relief. Photogrammetric Engineering & Remote Sensing, Journal 2006. Vol.72, 3: 287–298.

[12] Salamonowicz, P.. Comprehensive Assessment of the Shuttle Radar Topography Mission Elevation Data Accuracy. In: Gesh D., Muller J.-P., Farr T.G.. (eds.) The Shuttle Radar Topography Mission – Data Validation and Applications, Workshop, June 14–16 2005, Reston, Virginia, USA. ASPRS.

[13] Gorokhovich, Y. & Voustianiouk, A.. Accuracy assessment of the processed SRTM-based elevation data by CGIAR using field data from USA and Thailand and its relation to the terrain characteristics. Remote Sensing of Environment, An Interdisciplinary Journal 2006. Vol.104, 4: 409–415.

[14] Rodríguez, E., Morris, C.S. & Belz, J.E. 2006. A Global Assessment of the SRTM Performance. Photogrammetric Engineering & Remote Sensing, Journal 2006. Vol.72, 3: 249–260.

[15] Berry, P.A.M., Garlick, J.D. & Smith, R.G.. Near-global validation of the SRTM DEM using satellite radar altimetry. Environment, An Interdisciplinary Journal 2007. Vol. 106, 1: 17–27.

[16] Marschalk, U., Roth, A., Eineder, M. & Suchandt, S.. Comparison of DEMs derived from SRTM/X-and C-band. In: Geoscience and Remote Sensing Symposium, 2004. IGARSS '04. September 20-24, 2004. Anchorage, Alaska, USA. vol.7: 4531–4534. IEEE International.

[17] Hoffmann, J. & Walter, D.. How Complementary are SRTM-X and-C Band Digital Elevation Models? Photogrammetric Engineering & Remote Sensing, Journal 2006. Vol. 72, 3: 261–268.

[18] Ham, A., Rupe, C., Kuuskivi, T. & Dowding, S.. A Standardized Approach to Phase Unwrap Detection/Removal and Void Fill of the Shuttle Radar Topography Mission (SRTM) Data. In: Gesh D., Muller J.-P., Farr T.G.. (eds.) The Shuttle Radar Topogra-

phy Mission – Data Validation and Applications, Workshop, June 14–16 2005, Reston, Virginia, USA. ASPRS.

[19] Kuuskivi T., Lock, J., Li X., Dowding, S. & Mercer, B. 2005. Void Fill of SRTM Elevation Data: Performance Evaluations. In: ASPRS Annual Conference Proceedings, "Geospatial Goes Global: From Your Neighborhood to the Whole Planet", 2005. March 7-11. Baltimore, Maryland, USA.

[20] Hofton, M., Dubayah, R., Bryan Blair, J. & Rabine, D.. Validation of SRTM Elevations Over Vegetated and Non-vegetated Terrain Using Medium-Footprint Lidar. Photogrammetric Engineering & Remote Sensing, Journal 2006. Vol.72, 3: 279–286.

[21] Grohman, G., Kroenung., G. & Strebeck, J. 2006. Filling SRTM voids: The delta surface fill model. Photogrammetric Engineering & Remote Sensing, Journal 2006. Vol. 72, 3: 213–216.

[22] Ling, F., Zhang, Q.-W. & Wang, C.. Comparison of SRTM Data with other DEM sources in Hydrological Researches. In: Proceedings of the 31st ISPRS International Symposium on Remote Sensing of Environment, 2005. May 20–24. Saint Petersburg, Russian Federation. ISPRS.

[23] Luedeling, E., Siebert, S. & Buerkert, A.. Filling the voids in the SRTM elevation model — A TIN-based delta surface approach. ISPRS Journal of Photogrammetry and Remote Sensing 2007. Vol.62, 4: 283–294.

[24] Reuter H.I, A. Nelson, A. Jarvis, An evaluation of void filling interpolation methods for SRTM data, International Journal of Geographic Information Science, 2007. 21:9, 983-1008.

[25] CGIAR: Consortium for Spatial Information (CSI): SRTMV3/V4. http://srtm.csi.cgiar.org/ (accessed 24 August 2013).

[26] Guth, P.. Geomorphometry from SRTM: Comparison to NED. Photogrammetric Engineering & Remote Sensing, journal 2006. Vol.72, 3: 269–278.

[27] Yastikli, N., Kocak, G. & Buyuksalih, G.. Accuracy And Morphological Analyses of GTOPO30 and SRTM X-C Band DEMs in the Test Area Istanbul. In: Proceedings of the ISPRS Ankara Workshop on Topographic Mapping from Space, 2006. February 14–16. Ankara, Turkey. ISPRS.

[28] Wright, R., Garbeil, H., Baloga, S.M. & Mouginis-Mark, P.J.. An assessment of shuttle radar topography mission digital elevation data for studies of volcano morphology. Remote Sensing of Environment, An Interdisciplinary Journal 2006. Vol.105, 1: 41–53.

[29] Kiel, B., Alsdorf, D. & LeFavour, G.. Capability of SRTM C and X Band DEM Data to Measure Water Elevations in Ohio and the Amazon. Photogrammetric Engineering & Remote Sensing, Journal 2006. Vol.72, 3: 313–320.

[30] Ludwig, R. & Schneider, P.. Validation of digital elevation models from SRTM X-SAR for applications in hydrologic modelling. ISPRS Journal of Photogrammetry and Remote Sensing 2006. 60, 5: 339–358.

[31] Valeriano, M.M., Kuplich, T.M., Storino, M., Amaral, B.D., Mendes, J.N. & Lima, D.J.. Modeling small watersheds in Brazilian Amazonia with shuttle radar topographic mission-90 m data. Computers & Geosciences 2006. 32, 8: 1169–1181.

[32] Kellndorfer, J., Walker, W., Pierce, L., Dobson, C., Fites, J.A, Hunsaker, C., Vona, J. & Clutter, M. 2004. Vegetation height estimation from Shuttle Radar Topography Mission and National Elevation Datasets. Remote Sensing of Environment, An Interdisciplinary Journal 2004. 93, 3: 339–358.

[33] Simard, M., Zhang, K., Rivera-Monroy, V.H., Ross, M.S., Ruiz, P.L., Castañeda-Moya, E., Twilley, R.R. & Rodriguez, E.. Mapping Height and Biomass of Mangrove Forests in the Everglades National Park with SRTM Elevation Data. Photogrammetric Engineering & Remote Sensing, Journal 2006. 72, 3: 299–312.

[34] Walker, W.S., Kellndorfer, J.M. & Pierce, L.E.. Quality assessment of SRTM C-and X-band interferometric data: Implications for the retrieval of vegetation canopy height. Remote Sensing of Environment, An Interdisciplinary Journal 2007. 106, 4: 428–448.

[35] Menze, B.H., Ur, J.A. & Sherratt, A.G.. Detection of Ancient Settlement Mounds – Archaeological Survey Based on the SRTM Terrain Model. Photogrammetric Engineering & Remote Sensing, Journal 2006. 72, 3: 321–330.

[36] Surazakov, A. & Aizen, V.. Estimating Volume Change of Mountain Glaciers Using SRTM and Map-Based Topographic Data. IEEE Transactions on Geoscience and Remote Sensing 2006. 44, 10, Part 2: 2991–2995.

[37] Moccia, A., Chiacchio, N. & Capone, A.. Spaceborne bistatic synthetic aperture radar for remote sensing applications. International Journal of Remote Sensing 2000. 21, 18: 3395–3414.

[38] Massonnet, D.. The interferometric cartwheel: a constellation of passive satellites to produce radar images to be coherently combined. International Journal of Remote Sensing 2001. 22, 12: 2413–2430.

[39] Krieger, G., Fiedler, H., Mittermayer, J., Papathanassiou, K. & Moreira, A. 2003. Analysis of multistatic configurations for spaceborne SAR interferometry. In: IEE Proceedings-Radar, Sonar and Navigation 2003. June. 150, 3: 87–96.

[40] Moccia, A. & Fasano, G.. Analysis of Spaceborne Tandem Configurations for Complementing COSMO with SAR Interferometry, EURASIP. Journal on Applied Signal Processing 2005. 20: 3304–3315.

[41] Krieger, G., Moreira, A., Hajnsek, I., Werner, M., Fiedler, H. & Settelmeyer, E.. The TanDEM-X Mission Proposal. In: Heipke C., Jacobsen K. Gerke M. (eds.). Proceed-

ings of the ISPRS Hannover Workshop 2005: High-Resolution Earth Imaging for Geospatial Information, 2005. May 17-20. Hannover, Germany. ISPRS.

[42] Eineder, M., Krieger, G. & Roth, A. 2006. First Data Acquisition and Processing Concepts for the TanDEM-X Mission. In: Baudoin A., Paparoditis N.. (eds.). Proceedings of: From Sensors to Imagery, ISPRS Commission I Symposium 2006. July 3-6. Paris, France. ISPRS.

[43] Krieger, G., Moreira, A., Fiedler, H., Hajnsek, I., Werner, M., Younis, M. & Zink, M.. TanDEM-X: A Satellite Formation for High-Resolution SAR Interferometry. IEEE Transactions on Geoscience and Remote Sensing 2007. 45(11, Part 1): 3317–3341.

[44] Zink, M., Fiedler, H., Krieger, G. & Moreira, A. 2007. The TanDEM-X Mission. In: Lacoste H., Ouwehand L.. (eds.). Proceedings of Fringe 2007, November 26–30. Frascati, Italy. ESA Communication Production Office, ESTEC, Noordwijk, The Netherlands.

[45] Coletta, A., Galeazzi, C., Caltagirone, F., Scorzafava, E., Amorosi, L., Caliò, E., Notarantonio, A. & Moccia, A.. Interferometric Missions: Applications within ASI Space Programs. Proceedings of SpaceOps 2006. June 2006. Rome, Italy.

[46] ASTRIUM Geoinformation Services. ASTRIUM: WorldDEM™. http://www.astrium-geo.com/worlddem/ (accessed 3 September 2013).

[47] Eineder, M., Adam, N. & Yague-Martinez, N. 2007. First TerraSAR-X Interferometry Evaluation. In: Lacoste H., Ouwehand L.. (eds.). Proceedings of Fringe 2007, November 26–30. Frascati, Italy. ESA Communication Production Office, ESTEC, Noordwijk, The Netherlands.

[48] Meta, A., Prats, P., Steinbrecher, U., Scheiber, R. & Mittermayer, J. 2007. First TOPSAR image and interferometry results with TerraSAR-X. In: Lacoste H., Ouwehand L.. (eds.). Proceedings of Fringe 2007, November 26–30. Frascati, Italy. ESA Communication Production Office, ESTEC, Noordwijk, The Netherlands.

[49] Pasquali, P., Monti Guarnieri, A., D'Aria, D., Costa, L., Small, D., Jehle, M. & Rosich, B.. ALOS PALSAR Verification Processor. In: Lacoste H.,Ouwehand L.. (eds.) Proceedings of the Envisat Symposium 2007. April 23–27. Montreux, Switzerland. ESA Communication Production Office ESTEC, Noordwijk, The Netherlands.

[50] Shimada, M. 2007. Coherence dependency of the PALSAR POLinSAR on forest in Japan and Amazon. In: Proceedings of Geoscience and Remote Sensing Symposium 2007, IGARSS '07. July 23-28. 2636–2639. Barcelona, Spain. IEEE International.

[51] Deilami K, Hashim M. Very high resolution optical satellites for DEM generation: A review. European Journal of Scientific Research 2011, 49(4), 542-554.

[52] Grodecki J, Dial, G. Block adjustment of high-resolution satellite images described by Rational Polynomials. Photogrammetric Engineering and Remote Sensing 2003, 69 (1), 59-68.

[53] Cheng Ph, Chaapel C. Using WorldView-1 Stereo Data with or without Ground Control Points. GEOinformatics 2008, 11 (7), 34-39.

[54] Jacobsen K. Characteristics of very high resolution optical satellites for topographic mapping. International Archives of the Photogrammetry, Remote Sensing and Spatial Information Sciences 2011, XXXVIII-4/W19, on CDROM.

[55] Poli D, Caravaggi I. 3D information extraction from stereo VHR imagery on large urban areas: lessons learned. Natural Hazards 2012, Volume 68, Issue 1 (2013), page 53-78.

[56] Poli, D. A Rigorous Model for Spaceborne Linear Array Sensors. Photogrammetric Engineering & Remote Sensing 2007, 73(2), 187-196.

[57] Toutin, T. Review article: Geometric processing of remote sensing images: models, algorithms and methods. International Journal of Remote Sensing 2004, 25(10), 1893-1924.

[58] Poli D, Toutin T. Developments in geometric modelling for HR satellite push-broom sensors. The Photogrammetric Record 2012, 27: 58–73.

[59] Toutin T. Fine spatial resolution optical sensors. Chapter 8 in SAGE Handbook of Remote Sensing (Eds. T. A. Warner, M. D. Nellis & G. M. Foody) 2009. SAGE, London, UK. 568 pages: 108–122.

[60] Ebner H, Kornus W., Ohlhof, T. A simulation study on point determination for the MOMS-02/D2 space project using an extended functional model. International Archives of Photogrammetry and Remote Sensing 1992, 29(B4): 458–464.

[61] Kornus W. MOMS-2P Geometric Calibration Report. Results of laboratory calibration (version 1.1). DLR, Institute of Optoelectronics, Wessling, Germany. 16 pages, 1996.

[62] Kratky V. On-line aspects of stereophotogrammetric processing of SPOT images. Photogrammetric Engineering & Remote Sensing 1989, 55(3): 311–316.

[63] Jeong I.S. and Bethel J. Trajectory modeling for satellite image triangulation. International Archives of Photogrammetry, Remote Sensing and Spatial Information Sciences, 2001. Vol.37(1), pp. 901-907

[64] Poli D. Modelling of spaceborne linear array sensors. Doctoral thesis, IGP Mitteilungen Nr. 85, ETH Zurich, Switzerland. 204 pages, 2005.

[65] Fraser C. S., Baltsavias E, Gruen A. Processing of Ikonos imagery for submetre 3D positioning and building extraction. ISPRS Journal of Photogrammetry and Remote Sensing 2002, 56(3): 177–194.

[66] Grodecki J, Dial G. Block adjustment of high-resolution satellite images described by rational polynomials. Photogrammetric Engineering & Remote Sensing 2003, 69(1): 59–68.

http://dx.doi.org/10.5772/57584

[67] Gruen A. Development and Status of Image Matching in Photogrammetry. The Photogrammetric Record 2012, Special Issue: IAN DOWMAN RETIREMENT SYMPOSIUM, Volume 27, Issue 137, pages 36–57, March 2012.

[68] Remondino F., Spera M.G., Nocerino E., Menna F., Nex F., Gonizzi S. Dense Image Matching: comparisons and analyses. IEEE Proc. Digital Heritage Conference, Marseille, France, 2013

[69] Remondino F, El-Hakim S, Gruen A, Zhang L. Turning images into 3D models-Development and performance analysis of image matching for detailed surface reconstruction of heritage objects. IEEE Signal Processing Magazine 2008, Vol. 25(4), pp. 55-65.

[70] Agugiaro G., Poli D., Remondino F. Testfield Trento: geometric evaluation of very high resolution satellite imagery. Int. Archives of Photogrammetry, Remote Sensing and Spatial Information Sciences 2012, Vol. 39(1), pp. 191-196.

[71] Poli D. Remondino F, Angiuli E, Agugiaro G., 2013. Evaluation of Pleiades-1a triplet on Trento testfield. Int. Archives of Photogrammetry, Remote Sensing and Spatial Information Sciences, Vol. XL-1/W1, pp. 287-292.

[72] Gleyzes M A, Perret L, Philippe Kubik P. Pleiades system architecture and main performances. Int. Archives of Photogrammetry, Remote Sensing and Spatial Information Sciences 2012, Vol. 39(1), pp. 537-542.

[73] 4DiXplorer A.G.. SAT-PP: Satellite Precision Processing. http://www.4dixplorer.com/ (accessed 24 October 2013).

[74] Poli D, Wolff K, Gruen A. Evaluation of Worldview-1 stereo scenes and related 3D products. ISPRS Hannover Workshop 2007, Int. Archives of the Photogrammetry, Remote Sensing and Spatial Information Sciences 2009, Vol. 38(1-4-7/W5).

[75] H. Papasaika, E. Kokiopoulou, E. Baltsavias, K. Schindler, D. Kressner, Fusion of Digital ElevationModels Using Sparse Representations. In: Stilla U., Rottensteiner F., Mayer H., Jutzi B, Butenuth M.. (eds.). Proceedings of Photogrammetric Image Analysis 2011 October 5-7. Munich, Germany.

[76] Kanevski M. (eds.) *Advanced Mapping of Environmental Data. Geostatistics, Machine Learning and Bayesian Maximum Entropy*. ISTE, Wiley, London, 2008.

[77] William H.; Teukolsky, Saul A.; Vetterling, William T.; Flannery, Brian P.. Numerical Recipes: The Art of Scientific Computing (3rd ed.). New York: Cambridge University Press. 2007.

[78] Lazzaro D., Montefusco L.B.. Radial Basis Functions for the multivariate interpolation of large scattered data sets, Journal of Computational and Applied Mathematics, 140, pages 521-536, 2002.

Section 4

Land Motion Applications

Chapter 8

Mapping of Ground Deformations with Interferometric Stacking Techniques

Paolo Pasquali, Alessio Cantone, Paolo Riccardi,
Marco Defilippi, Fumitaka Ogushi,
Stefano Gagliano and Masayuki Tamura

Additional information is available at the end of the chapter

http://dx.doi.org/10.5772/58225

1. Introduction

Interferometric stacking techniques emerged in the last decade as methods to obtain very precise measurements of terrain displacements, and especially of subsidence phenomena. In particular, the so-called Persistent Scatterers [1] and Small BASeline [2] methods can be considered as the two most representative stacking approaches.

In both cases, the exploitation of 20 or more satellite Synthetic Aperture Radar (SAR) acquisitions obtained from the same satellite sensor with similar geometries on the interest area allows to measure displacements with an accuracy in the order of a few mm / year, and to derive the full location history of "good" pixels with an accuracy of 1cm or better for every available date.

This chapter is presenting an extensive analysis of the two methods, a validation of the results obtained in same geographical areas with the different techniques and an evaluation of the suitability of these techniques for different applications.

All results shown in this chapter have been generated with the SARscape® software package.

2. Interferometric stacking techniques

While trying to provide an answer to the same problem, i.e. how to measure small land displacements out of a series of SAR images acquired on a same geographical area under a same geometry, the PS and SBAS approaches designed two algorithms that focus each on a different type of objects and land cover to favour in the analysis: the PS technique focuses on so-called Point Targets, i.e. objects possibly of small size and with a very well characterized geometry like corner reflectors (e.g. buildings, rocks) and with a high temporal stability of the backscattered signal; the SBAS technique vice-versa is concentrating the analysis on so-called distributed targets, like open fields and not very geometrically characterized objects. Both approaches, exploiting a reference Digital Elevation Model in input of the workflow, also estimate its errors as residual height component together with the displacement series.

As shown in Figure 1 on the left, the original PS approach is selecting one of the images of the input stack as reference and generating all differential interferograms between this and the other acquisitions. Focusing on Point Targets, no limits of critical baseline shall be considered, and no spectral shift or other filtering are performed [3]; the algorithm is then operating on the phase time series of each of the pixels separately, without performing any phase unwrapping. Both these features allow to obtain the maximum spatial resolution of the final results i.e. total independency of the measures for adjacent pixels, as well as eliminate the possibility of unwrapping errors propagation.

The estimation of the average linear displacement rate and of the residual height correction factor is performed through a kind of frequency analysis: a range of linear displacement rates and of residual heights is explored and, for each of the explored values, the phase trend corresponding to the tested values is subtracted from the measured phase time series, and a so-called *temporal coherence* is estimated by normalising the summation of the de-trended complex time series.

This analysis is in the original algorithm performed on selected pixels (*PS candidates*) that show no significant variations of the backscatter amplitude over the time; in any case, only pixels showing high values of the temporal coherence are finally considered as true PSs.

It shall be noticed that the temporal coherence should more appropriately be considered as a measure of linearity of the phase time series; points characterised by stable radar backscatter and very low phase noise and temporal decorrelation, but characterised by significantly non-linear displacement regimes will not be recognised as PSs by such analysis.

Being the PS estimation performed independently on each pixel, large variations of phase from one date to the next in the time series due to large displacements or irregular temporal sampling will also make the analysis difficult, resulting in such cases in wrong estimations of the displacement rates and / or difficulties in the identification of the corresponding PSs.

The SBAS approach, as shown in Figure 1 on the right, is generating all differential interferograms from the input images stack that fulfil criteria of temporal and geometric baseline within a given interval in time or respectively normal baseline respect to the critical one. The second condition is trying to limit the impacts of volume decorrelation over natural distributed targets; to re-enforce this, spectral shift and adaptive filtering steps (typical of standard SAR interferometry) are included in the workflow.

The interferograms (coregistered on a common geometry) are then unwrapped, either with a conventional 2D (interferogram-by-interferogram) or with a combined 3D approach [4]. This

http://dx.doi.org/10.5772/58225

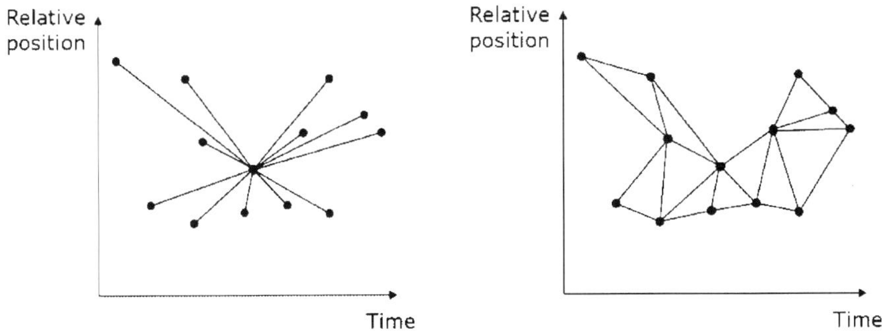

Figure 1. Typical processing schemes for the PS (on the left) and SBAS (on the right) techniques each circle corresponds to one acquisition, each connecting segment one interferogram.

step, surely the most challenging in SAR interferometry, relies on some spatial continuity of the interferometric phase, and hence is expecting some form of spatial correlation of the displacement phenomena happening in the region of interest, and it is a potential source of errors propagation. On the other side, this step makes the approach more robust toward irregular series and fast displacements.

The average displacement rate and the residual height correction factor are then estimated by inverting a linear system, possibly with a robust SVD approach, that includes all measures (one for each interferogram) together with their proportionality coefficients depending on the temporal and geometric baseline of each pair.

The final identification of reliable results is derived for example from the average (spatial) coherence of each pixel; no assumption is made in particular on the linearity of the true underlying displacement, and average displacement rates can also be estimated for pixels characterised by strongly non-linear behaviours, given that they remain coherent in most of the interferograms of the network.

Different models (e.g. quadratic or cubic) can be exploited during the inversion phase, and the reconstruction of the LMS time series of the displacements is also possible with this approach without the need of assuming any type of model for the true temporal deformation.

Both algorithms, after a first estimate of the average displacement rates and height correction factors, perform an estimated of the so-called "Atmospheric Phase Screen". The impact of atmospheric heterogeneities on the propagation of the SAR signal is considered as a spatially low frequency signal, with very short temporal correlation. Appropriate space – time filtering is performed on the phase time series after removing the results of the first estimate to evaluate the pixel-by-pixel and date-by-date impact of the APS and remove it from the original phase time series. A second and final iteration is then performed of the APS-corrected data with either of the PS or SBAS approaches to obtain a refined final estimate of the displacement rate and height correction.

As final step, the time series of the displacement for each pixel and date is obtained in both approaches.

A synthetic comparison of the characteristics of the two approaches is given in the following Table.

PS	SBAS
Analyse independent, uncorrelated motions	Monitor at best spatially correlated motions
Expect pixel-wise continuous time series	Capable of handling time series with temporal holes
Time interval between two acquisitions limited by displacement rate	Time interval between two acquisitions limited by temporal decorrelation
Very accurate on PSs	Slightly less accurate
Linear displacements favoured	Larger variety of parametric models possible. Non-parametric modelling possible

Table 1. Main characteristics from the exploitation point of view of the PS and SBAS approaches.

3. Factors affecting the final expected precision

As for standard interferometry, temporal decorrelation can be considered as the most important factor affecting the precision obtainable from the processing of an interferometric stack of SAR images.

An estimation of the expected precision of the measured displacements can be derived from the interferometric coherence. Known its value for a certain pixel, well known relationships [6] allow to obtain an estimate of the interferometric phase standard deviation, then scaled by the system wavelength to obtain a correspondent estimation of the expected displacement precision. By exploiting this approach, it is possible to evaluate for example which is the level of coherence necessary to obtain the same precision with a certain system wavelength as with another wavelength at a given coherence level, as shown in Figure 2.

Here a C-band system (as ERS-1/2, ENVISAT ASAR, Radarsat-1/2 and the forthcoming Sentinel-1) has been selected as reference: the red curve provides an estimate of the coherence level that is necessary, for every C-band coherence value, to obtain the same displacement precision with a L-band system (as PALSAR-1 and the forthcoming PALSAR-2), while the blue curve presents a similar comparison between a C-Band and a X-band (as COSMO-Skymed and TerraSAR-X / TanDEM-X) system.

It can here be seen how, assuming a minimum acceptable value of 0.2 for the C-Band coherence, comparable accuracies can be obtained from L-band measurements only for coherence levels of 0.6 or more. This corresponds to the L-band longer wavelength (if compared to C-band) and hence poorer displacement sensitivity.

http://dx.doi.org/10.5772/58225

On the other hand, as expected due to the shorter wavelength and hence better displacement sensitivity, X-band systems provide, with coherence 0.2, displacement precisions that are comparable with what can be obtained from a C-band system with coherence 0.4 or more.

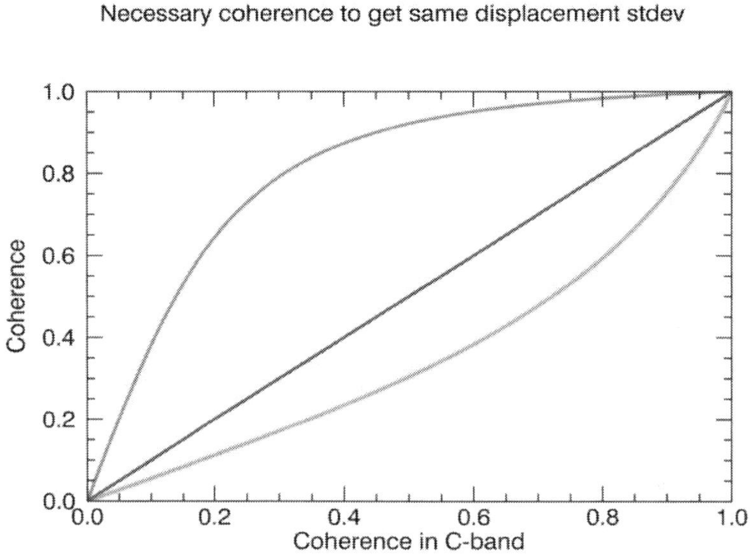

Figure 2. Interferometric coherence necessary to obtain the same measurement precision in mm as depending from the different system frequency. Red: L-band; Blue: C-band (reference); Green: X-band.

The amount of temporal decorrelation and hence of phase noise depends itself of the characteristics of the observed areas in relationship with the system wavelength. It is well known that repeat-pass SAR acquisitions are showing complete decorrelation over water bodies, and very high correlation over point targets and man-made features, independently from the system wavelength. On the other hand natural, distributed objects show temporal correlation that depend on the observation frequency: in general the higher the system frequency (and hence the shorter the wavelength), the shorter the penetration of the transmitted signal into vegetation layers, and hence the stronger the effects of temporal decorrelation.

Summarising, it can be expected that lower frequency (e.g. L-band) systems show more extensive coverage over natural areas than systems with higher (e.g. C-and X-band) frequency; on the other hand, where measurements can be obtained with systems with different wavelength on a same area, better precisions can be expected from the systems with shorter wavelength.

One example of these effects is shown in Figure 3, where average displacement rates obtained from stacks of PALSAR-1 (on the left) and of ENVISAT ASAR data (on the right) spanning a period between 2006 and 2010 are presented for a region in Japan between the cities of Tokyo and Chiba. It is evident here how many black (no value) areas are visible in the ASAR data, while the same regions show reliable measurements obtained with PALSAR data. On the other hand, areas that are covered by both systems show higher variability and hence most likely poorer precision in the PALSAR dataset.

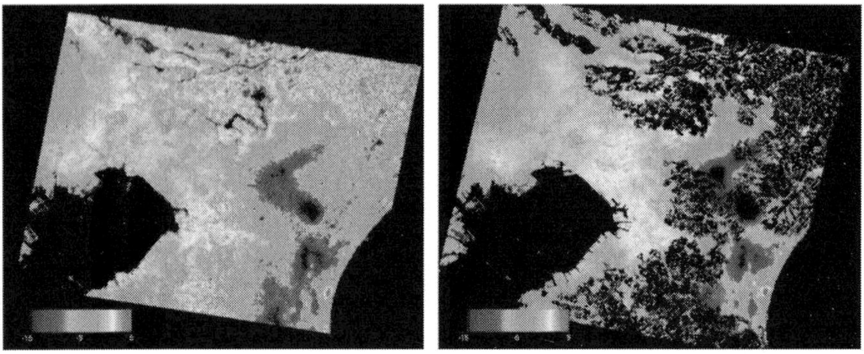

Figure 3. Comparison of average displacement rates (color scale between-15 mm/year in blue and+5mm/year in red) as derived with SBAS from ALOS PALSAR (left) and ENVISAT ASAR (right) data over the Tokyo – Chiba area (Japan).

It shall be noticed that models have been proposed to help quantifying the expected temporal evolution of the interferometric coherence (as for example in [7]); a limitation of these approaches is that they focus more on a general, statistical point of view, without considering phenomenological aspects like spatial variability of the land cover [8] and seasonal and long-term effects [9] [10]. It has been seen how these last aspects have finally effects that can be foreseen only in a qualitative way, but of such importance on the decorrelation to significantly decrease the applicability of the theoretical models.

One example is the temporal coherence obtained with very-high resolution X-band systems like TerraSAR-X and COSMO-Skymed: the values measured from real data are often surprisingly higher than that forecasted by models. This is not actually due to the system frequency but more by their very-high spatial resolution: if the backscatter signal measured by coarser resolution system (e.g. ~25m for ERS-1/2 and ASAR) is a mixture of returns from different land cover types (vegetation, bare soil, rocks etc.), and the contribution of stable scatterers is often hidden by the other components, the very-high resolution data (e.g. ~3m for TerraSAR-X and COSMO-Skymed) have higher likelihood of keeping the different contributions separate in different pixels, hence stable scatterers can be better identified as highly coherent pixels.

Some phenomenology may also have an effect on the temporal variability of decorrelation: on one hand, it is commonly assumed that interferometric coherence decreases with time; on the other hand, long time, such as winter-winter interferograms (hence computed for seasons where vegetation changes are typically negligible), often show coherence higher than short-time interferograms computed within periods of strong vegetation changes (e.g. spring).

3.1. Validation

Two main approaches may be identified for the validation of the measurements obtained with interferometric stacking techniques: absolute methods for evaluating the accuracy of the measures based on external, reference measurements obtained for example with GNSS, levelling or other systems (e.g. [11] [12] [13]), and more relative methods, evaluating the precision of the SAR-based measurements by comparing results obtained from different sensors (e.g. [13]) and / or different processing approaches (e.g. PS and SBAS).

This Section presents results obtained with both approaches about the validation of interefrometric stacking PS and SBAS measurements from ASAR and PALSAR data over the area in Japan shown in Figure 3, to highlight the different issues relative to validation and typical results in terms of accuracy and precision obtainable with these techniques.

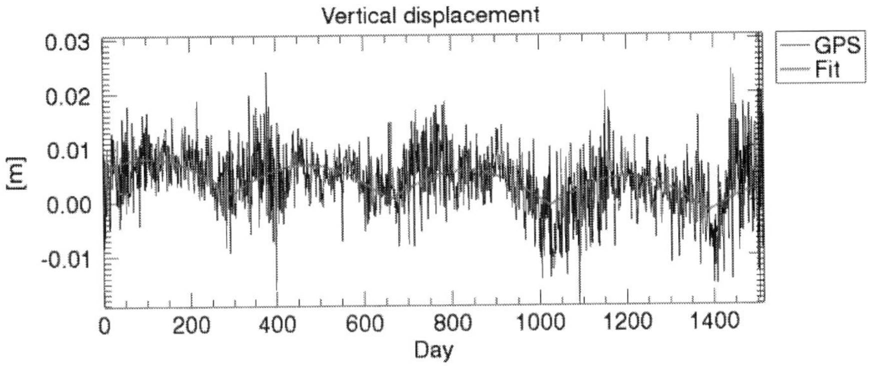

Figure 4. Vertical displacement of one of the permanent GPS stations of the GEONET network. Black: original GPS measure; red: fitted values. Data courtesy of GSI.

Permanent GPS station measurements from the GSI GEONET nation-wide network [14] have been exploited to perform an absolute validation of the SBAS data. A first investigation of the GPS data highlighted two main effects, easily detectable from Figure 4: GPS data have themselves a certain dispersion, as expected, and systematic trends (most likely due to uncompensated tidal effects [15] [16]) are present, in form of yearly periodical cycles. A straightforward date-by-date difference between the SAR and GPS (projected along the Line Of Sight direction) measurements is therefore not feasible, and a different strategy shall be adopted. A simple model, composed by a linear displacement combined with a half-cycle sinusoidal oscillation with yearly period, has been selected to describe the GEONET data, and the corresponding parameters have been estimated independently for each of the stations present in the area of study.

Finally, the station-by-station difference between the GENOET fitted linear displacement (re-projected along LOS) and the SBAS average linear displacement has been computed, as shown in Figure 5.

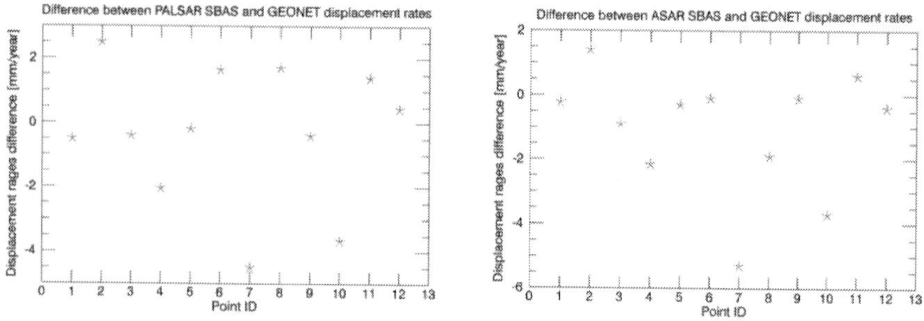

Figure 5. Difference between GPS and SBAS PALSAR (on the left) or ASAR (on the right) linear displacement rates.

A summary of the statistics of the differences between the average displacement rates as measured from SAR and GPS is shown in Table 2, confirming the very high accuracy that can be obtained with the interferometric stacking techniques.

	GPS - PALSAR	GPS - ASAR
Mean difference [mm/year]	-0.28	-1.1
Difference Standard Deviation [mm/year]	1.9	1.8

Table 2. Difference between GPS and SBAS average displacement rates.

The validation has then been continued by comparing the results obtained from SAR data only, by exploiting independent processing applied either to different input time series, derived from different sensors, or to the same time series but exploiting different (PS and SBAS) approaches.

Figure 6 shows the difference between the average displacement rates presented in Figure 3, showing in general a very good agreement between the measures derived from L-band and C-Band data.

Figure 7 shows the differences between average displacement rates obtained with the same sensors but with different (PS and SBAS) methodologies. It can be seen here how in both cases the two approaches provide very similar results but, as expected, the PALSAR (L-band) data show higher variability, due to the longer wavelength and hence smaller sensitivity to displacements.

The statistics summarising the results shown in these figures are presented in Table 3, confirming the comments made on the single plots.

As a final, qualitative cross-validation, displacement time series are shown in Figure 8 for a same pixel as obtained from different data and with different algorithms. Here again, as for the average displacement rates, it can be seen how longer wavelength data provide higher variability of the measures; on the other hand, SBAS time series show consistently less dispersions when compared with PS time series, mainly due to the filtering performed during the SBAS interferometric processing.

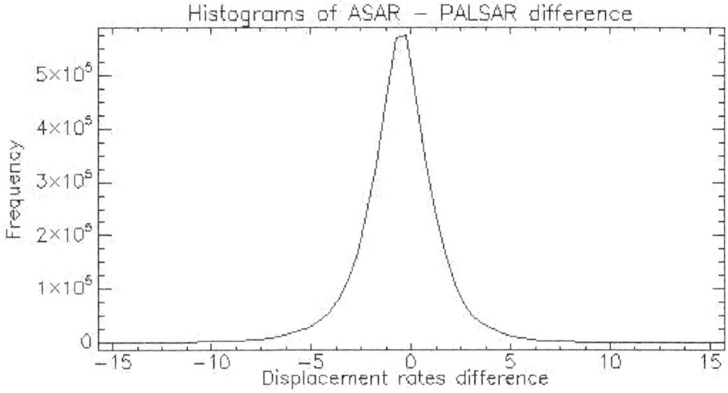

Figure 6. Difference between SBAS PALSAR and ASAR linear displacement rates.

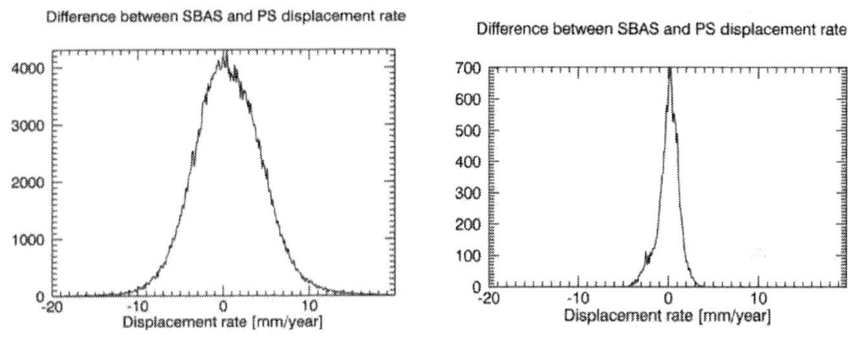

Figure 7. Difference between SBAS and PS displacement rates as derived from PALSAR (on the left) and ASAR (on the right) data.

	SBAS ASAR -PALSAR	PALSAR SBAS – PS	ASAR SBAS – PS
Mean difference [mm/year]	-0.33	0.7	0.1
Difference Standard Deviation [mm/year]	2.1	4.1	1.2

Table 3. Difference between average displacement rates obtained from PS and SBAS from ASAR and PALSAR data.

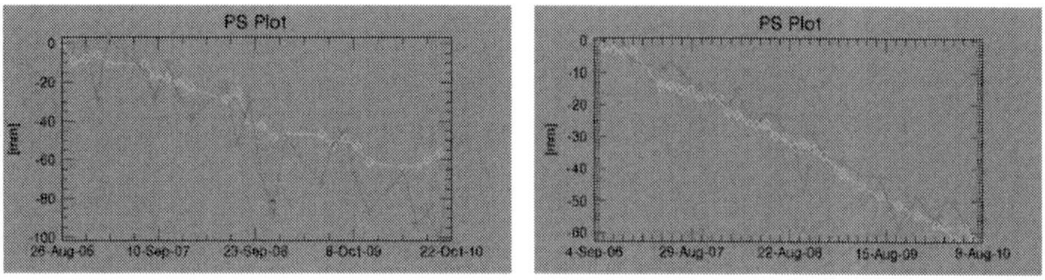

Figure 8. Comparison of displacement time series as derived from ALOS PALSAR (on the left) and ENVISAT ASAR (on the right) data exploiting the SBAS (green diamonds) and PS (red crosses) techniques.

3.2. Application examples

Some examples are shown in this Section to discuss the applicability of the major interferometric stacking techniques in areas characterised by different land cover, different types of displacements and in view of various land applications.

Figure 9 shows first the average displacement rates as obtained from very-high resolution TerraSAR-X data through PS processing for an area of the city of Budapest, where a new Metro line was under construction, and generally a high PS density could be obtained. A trace of the line is drawn over the area, and the average displacement rate shows in general measurable displacements of subsidence around this trace. The red square indicates the area where the new Metro station of Szent Gellért had been built: a poor PS density is unfortunately achievable here. The middle image in this Figure shows a close-up of the average displacement rate as obtained for the same small area through SBAS processing; the spatial density of measures is much higher in this case, in particular in the area of more severe subsidence.

As the time series shows in the lower image, the displacement is strongly non-linear in this area, characterized in particular by a sudden descent, in the order of magnitude of half of the system wavelength, in a period of about 3 months. This kind of step cannot be resolved well by the PS approach, that tends to fit a linear velocity to the other parts of the time series, but obtaining large residuals, and hence low temporal coherence, discarding the pixels as no PSs. The SBAS approach is on the other hand, also thanks to the very-high resolution of the imagery, capable of obtaining reliable results, and well reconstructing the temporal behaviours of the displacement, that shows good spatial correlation.

Figure 10 shows the historical evolution of the landfills close to Urayasu (Japan). The average displacement rate in the period 2006 – 2010 in the same region, as obtained through PS and SBAS processing of ASAR and PALSAR data is shown in Figure 11. It shall be noticed that these data are a portion of the data shown in Figure 3.

The coverage obtained with the two methods from different data is in this urban area very similar. As expected, the SBAS data show smoother results, while the spatial resolution obtained in the PS case is better and many spatial details are more accurately preserved; PALSAR data show higher spatial variability, while ASAR data provide higher accuracy.

Nevertheless, all 4 approaches provide very consistent results, able to well delineate for example areas of terrain compaction at the borders of different land-fills. It is furthermore interesting to notice how the areas undergoing large subsidence correspond to those having thick layers of soft soil over stiff basement, as shown in Figure 12.

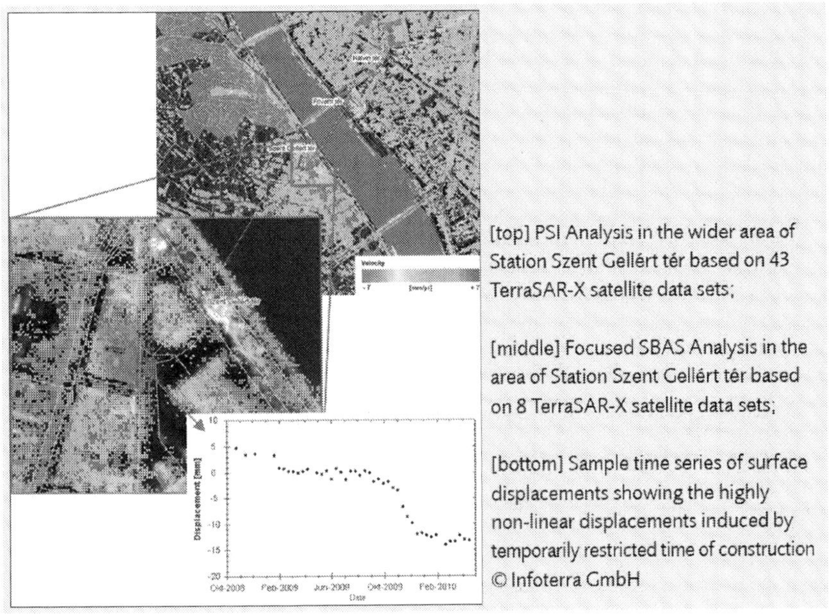

Figure 9. Average displacement rate as obtained through PS (above) and SBAS (middle) processing; SBAS displacement time series (below) for an area in Budapest (Hungary). ©Airbus Defence and Space

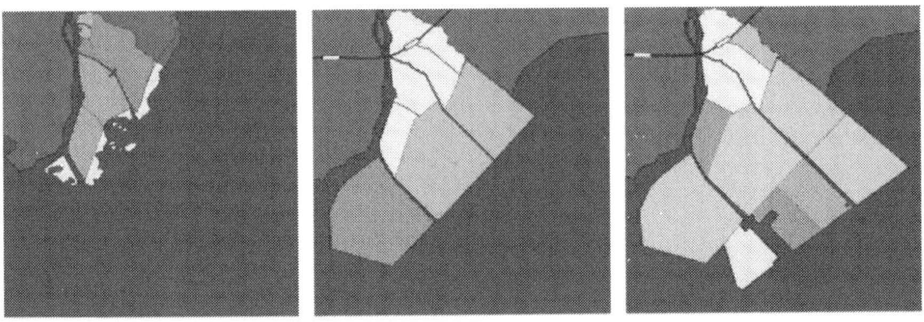

Figure 10. Landfill area around Urayasu (Japan) in years 1950, 1975, 1980 from left to right [17].

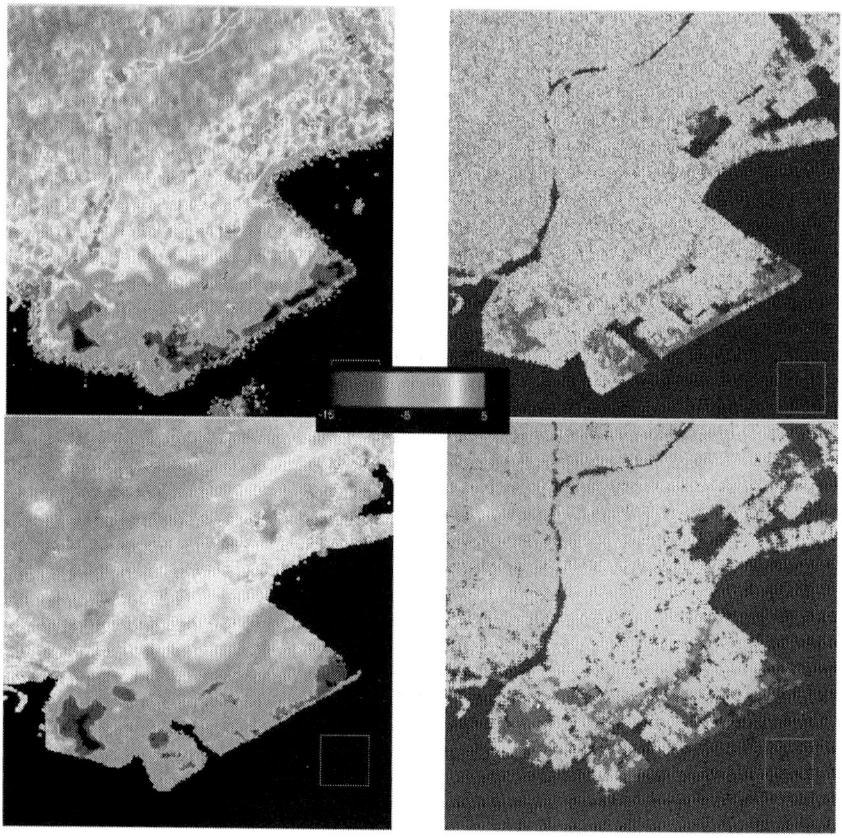

Figure 11. Average displacement rate in the 2006-2010 period in the area of Urayasu, as obtained from PALSAR (above) and ASAR (below) data through SBAS (on the left) and PS (on the right) processing.

http://dx.doi.org/10.5772/58225

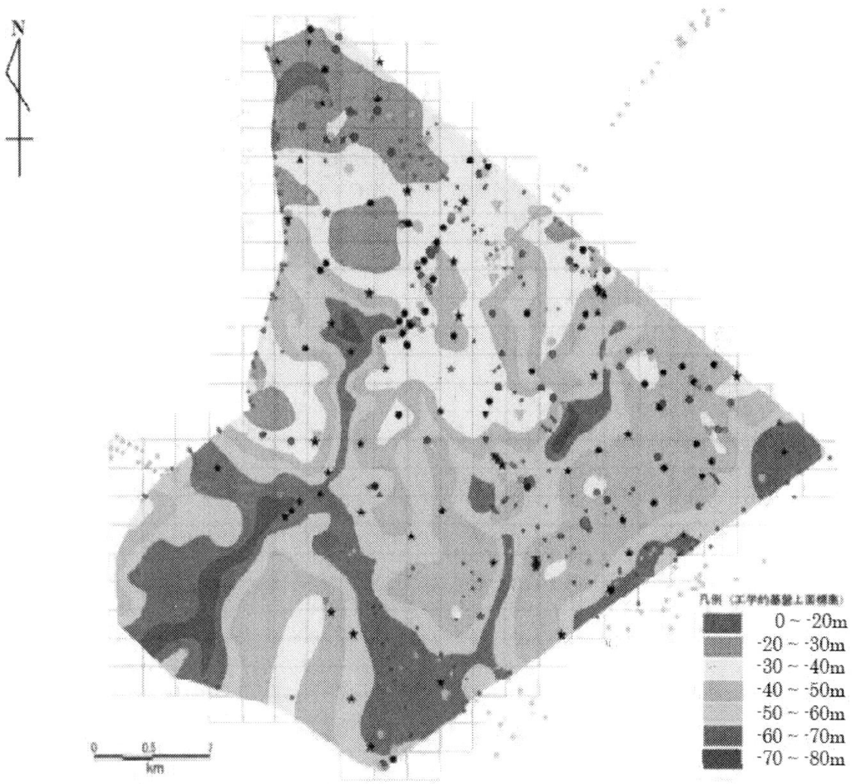

Figure 12. Depth of the upper surface of the solid geological stratum (Pleistocene sand stratum with the standard penetration test N value > 50) in Urayasu city [18].

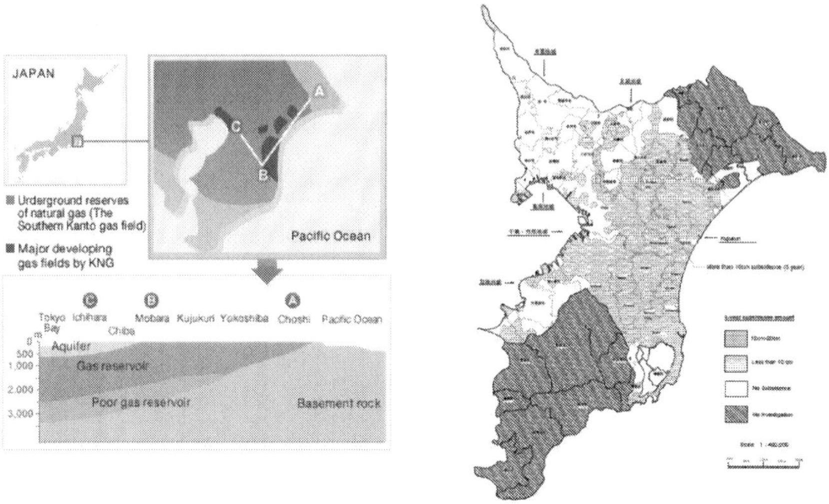

Figure 13. Southern Kanto gas field, Kanto Natural Gas Development Co Ltd. [19] (on the left); known 5-years displacement in the region of Togane, eastern part of the same area [20] (on the right).

The whole region of Figure 3 is located over the Southern Kanto gas field [19], shown in Figure 13. Various phenomena of subsidence are known in this region [20], in particular on its eastern side, as shown in the same Figure. This information provide an additional clear indication, although in this case from a qualitative point of view since too coarse respect to the SAR data, about the reliability of the results obtained with the SBAS approach, both in terms of order of magnitude of the average displacement rate and of its spatial distribution.

A closer look at the region on the east coast is shown in Figure 14 as obtained through PS and SBAS processing of ASAR and PALSAR data. An area of uplift is present here, resulting from strong water injection as compensation of the gas extraction. Appropriate coverage can be obtained here only with the SBAS approach, or exploiting PALSAR data and PS processing, while ASAR data and PS processing are providing no reliable results and loosing most of the spatial variation of the displacement. As expected, lower frequency data show rougher spatial distribution; nevertheless, the three applicable approaches show very consistent results both in terms of value and spatial distribution of the obtained results.

Finally, Figure 15 shows the average displacement rate obtained from ASAR imagery over the Lisan peninsula, an area in the Dead Sea characterised by diapirism and severe subsidence phenomena related to the lake level lowering (1m/year), with the PS and the SBAS approaches.

Although the area is very dry and stable from the radar backscattering point of view, the complex land cover and the very complex displacement regimes result in a significantly different spatial coverage obtained with the PS and SBAS methods, making only the second one suitable for reliable investigations in such cases.

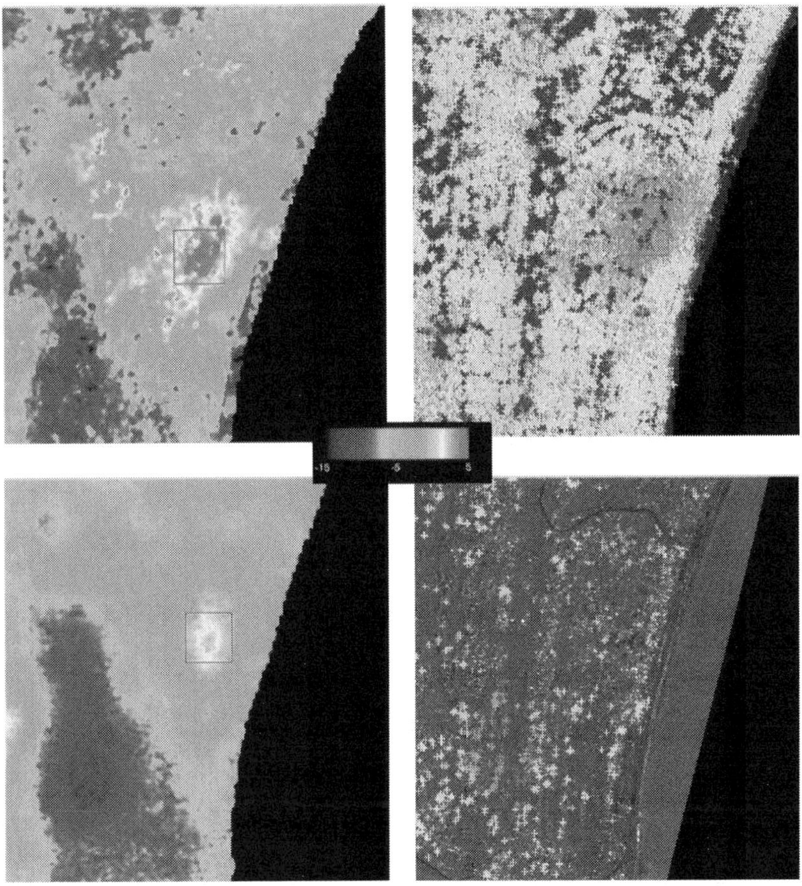

Figure 14. Average displacement rate in the 2006-2010 period in the area of Chosei (Japan) obtained from PALSAR (above) and ASAR (below) data through SBAS (on the left) and PS (on the right) processing.

More in-depth analysis of SBAS results for ground deformation monitoring over the Dead Sea region are presented in the Chapter *„Dikes stability monitoring versus sinkholes and subsidence, Dead Sea region, Jordan"* of this book.

4. Analysis of time series non-linearity

The exploration performed in the previous Section concerning the exploitability of Interferometric Stacking techniques mainly focused on the analysis of the average displacement rate as measured on a stack of SAR imagery. On the other hand, as already mentioned in the previous

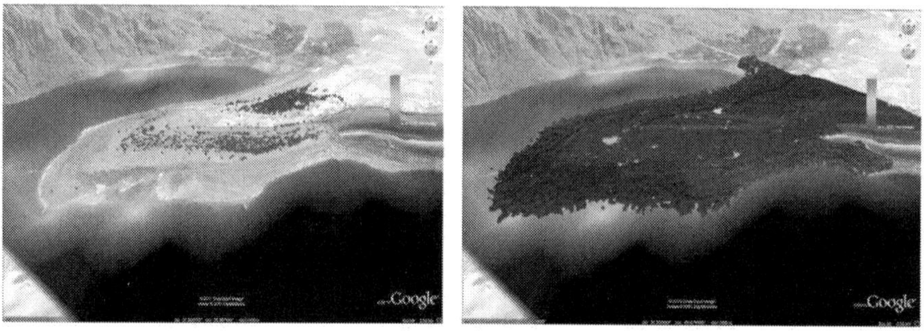

Figure 15. Comparison of average displacement rates in the period 2003 – 2010 as derived with PS (left) and SBAS (right) processing from ENVISAT ASAR data over the Lisan peninsula, Dead Sea, Jordan.

Sections, the information that can in principle be extracted from such input data is far more, not only limited to a simple average value, but potentially describing the full temporal evolution of phenomena that vary with the time.

One evident advantage of exploiting the average displacement rate is that it can be easily displayed, and regions showing different average behaviours can be easily identified with a simple visual analysis. These observations start to show their limitation as soon as more complex, non-linear behaviours are to be expected (as natural) in a certain region, and different parameters shall be sought to provide a synthetic way to visualise and identify areas with similar, non-linear characteristics.

Figure 16 in the upper left part shows an excerpts of a South-Eastern region of Figure 3 right, for which the average displacement rate is displayed, as estimated through the SBAS algorithm. As previously described, the SBAS algorithm obtains this information through a Lest Mean Square fit / inversion, whose statistical significance is often estimated by measuring the corresponding χ^2 value. When a simple linear model is assumed during the SBAS fit / inversion for this area, the corresponding χ^2 resembles what is shown in Figure 16 upper right; two regions can be easily identified in this image, showing systematically high values of χ^2, hence good candidates for a further investigation, focused on the identification of non-linear displacement behaviours. Figure 16 shows in the lower row temporal plots for the two regions highlighted in the χ^2 image; the plot on the left in particular shows in red the displacement time series of the point in region 1 showing highest χ^2 values, while in blue (as comparison) the time series of a neighbouring pixel that shows low values of χ^2. The same Figure shows in the lower right part in red the displacement time series of the point in region 2 showing highest χ^2 values, while in blue (as comparison) the time series of the neighbouring pixel that shows largest average displacement rate, as in the upper lower image of the same Figure, but linear variation.

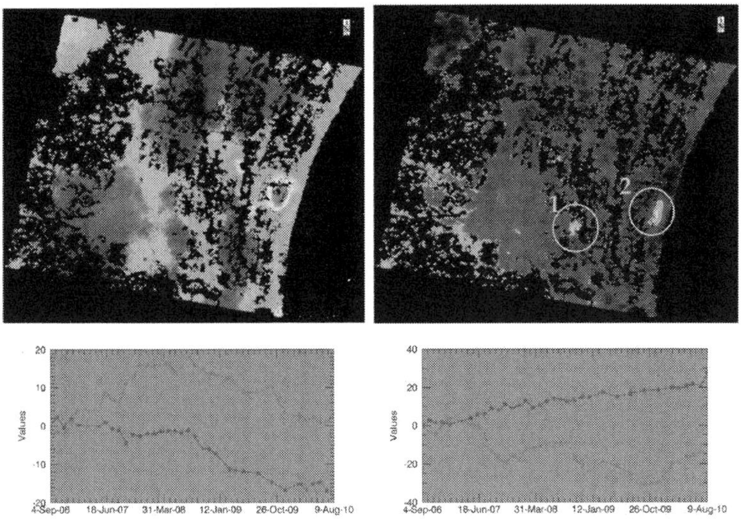

Figure 16. ENVISAT ASAR SBAS results over the Togane region (Japan). Left to right and top to bottom: average displacement rate, χ2 from linear model, location time series in [mm] from areas 1 and 2 in the χ2 image.

The plots in this Figure show very clearly two different non-linear behaviours: area 1 is characterised by a parabolic (first increasing an then decreasing) trend, while area 2 is characterised by a periodical trend, with a yearly cycle. As it can easily be seen, these two regions could not at all be identified from the average displacement rate image, while their temporal behaviours significantly deviate from a linear one, that characterizes most neighbouring regions.

The χ^2 value seems hence a good candidate for a simple yet efficient identification of areas showing non-linear but consistent temporal displacement behaviours, as identified by the SBAS processing.

As a further trail, the χ^2 value has been evaluated again after performing a new SBAS processing that considers a 3rd order polynomial model during the inversion steps; the corresponding result is shown in Figure 17. Here it can be easily seen how the χ^2 value is significantly reduced in region 1, that originally showed parabolic trends, hence easily fitted with a third order model; the same cannot be said for region 2, where the periodical trend cannot be easily described by a polynomial function.

Another non-linearity analysis example is presented in Figures 18-24 for a region in Jordan close to Madaba city, where land subsidence relates with water extraction for agricultural practices.

Figure 18 shows the average displacement rate estimated with a liner model and the corresponding χ^2. It is interesting to notice that, although the two measures show large values

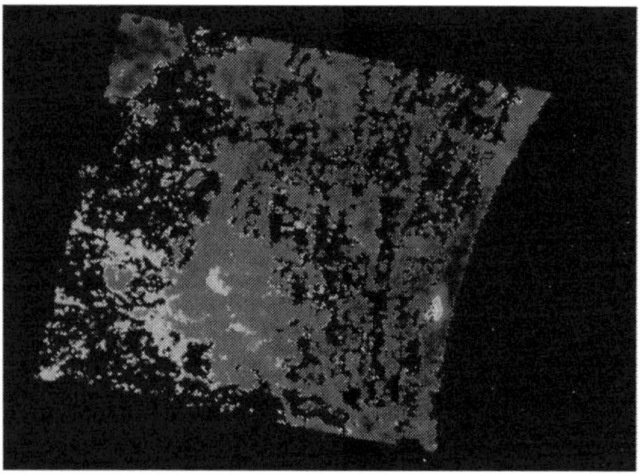

Figure 17. ENVISAT ASAR SBAS results over the Togane region (Japan), χ2 from cubic model.

generally in the same geographic area, their spatial distribution is quite different; most of large displacements are characterised by strongly non-linear motions, but not all and not in the same way, hence the average displacement rate layer is not showing all the information.

Figure 19 is presenting results obtained on the same area by exploiting a cubic deformation model. As to be expected, the accuracy of the fit of the SAR time series increased (and hence the χ^2 values decreased), and most of the displacements can be statistically well described by a 3rd order polynomial.

Figure 20 shows the difference between the χ^2 values as resulting from the linear and respectively cubic polynomial fit. As expected, most of the image shows no significant change, i.e. the linear model is good enough to describe the displacement behaviour. On the other hand, a few areas show significant decrease of the χ^2 values when increasing the model complexity. Some of them have been highlighted in the same Figure, and their displacement time series is shown in Figure 21, together with that of an area with large displacement rate but already low χ^2 values, highlighted in Figure 18.

Here it can be seen how the non-linear displacement in area 1 is characterised by a first almost linear regime, followed by a significant acceleration and finally a possible little deceleration. Areas 2 and 3 show similar behaviours, where the initial linear trend is almost constant (no displacement) and the final deceleration is possibly less pronounced. All three cases seems nevertheless well suitable for a cubic polynomial representation; a further information that could be of interest for the exploitation is the date of change between the linear (or constant) and the accelerated regime.

Area 4 shows a quite clear quadratic behaviour, that could have equally well fitted with a second order model.

http://dx.doi.org/10.5772/58225

Area 5 presents, on the contrary, a simple linear regime that, even if the average displacement rate is quite large, could be well fitted with a first order model, showing no significant improvement (reduction) in the χ^2 values when increasing the model complexity.

It might be of interest at this point to comment the average acceleration and acceleration variation images shown in Figure 22, as estimated with the third order model.

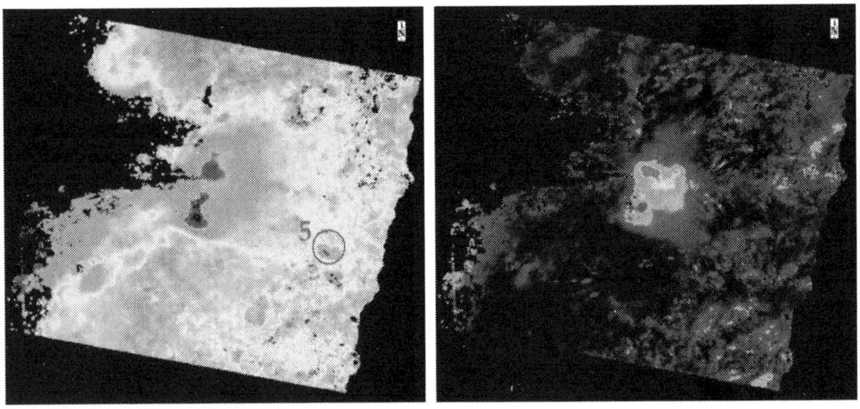

Figure 18. ENVISAT ASAR SBAS results over the Madaba region (Jordan). Average displacement rate from linear model (on the left) and corresponding χ2 (on the right).

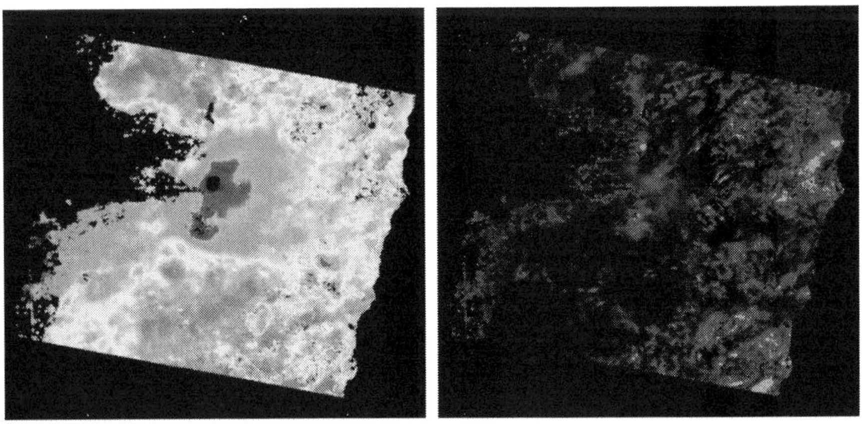

Figure 19. ENVISAT ASAR SBAS results over the Madaba region (Jordan). Average displacement rate from cubic model (left) and corresponding χ2 (right).

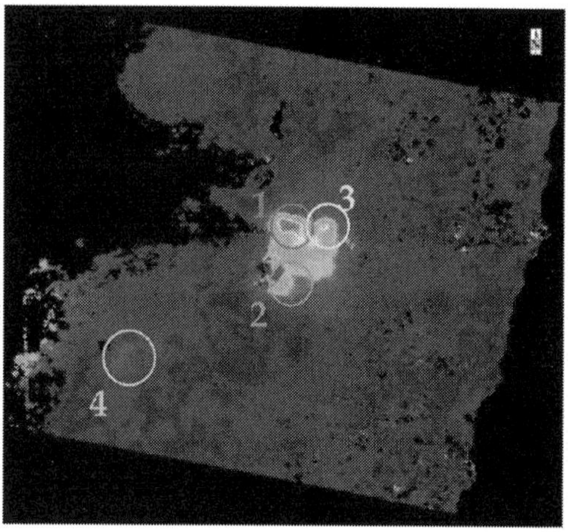

Figure 20. ENVISAT ASAR SBAS results over the Madaba region (Jordan). Difference of χ2 between linear and cubic model.

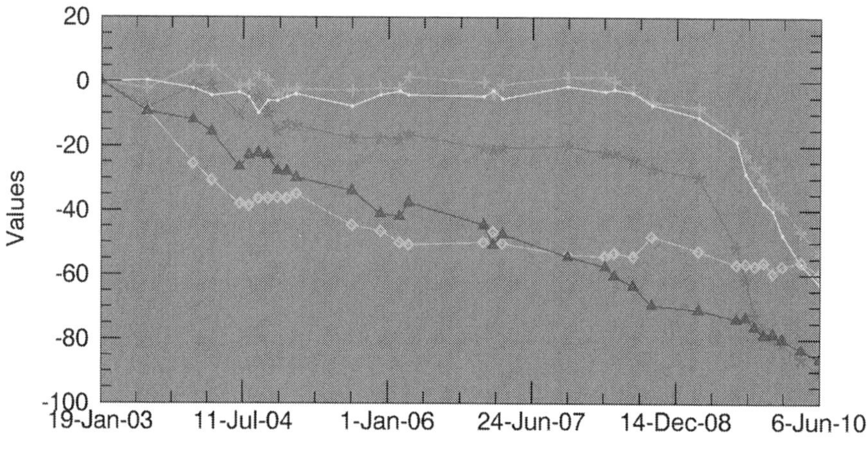

Figure 21. Displacement time series in [mm] from areas 1 to 5 of Figure 20 and Figure 18.

The average acceleration provides a measure of the main curvature of the time series: a big part of the area shows very small curvature (green – yellow areas, hence characterised by mostly linear displacements, as for curve 5); areas in red correspond to positive curvatures, with mainly a decrease of the displacement rate with the time, as in curve 4 of Figure 20-21, while blue areas show negative curvature, hence corresponding to an increase of the displacement rate (in absolute value), as in curves 1 to 3 in the same Figures.

It shall be noticed again how these images have again a different although overall similar spatial distribution with respect to the average displacement rate one, showing the amount of additional information on the discrimination of different displacement regimes that they can bring.

Figure 22. ENVISAT ASAR SBAS results over the Madaba region (Jordan). Average displacement acceleration from cubic model (left) and corresponding acceleration variation (right).

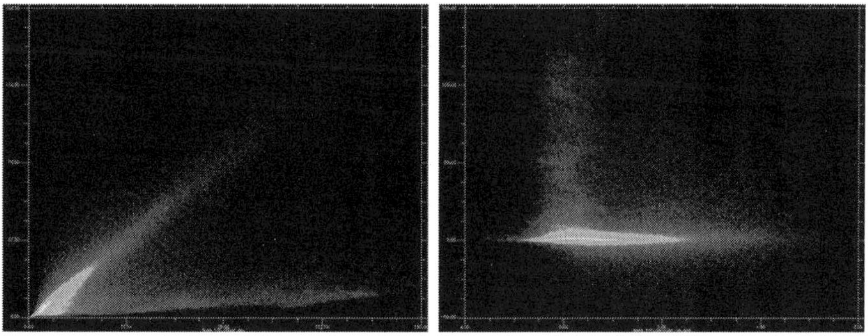

Figure 23. ENVISAT ASAR SBAS scatter plots over the Madaba region (Jordan). On the left: χ2 from linear model (x axes) against χ2 from cubic model (y axes). On the right: scatter plots of expected displacement rate precision (x axes) against χ2 decrease from linear to cubic model (y axes).

Figure 23 presents two scatter plots for the same case, where the χ^2 values obtained with the two model types are plotted together (on the left), and their difference is plotted against the value of expected displacement accuracy, as estimated from the original pixels coherence.

The first plot shows, as expected, mainly two groups of pixels, one for which the increase of model order is not making a real difference (mostly linear motions, and some non-linear non-polynomial motions, or noise), and the other for which a higher order model is more appropriate. Mainly no pixel shows significant decrease of the fit accuracy at the increase of the model order.

The scatter plot on the right is on the other hand showing how the increase of the fit accuracy can be obtained only for pixels that, a priori, can be assumed as reliable and for which any analysis or fit of the displacement time series can provide meaningful results.

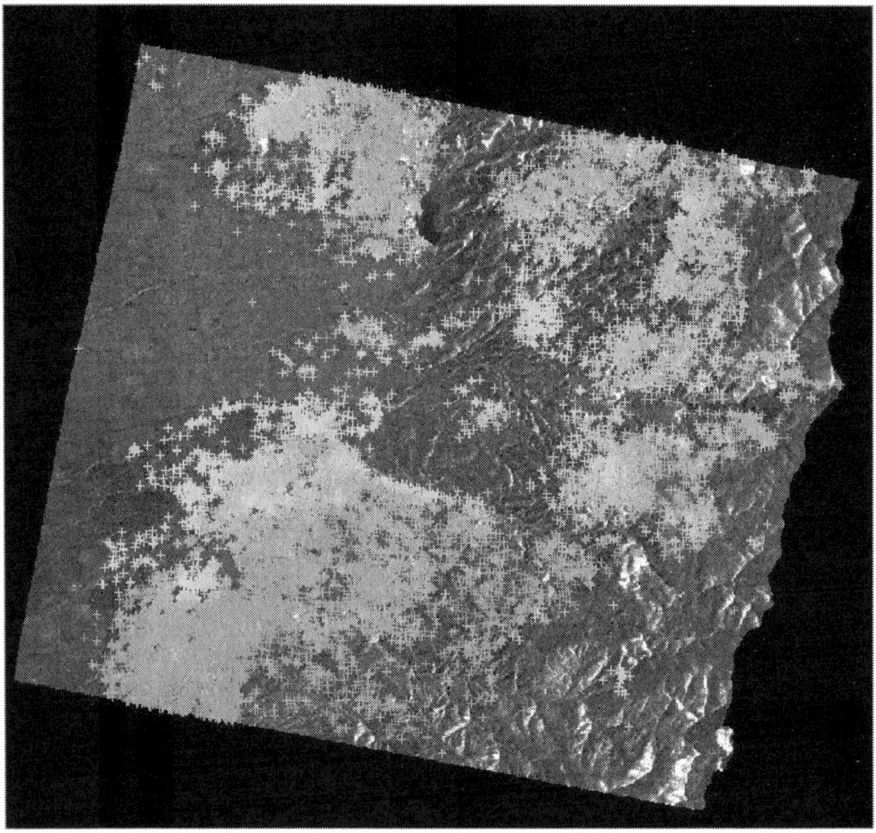

Figure 24. ENVISAT ASAR PS results over the Madaba region (Jordan).

Figure 24 shows the results of the PS processing over the same area. It is well visible that reliable results can, with this approach, be obtained only for areas characterised by a well linear displacement behaviours; area showing large χ^2 values in case of linear model do not provide any result when processed with the PS technique.

The following Table summarises the findings obtained in the analysis of the χ^2 layers of the two previous cases and the information about discriminating different temporal behaviours of the displacements.

	Small linear χ^2	**Large linear χ^2**
Small cubic χ^2	Linear motion	Non-linear polynomial motion
Large cubic χ^2	Noisy pixel	Non-linear non-polynomial motion (e.g. periodical – seasonal)

Table 4. Discrimination of different displacement temporal regimes based on χ^2 analysis.

5. Conclusions and outlook

The previous Sections presented a number of validated examples, showing how interferometric stacking techniques can provide accurate and reliable information concerning ground deformations that can be very valuable for different applications. The PS and SBAS approaches can be considered as complementary, each with specific unique features.

The huge amount of measures, both in terms of spatial and temporal density, that these techniques can provide, when for example compared with other systems like GNSS or levelling, can be on one hand considered as a wealth of information. On the other hand, such detail can often make the interpretation very complex.

As for example discussed in the previous Section, the average displacement rate is often just one of the synthetic descriptors that can be used to represent in a compressed way the complexity of the whole displacement time series. When the displacement regimes are more complex, robust algorithms as the SBAS one shall be exploited, or adaptations of the original PS approach shall be considered that could extend (even if not completely) its applicability toward more non-linear [21] and / or non-continuously coherent displacements [22]. Other indicators and analysis approaches shall also be developed to help the identification and extraction of the different regimes.

Different physical factors start to play a role in the interpretation of the obtained measurements, as soon as their accuracy start to be very high. It is for example well known that the temperature variations start to be clearly recognisable in interferometric stacking measurements, in particular when performed over large metallic structures (bridges, buildings with big metallic components etc.). One example of such an effect is shown in Figure 25 for an area of the city of Beijing as obtained through PS processing of very-high resolution COSMO-Skymed data. Piles of PSs are visible, corresponding to the reflections of the different floors of single buildings. One of them, in the lower left part of the image is showing average displacement rates that well correlate with its height: higher locations have larger thermal expansion, as to be expected for this tall metallic building.

This component cannot be neglected, and should for example be included within the inversion steps by opportunely modifying the corresponding algorithms [23].

Figure 25. Average displacement rate estimated with the PS approach over an area of Beijing from COSMO-Skymed data. Each circle correspond to one identified PS.

As recalled in the previous Sections, all interferometric stacking algorithms include a step of estimation and subtraction of the APS, based on some statistical assumptions on the spatial and temporal distribution of the atmospheric artefacts. These assumptions may be often in contrast with the spatial distribution of the displacement, also considering sometime a similar dependency of the displacement and of the atmosphere parameters with the terrain topography, as for volcanoes and for other specific cases and morphologies. It is hence very interesting to explore alternative approaches, based on additional independent measurements, to estimate and mitigate the impact of atmospheric heterogeneities, as for example as suggested in [24]-[30] with different approaches.

Last but not least, it has been shown how both PS and SBAS provide very interesting and complementary information; approaches that somehow merge the best of the two worlds can hence only considered with big interest [31] [32].

http://dx.doi.org/10.5772/58225

Author details

Paolo Pasquali[1*], Alessio Cantone[1], Paolo Riccardi[1], Marco Defilippi[1], Fumitaka Ogushi[2], Stefano Gagliano[3] and Masayuki Tamura[4]

*Address all correspondence to: paolo.pasquali@sarmap.ch

1 Sarmap SA, Purasca, Switzerland

2 Exelis Visual Information Solutions K.K., Tokyo, Japan

3 Exelis Visual Information Solutions Italia, Concorezzo, Italy

4 Department of Civil and Earth Resouces Engineering, Kyoto University, Kyoto, Japan

References

[1] Ferretti, A., Prati, C., & Rocca, F.: *"Permanent scatterers in SAR interferometry"*, IEEE Transactions on Geoscience and Remote Sensing, 39, 8-20, 2001.

[2] Berardino, P., Fornaro, G., Lanari, R., & Sansosti, E.: *"A new algorithm for surface deformation monitoring based on small baseline differential SAR interferograms"*, IEEE Transactions on Geoscience and Remote Sensing, 40, 2375-2383, 2002.

[3] Gatelli, F., Monti Guarnieri, A., Parizzi, F., Pasquali, P., Prati, C., Rocca, F., "*The wavenumber shift in SAR interferometry*", *Geoscience and Remote Sensing, IEEE Transactions on*, vol.32, no.4, pp.855,865, Jul 1994.

[4] Pepe, A.; Lanari, R., "On the Extension of the Minimum Cost Flow Algorithm for Phase Unwrapping of Multitemporal Differential SAR Interferograms," *Geoscience and Remote Sensing, IEEE Transactions on*, vol.44, no.9, pp.2374,2383, Sept. 2006.

[5] Lanari, R.; Mora, O.; Manunta, M.; Mallorqui, J.J.; Berardino, P.; Sansosti, E., "*A small-baseline approach for investigating deformations on full-resolution differential SAR interferograms,*" *Geoscience and Remote Sensing, IEEE Transactions on*, vol.42, no.7, pp.1377,1386, July 2004.

[6] Zebker, H.A.; Villasenor, J., "*Decorrelation in interferometric radar echoes*", *Geoscience and Remote Sensing, IEEE Transactions on*, vol.30, no.5, pp.950,959, Sep 1992.

[7] Rocca, F., "*Modeling Interferogram Stacks*", *Geoscience and Remote Sensing, IEEE Transactions on*, vol.45, no.10, pp.3289,3299, Oct. 2007.

[8] A. Parizzi, X. Y. Cong, and M. Eineder, *"First results from multifrequency interferometry-a comparison of different decorrelation time constants at X, C and L-band"*, *Fringe 2009 Workshop Proceedings*, 2009.

[9] Meng Wei; Sandwell, D.T., "Decorrelation of L-Band and C-Band Interferometry Over Vegetated Areas in California," *Geoscience and Remote Sensing, IEEE Transactions on*, vol.48, no.7, pp.2942,2952, July 2010

[10] Parizzi, A., "Speckle statistics and long-term coherent SAR interferograms," *Geoscience and Remote Sensing Symposium (IGARSS), 2012 IEEE International*, vol., no., pp. 5590,5593, 22-27 July 2012.

[11] Persistent Scatterer Interferometry Codes Cross-Comparison And Certification (PSIC4) Project. *http://earth.esa.int/psic4/* (accessed 25 September 2013).

[12] D. Raucoules, B. Bourgine, M. de Michele, G. Le Cozannet, L. Closset, C. Bremmer, H. Veldkamp, D. Tragheim, L. Bateson, M. Crosetto, M. Agudo, M. Engdahl, *"Validation and intercomparison of Persistent Scatterers Interferometry: PSIC4 project results"*, Journal of Applied Geophysics, Volume 68, Issue 3, July 2009, Pages 335-347

[13] Ferretti, A.; Savio, G.; Barzaghi, R.; Borghi, A.; Musazzi, S.; Novali, F.; Prati, C.; Rocca, F., "Submillimeter Accuracy of InSAR Time Series: Experimental Validation," *Geoscience and Remote Sensing, IEEE Transactions on*, vol.45, no.5, pp.1142,1153, May 2007.

[14] Miyazaki, S., Y. Hatanaka, T. Sagiya and T. Tada: *"The Nationwide GPS Array as an Earth Observation System"*, Bull. Geographical Survey Institute, 44, 11-22, 1998.

[15] Hatanaka, Y., A. Sengoku, T. Sato, J.M. Johnson, C. Rocken and C. Meertens: *„Detection of Tidal Loading Signals from GPS Permanent Array of GSI Japan"*, J. Geod. Soc. Japan, 47, 187-192, 2001.

[16] Hatanaka, Yuki, Toyohisa Iizuka, Masanori Sawada, Atsushi Yamagiwa, Yukie Kikuta, James M. Johnson, and Christian Rocken, *"Improvement of the analysis strategy of GEONET." Bull. Geogr. Surv. Inst* 49 (2003): 11-37.

[17] Urayasu City Government Homepage: http://www.city.urayasu.chiba.jp/, (accessed 25 September 2013).

[18] The Japanese Geotechnical Society, Japan Society of Civil Engineering, and Architectural Institute of Japan, *"Report of the Urayasu city review and research committee on liquefaction countermeasure techniques"*, March 2012.

[19] Kanto Natural Gas Development Co Ltd., Abundant Reserves. http://www.gasukai.co.jp/english/gas/index4.html, (accessed 25 September 2013).

[20] Chiba Prefecture Homepage, http://www.pref.chiba.lg.jp/, (accessed 25 September 2013).

[21] Ferretti, A.; Prati, C.; Rocca, F., *"Nonlinear subsidence rate estimation using permanent scatterers in differential SAR interferometry," Geoscience and Remote Sensing, IEEE Transactions on*, vol.38, no.5, pp.2202,2212, Sep 2000.

[22] Colesanti, C.; Ferretti, A.; Novali, F.; Prati, C.; Rocca, F., "SAR monitoring of progressive and seasonal ground deformation using the permanent scatterers technique,"

Geoscience and Remote Sensing, IEEE Transactions on, vol.41, no.7, pp.1685,1701, July 2003.

[23] Monserrat, O.; Crosetto, M.; Cuevas, M.; Crippa, B., "The Thermal Expansion Component of Persistent Scatterer Interferometry Observations," *Geoscience and Remote Sensing Letters, IEEE,* vol.8, no.5, pp.864,868, Sept. 2011.

[24] S. Williams, Y. Bock, and P. Fang, "Integrated satellite interferometry: Tropospheric noise, GPS estimates and implications for interferometric synthetic aperture radar products," J. Geophys. Res., vol 103(B11), pp. 27051-27068, 1998.

[25] F. Onn and H. A. Zebker, "Correction for interferometric synthetic aperture radar atmospheric phase artifacts using time series of zenith wet delay observations from a GPS network," J. Geophys. Res., vol. 111, B09102, 16 pp, 2006.

[26] Li, Z., E. J. Fielding, P. Cross, and J.-P. Muller, *"Interferometric synthetic aperture radar atmospheric correction: Medium Resolution Imaging Spectrometer and Advanced Synthetic Aperture Radar integration", Geophys. Res. Lett., 33.* 2006.

[27] Li, Z., E. J. Fielding, P. Cross, and J.-P. Muller, *"Interferometric synthetic aperture radar atmospheric correction: GPS topography-dependent turbulence model", J. Geophys. Res., 111, 2006.*

[28] Doin M.-P., C. Lasserre, G. Peltzer, 0. Cavalié, C. Doubre, *"Corrections of stratified tropospheric delays in SAR interferometry: validation with global atmospheric models"*, J. of Applied Geophysics, 69 (1), 2009.

[29] Löfgren, J.S.; Björndahl, F.; Moore, A. W.; Webb, F.H.; Fielding, E.J.; Fishbein, E.F., *"Tropospheric correction for InSAR using interpolated ECMWF data and GPS Zenith Total Delay from the Southern California Integrated GPS Network," Geoscience and Remote Sensing Symposium (IGARSS), 2010 IEEE International,* vol., no., pp.4503,4506, 25-30 July 2010.

[30] Walters, R. J., J. R. Elliott, Z. Li, and B. Parsons, *"Rapid strain accumulation on the Ashkabad fault (Turkmenistan) from atmosphere-corrected InSAR"*, J. Geophys. Res. Solid Earth, 118,3674–3690. 2013.

[31] Hooper, *"A multi-temporal InSAR method incorporating both persistent scatterer and small baseline approaches"*, Geophysical Research Letters 35, 2008.

[32] Ferretti, A.; Fumagalli, A.; Novali, F.; Prati, C.; Rocca, F.; Rucci, A., *"A New Algorithm for Processing Interferometric Data-Stacks: SqueeSAR", Geoscience and Remote* Sensing, IEEE Transactions on, vol.49, no.9, pp.3460,3470, Sept. 2011.

Chapter 9

SAR Data Analysis in Solid Earth Geophysics: From Science to Risk Management

S. Atzori and S. Salvi

Additional information is available at the end of the chapter

http://dx.doi.org/10.5772/57479

1. Introduction

One of the scientific disciplines in which SAR data have generated a real breakthrough is Solid Earth Geophysics. As soon as the first differential SAR interferograms appeared on scientific journals, describing the surface crustal changes due to earthquakes or inflation of volcanic edifices, geophysicists realized that their science was to greatly benefit from these new data.

The reasons were clear: before the InSAR era crustal deformation was measured using very time-consuming, ground based methods, as leveling, triangulation, trilateration; all requiring costly networks and measurement campaigns. The same was true for the GPS technique, appeared only few years before. SAR interferometry made things simpler and cheaper, although with some limitations; probably the most appealing features of InSAR deformation measurements were their spatial imaging capacities and the high ground resolution. These allowed an unprecedented way to look at all the natural phenomena causing, or derived from, surface deformation.

In this chapter we start describing technical aspects related to the way InSAR data are exploited to derive the parameters of the sources of an observed phenomenon, focusing the attention on the seismic and volcanic activity. We present the state-of-the-art techniques to this inference problem, describing the most common inversion algorithms and analytical models, especially those implemented for the description of the seismic and volcanic cycles. Other technical aspects, as the role played by the data uncertainty, the existing strategies to reduce the data redundancy and the way sources of different nature interact with each other, are also presented to provide useful basics for the reader interested InSAR data modeling.

The second part of this chapter focuses on the practical use of InSAR data and derived models, describing their assimilation in the seismic risk assessment and prevention or during the disaster response. Examples of already existing operational procedures, research and development pilot projects, current and future SAR missions, including Sentinel-1, are also presented to complete this detailed overview on the role played by SAR and SAR-derived information in Solid Earth Geophysics.

2. Analytical source modeling

InSAR modeling is an inference process used to define the properties of a realistic geophysical source, or combination of sources, from a set of geodetic measurements. First aspect to clarify is the definition of "realistic", strongly related to *how* resulting models are exploited. Few hours after an earthquake, the availability of an approximated fault model can be more useful to a decision maker than a complex finite element solution obtained in a week. Conversely, the study of a tectonic setting in a seismogenic area might need a detailed source description, including mechanical discontinuities, the effects of topography, non-elastic rheology and a stratified crust, characteristics not considered by analytical models.

Accepting a simple approximations, as a free surface without topography overlying an homogeneous and isotropic half-space, ordinary geophysical models can be expressed as a set of equations in explicit form, where the modeled surface displacement $\mathbf{d}_{\mathbf{mod}}$ for a given point located in (x,y) is obtained as a function

$$\mathbf{d}_{\mathbf{mod}}(x,y) = \boldsymbol{func}(m_1,m_2, \ldots,m_n, \ldots) \tag{1}$$

being m_1, m_2,... m_n the model parameters.

Though the above approximations might appear excessive, in most cases they provide very good description of the sources, a fact that is indeed reflected in their massive use in Geophysics. Commonly used sources in the literature are:

elastic dislocation in a finite rectangular source [1]: it is definitely the most used model to predict the surface displacement due to an earthquake, represented as a shear dislocation over a finite rectangular fault. The Okada model, however, can be also used to describe magma intrusion like sills or dykes [2],[3],[4], interseismic and post-seismic deformations (see Section 4), landslides [5] and ground subsidence induced by fluid extraction [6]. Source parameters are: East and North position, depth, length, width, strike angle, dip angle, dislocation (or slip), dislocation angle (rake), opening (Figure 1);

point pressure source [7]: it is one of the simplest and effective source used in volcanology, as its description requires only 4 parameters: depth, east and north position, volume variation or pressure variation[1] (Figure 2)

1 The Mogi equations can be indifferently written in terms of pressure change or volume change; however, the volume/pressure conversion requires the definition of a virtual source radius.

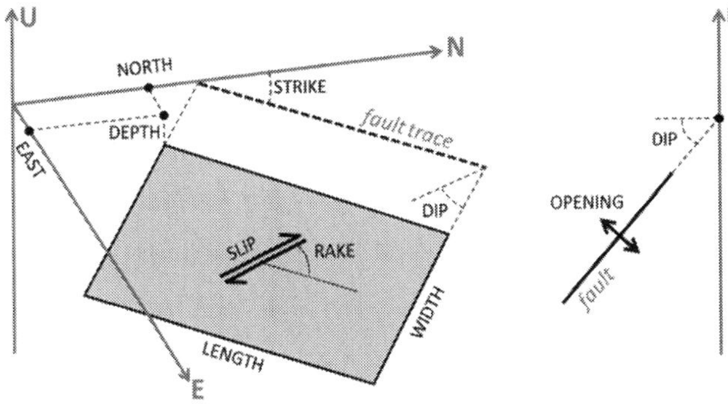

Figure 1. Parameters required by the Okada model.

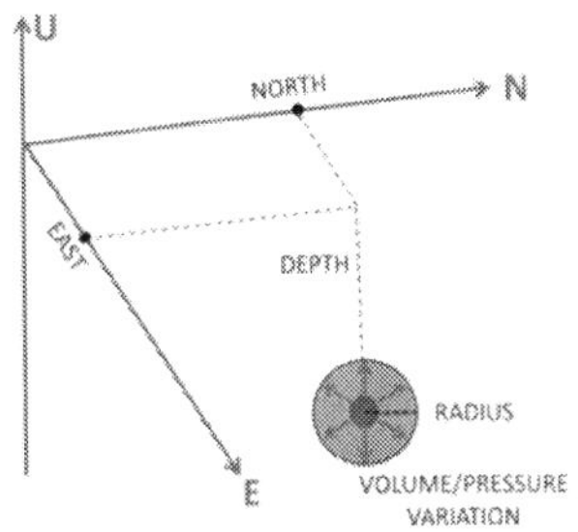

Figure 2. Parameters required by the Mogi model

Several other sources have been proposed in the literature, with the aim of providing more realistic solutions to describe geophysical phenomena: dislocation over a finite triangular source [8]; volume variation of a dipping finite prolate spheroid [9]; inflation of an arbitrarily oriented triaxial ellipsoidal cavity [10]; pressure change in a penny-crack source [11]; closed vertical pipe [12]; stress induced by a finite spherical source [13]. A description of the differences among all these sources is beyond the scope of this article, and we refer the reader to the cited literature.

Lastly, we remark the existence of semi-analytical solutions for a layered crust [14] [15], even with a visco-elastic rheology [16]; however they often require a considerable computational time, preventing their use in non-linear inversion schemes, where thousands of forward calculations are often needed. The same limitation affects finite-element models, unless used to build the Green's function matrix as explained in the next section.

For almost all the listed models, the geometric parameters (position, depth, dimension, orientation, etc...) are non-linearly related to the surface displacement. On the contrary, "intensity" parameters, as the dislocation for the Okada model or the pressure change for a Mogi model, have a linear dependency with the surface displacement. The retrieval of non-linear and linear parameters from geodetic data follows different inversion strategies, explained in the next section.

A further Okada-derived model, widely used to model earthquakes, consists of an array of single Okada sources, or *patches*, for which all the non-linear parameters are already constrained (Figure 3). This compound source is used to reproduce a fault rupture with a variable shear dislocation (or tensile opening), thus improving the modeling performances compared with the uniform slip Okada source.

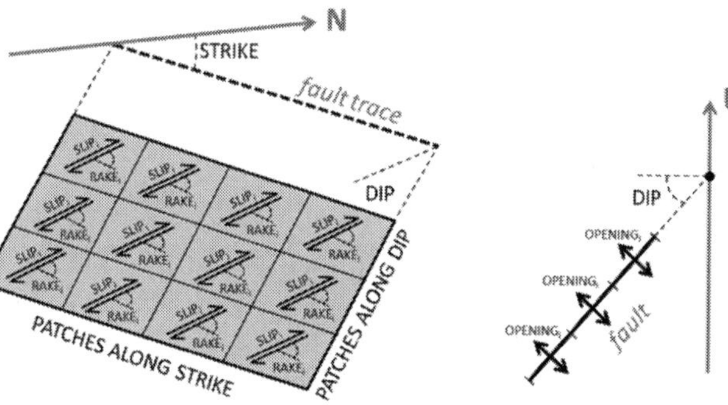

Figure 3. Okada-based, compound source for slip or opening distributed models

InSAR measurements are relative, therefore the displacement maps are often related to a reference point considered stable or with a known deformation. This point sometimes turns out to be inappropriate: it could fall in an area affected by atmospheric artifacts; the non-zero displacement field could be larger than the image frame; the GPS value used to tie the reference point might contain a very long wavelength tectonic signal. Furthermore, orbital inaccuracies might introduce artificial linear ramps or even quadratic surfaces. Thus a further non-geophysical source must be often considered, to account for possible improper reference points and/or possible orbital artifacts.

This model can be implemented as a second order surface of the type

$$\mathbf{d}_{\mathbf{mod}}(x,y) = A + B{\cdot}x + C{\cdot}y + D{\cdot}x^2 + E{\cdot}y^2 + F{\cdot}xy \qquad (2)$$

so that the apparent displacement $\mathbf{d}_{\mathbf{mod}}$ for a given point (x,y) is a linear combination of the coefficients A, B… F. Second order terms can be neglected when orbital artifacts are well reproduced only with a linear ramp.

We remark that all the geophysical and orbital sources[2] must always be solved simultaneously in the inversion, otherwise the first source used to model the data will tend to improperly reproduce the signal we might fit with other sources.

2 We adopt the misleading definition "orbital source" only because it is mathematically handled just like the Okada or Mogi equations.

3. Non-linear and linear inversions

Current approaches to the modeling of geodetic data are the evolution of pioneer studies where the amount of geodetic measurements where certainly lower than nowadays [17][18][19] and the availability of geophysical models were still limited: solutions for dislocation in an elastic medium were available only for some simple fault configurations and for point sources [20] [21]. However, non-linear and linear inversion strategies were essentially the same presently adopted. What deeply changed through the decades is the amount of observed data, with InSAR playing a lead role.

An inversion algorithm is a procedure to infer some source parameters so that the modeled, or predicted, surface displacement best reproduces the observed one. We must firstly provide a definition for "best fit"; almost every model shown in literature assumes that the best parameter estimate ($\mathbf{m}_{est}$) occurs when the square of the residuals between observed ($\mathbf{d}_{obs}$) and modeled ($\mathbf{d}_{mod}$) data, i.e. a cost function based on the L_2 norm, is minimized:

$$\mathbf{m}_{est} : \min\left(func\left(L_2\right)\right), \text{where } L_2 = \sqrt{\left(\mathbf{d}_{obs} - \mathbf{d}_{mod}\right)^2} \tag{3}$$

This choice implies that $\mathbf{d}_{obs}$ data obey to a Gaussian statistics, so that the presence of large outliers is strongly unlikely [22]. However, uncertainties in InSAR data may arise at different stages until the unwrapped and geocoded displacement map is generated, and large spurious artifacts are not so unlikely (see [23] and references therein). Thus the assumption of an L_2 norm criterion should be taken with care, since it may not be the best choice to build the cost function to minimize.

The inversion strategy must be chosen according to the source parameters $\mathbf{m}$ linearity with respect to the surface displacement $\mathbf{d}_{mod}$. In this section, we assume that the surface displacement is a non-linear combination of the source parameters we want to infer from InSAR data.

Any non-linear inversion scheme is based on the realization of a sequence of forward calculations as in (1) until the condition (3) is satisfied or, in other terms, a set of rules to iteratively change the $\mathbf{m}$ parameters until the best estimate $\mathbf{m}_{est}$ is obtained. A peculiar aspect of InSAR data is that displacement measurements are provided in the line-of-sight direction, while every modeled dataset $\mathbf{d}_{mod}$ is calculated in a Cartesian reference system; therefore the calculation of the L_2 norm must be preceded by projection of modeled data into the line-of-sight.

A typical issue to face in a non-linear inversion is the presence of an unpredictable number of local minima, corresponding to unsatisfactory solutions; a robust algorithm should be able to take a cost function out from local minima. Unfortunately, only few strategies guarantee the achievement of this goal, for instance Simulated Annealing [24]; however, they are incompatible with a reasonable processing time[3]. We also remark that the level of non-linearity depends

3 Simulated Annealing is largely used in non-linear inversions due to the ease of its implementation, but never with the cooling schedule that guarantees to find the global minimum, that would require a nearly endless computation time.

on the parameters to invert: for instance, when periodic parameters like rake or strike angles in the Okada model are fixed or forced to vary in a narrow range, the impact of local minima decreases.

In our experience, the Gauss-Newton or gradient descent methods, as the Levemberg-Marquardt algorithm which is a mix of both [25], provide excellent results with a short processing time; although this algorithm does not consider any stratagem to escape from local minima, its implementation with multiple restarts provides an efficient strategy to identify the global minimum.

We incidentally remark that the repeated calculation of forward models makes unfeasible the use of finite element models in any inversion scheme, because of the long time needed to setup and calculate even a single forward model.

A different approach is adopted when all the parameters to invert are linearly related to the surface displacement, as occurs with the strike-slip, dip-slip and opening components for an Okada model, with the volume variation in a Mogi source or with the coefficients of equation (2). In this case, the inversion can be set up in a matrix form of the type:

$$\mathbf{d}_{\text{obs}} = \mathbf{G} \cdot \mathbf{m} \tag{4}$$

where the vector collecting all the parameters $\mathbf{m}$ is related to the known displacement values $\mathbf{d}_{\text{obs}}$ through the Green's function matrix $\mathbf{G}$.

It is worth noting that the number of unknowns $\mathbf{m}$ is almost always lower than the number of equations in the linear system (4), given the abundance of InSAR measurements $\mathbf{d}_{\text{obs}}$. Therefore the problem is generally over-determined and the solution, in the simplest case, is calculated in the least square sense on the data, as follows

$$\mathbf{m}_{\text{est}} = \mathbf{G}^{-g}\mathbf{d}_{\text{obs}} \text{ where } \mathbf{G}^{-g} = \left[\mathbf{G}^{\mathrm{T}}\mathbf{G}\right]^{-1}\mathbf{G}^{\mathrm{T}} \tag{5}$$

being $\mathbf{G}^{-g}$ the generalized inverse of $\mathbf{G}$ [22].

However, if we consider the practical problem of finding the coseismic slip distribution as in Figure 6e, solutions found via (4) and (5) are generally not acceptable because of the weak control that InSAR surface measurements have on deep parameters, leading to highly scattered $\mathbf{m}_{\text{est}}$ values. This problem can therefore be treated as partially underdetermined, adding constraints on the model parameters to get a more reliable result. In this case, since the problem is not mathematically underdetermined, the model reliability is obtained at the expense of the data fit. The system of (4) can be arbitrarily extended to introduce almost any type of *a priori* constraint; in earthquake modeling, for instance, we can force the slip to gently vary across the fault and this can be achieved introducing a damping parameter in the form:

$$\begin{pmatrix} d_{obs} \\ 0 \end{pmatrix} = \begin{pmatrix} G \\ \varepsilon \bullet \nabla^2 \end{pmatrix} \bullet \mathbf{m} \tag{6}$$

where the Green's functions $\mathbf{G}$ are extended with a Laplacian operator ∇^2 of the model parameters, opportunely weighted with an empirical parameter ε [19][26]. In this *damped least square* solution, the ε coefficient is problem dependent and must be chosen via trial and error or through the construction of an empirical fit vs. smoothness curve, to find the desired compromise.

Other inequality constraints, such as the positivity (Non-Negative Least Square), can be introduced as well, to further increase the solution reliability; in the modeling of a coseismic displacement field, the constraint $\mathbf{m}>0$ is used to avoid unrealistic "back-slip" conditions. For a complete review of the strategies used to add conditions to the linear system (4), refer to [22].

Unlike non-linear inversion, in linear inversions the use of finite elements models (FEM) is not prohibitive in terms of computational time. In this stage, the heavy forward calculation must be done only once to build the kernel matrix **G** [27]. After the Green's function matrix is available, the problem is reduced to the solution of equations (4), (6) or equivalent.

4. Seismic cycle imaging and modeling

Based on the elastic rebound theory formulated by Harry Fielding Reid in 1910, after the 1906 San Francisco earthquake, a seismic cycle is formed by a slow accumulation of stress and deformation, as consequence of the forces acting from the underlying Earth mantle, followed by an impulsive release of stress and energy when the internal strength is exceeded such that the brittle crust breaks (Figure 4). These two phases, which we refer to as inter- and co-seismic, are completed by a post-seismic phase, where different phenomena may induce further deformations during a short- to mid-term period after the earthquake (Figure 5).

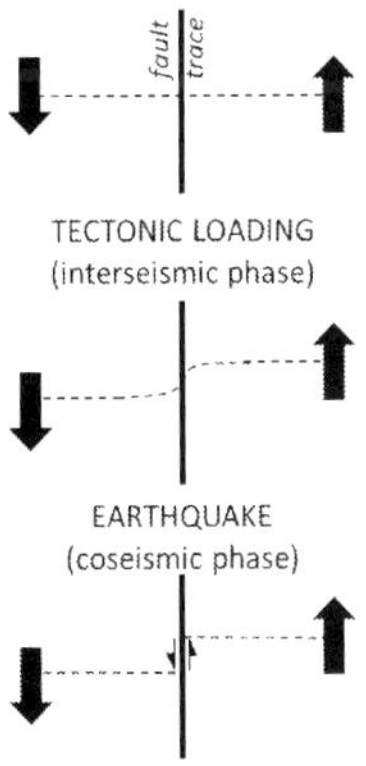

Figure 4. Simplified description of the elastic rebound theory: black arrows describe the steady interseismic forces acting on a locked fault until its sudden failure.

The co-, post- and inter-seismic phases have completely different characteristics in terms of duration and crust behavior. During the inter-seismic phase, faults are locked on the upper crust and the underlying forces act to deform the surface with relatively constant rates of few millimeters per year over large areas. As soon as the accumulated stress exceeds the locking frictional forces, the crust cracks and part of the deformation around the faulted area is elastically and permanently recovered in few seconds. Soon after the earthquake, for a period lasting from few minutes to some years, a further deformation occurs as a consequence of the sudden co-seismic stress release. Deformation rates roughly follow a decreasing exponential law (Figure 5) and can be explained in terms of one or a combination of the following factors: residual dislocation on the ruptured fault (after-slip models), viscoelastic relaxation of the lower crust driven by the coseismic stress change, poro-elastic rebound due to the migration of fluids in the crust ([28] and references therein).

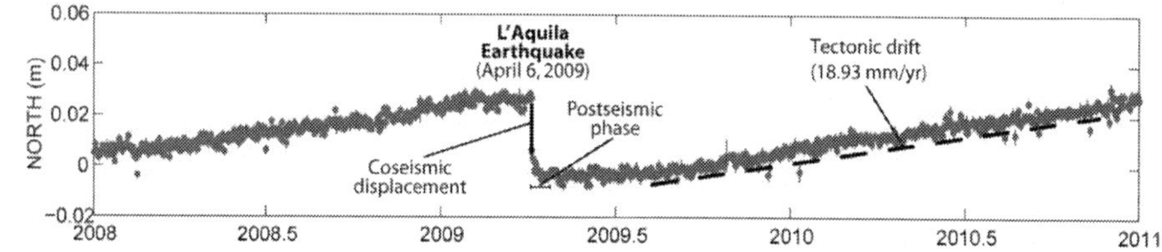

Figure 5. CGPS data related to the 2009 L'Aquila earthquake, showing the steady tectonic drift, the sudden coseismic displacement and the exponential post-seismic relaxation (by courtesy of Roberto Devoti, INGV)

This basic description allows to state how SAR derived data can play a crucial role in the understanding of a seismic cycle. In the co-seismic phase, the expected surface displacement, from tens of centimeters to several meters, and the nearly perfect elastic behavior due to the instantaneous deformation are perfect conditions to model the standard two-pass interferometry with the Okada solutions. [30][31][32] showed that since the 1992 Landers earthquake, the onshore deformation for all the significant earthquakes worldwide (M > 5.5) have been imaged with the standard InSAR interferometry and most of them have been modeled with the Okada model. A common problem found in this approach is that, in general, coseismic interferograms contain a contribution from the post-seismic deformation that can affect some fault parameters [31]. Isolating the co-seismic signal may not be straightforward, but the introduction of continuous measurements, such as CGPS, may be helpful in the post-seismic contribution removal.

During the inter-seismic phase, steady deformation rates are often assumed; however, expected values are of the order of some millimeters per year, so they can be hardly detected with two-pass interferometry: the time needed to accumulate a signal above ordinary InSAR artifacts would be too long to preserve the phase coherence; for this reason, time-series techniques like PS [33], SBAS [34], or image stacking are indicated to get mean velocity maps with millimetric accuracy [35], provided that a consistent number of images is available [36] [37]. Elastic models have also been used to fit inter-seismic data [38][39][40], however its use must be carefully considered because the assumption of elasticity can lack of realism.

Another important aspect is the apparent similarity between long-wavelength tectonic signals spreading through a whole SAR image frame and the orbital artifacts [41]. In this context, the use of GPS, unaffected by such artifacts, to constrain the velocity maps can be very effective [36][41].

Post-seismic relaxation has intensity and duration strongly dependent on the earthquake magnitude; short intervals can be easily encompassed between two SAR acquisition [42][43], often cumulated with the co-seismic effects. In this case the two contributions cannot be distinguished unless external continuous observations, as GPS, are introduced [44]. For large earthquakes, the post-seismic effects can last from months to years, and the InSAR time-series approach can be effectively used to describe the crustal displacement time evolution [28]. While the observed displacement is interpreted in terms of after-slip over the seismogenic fault, the Okada solutions can be used to model the signal [45]. However, for long-term deformation, visco-elastic models should be used [42].

Regardless from the seismic cycle phase, deformation modeling with the Okada solution is generally subdivided into two steps: a first non-linear inversion to retrieve all the unconstrained source parameters, followed by a linear inversion to get the distribution of the dislocation over the fault(s). The only measure to adopt before running the linear inversion is the widening of the fault plane obtained via non-linear inversion: the latter represents only a mean source with a mean dislocation value, therefore the fault plane must be enlarged to let the slip vanish to zero.

While most of the signal is already reproduced with uniform slip sources, high frequency spatial fluctuations are recovered adopting the Okada-derived distributed slip sources (Figure 6), solved with the linear system (4).

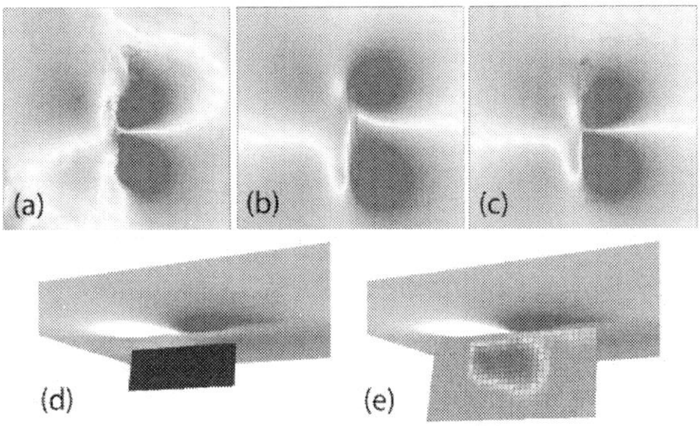

Figure 6. Coseismic displacement field, for the 2003 Bam (Iran) earthquake, retrieved with InSAR (a), modeled with a uniform slip Okada source (b) and with a distributed slip, Okada source (c). Models for (b) and (c) are shown in (d) and (e), respectively.

A last remark is about the possibility of obtaining a more realistic source by relaxing the condition of a flat half-space. The possibility of accounting for the topography in the overall system setup will be described in the next section, where this aspect plays an important role.

5. Volcanic modeling

The volcanic activity monitoring is conditioned by our ability to describe and understand the eruption cycle. Several steps can be identified in this cycle: magma generation, melting, storage and ascent, crustal assimilation, degassing, crystallization and surface eruption, not all of which can necessarily occur. In any case, evidences of the incoming eruption can be noticed only late in the cycle and InSAR data are crucial to provide information for the hazard mitigation, even for not erupting stages [46].

In the case of volcanic phenomena, several factors contribute to make the assumption of elasticity less reliable than the co-seismic case; magma intrusion starts below the seismogenic crust, thus involving ductile, high temperature layers able to deform aseismically. Furthermore, inflating and deflating phenomena involve times long enough to activate a visco-elastic behavior. Lastly, the half-space approximation is debatable as well, because of the inevitable significant topography of the investigated areas. Despite this limitations, the aforementioned elastic models have been largely applied in volcanology, with the aim of reproducing almost all surface deformations detected with InSAR [47][3][48][4][49][50].

Magma chamber inflation or deflation is generally modeled with the simple Mogi source, due to the ease of its implementation compared with its effectiveness [47][51]. For magma intrusion in vertical (dykes) or horizontal (sills) cracks, distributed opening sources based on the Okada model are instead adopted [48][50].

Sometimes, complex patterns revealed by InSAR suggest the implementation of a multiple-source system, where also seismic sources can have a role, as shown by the 2005 Afar dyking phase described in [52] (Figure 7). In such contexts, the double step non-linear/linear inversions is adopted to first constrain the source geometries, then to retrieve the slip and opening distributions.

For the long-term crustal deformation, time-series techniques have shown their effectiveness to describe the surface change, even when repeated inflation-deflation cycles are present [53]. In this case, modeling can be carried out by fixing the non-linear source parameters and then fitting the time dependent signal by only varying the linear parameter, i.e. the volume or pressure change.

A further remark, in this context, is about the important role played by topography, since strong elevation variations are expected in the investigated areas; for Mt. Etna, for instance, total relief difference is over 3000 m. To mitigate the assumption of flat half-space, topographic corrections can be applied to the analytical models. This correction consists in the calculation of the source depth not from the zero level of the free surface, but adding the real elevation of the point for which the predicted displacement is being calculated. Such compensation has

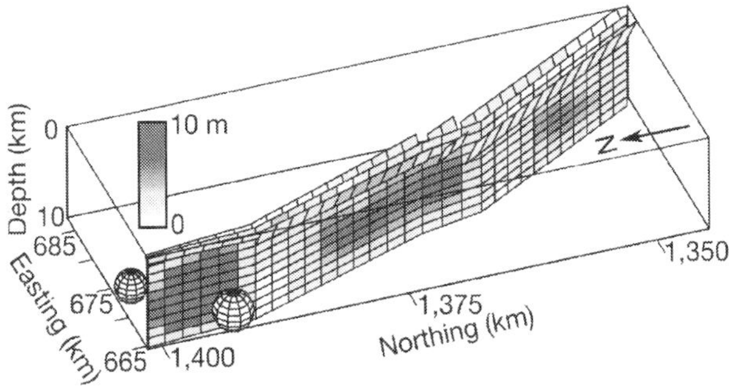

Figure 7. The complex system of sources (magma chambers, dykes and faults) used to model the Afar dyking phase (from [52])

been compared with finite element models, providing a satisfactory improvement on the modeled data [54][55], as shown in Figure 8.

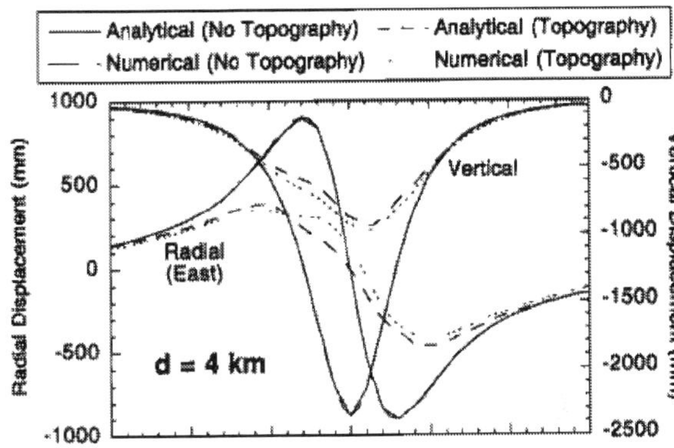

Figure 8. Comparison between displacement profiles from analytical and numerical models, with and without topographic corrections (from [54])

Finally, the frequent presence of stratified atmosphere, altering the interferogram with fringes due to topography-related radar delays, can be discriminated by comparing independent interferograms, assuming weather conditions uncorrelated in time.

6. Data downsampling

Since displacement maps derived from InSAR processing may contain millions of valid pixels, with an high degree of spatial correlation, a way to reduce the data must be adopted in any inversion strategy. Several criteria have been proposed in literature, among which we recall here three different approaches: Quadtree decomposition [56][57], resolution-based [69] and regular mesh [58][59].

A general rule to state which is the best method does not exist, though the Quadtree algorithm is the most used. It is a decomposition algorithm aimed at preserving the "amount of information" in the image, and is in general based on the spatial gradient of the signal; it follows that areas with higher displacement values are sampled at higher spatial frequencies (Figure 9b).

The resolution-based algorithm proposed by [69] is driven by an already known source; this allows to calculate the data resolution matrix [22] used to define where surface data must be sampled to constrain the linear source parameters.

Finally, regular sampling is also widely adopted, since its ease of implementation and its effectiveness in imposing a sampling density independent from displacement values. In fact, the InSAR data resolving power, i.e. the maximum detail level achievable on a source, strongly depends on the location of the observed points, as shown in [60] and not on the displacement field itself. The sampling can be manually customized by defining areas with different sampling density; this also allows to have a good control on the number of observed data to handle in the inversion (Figure 9b).

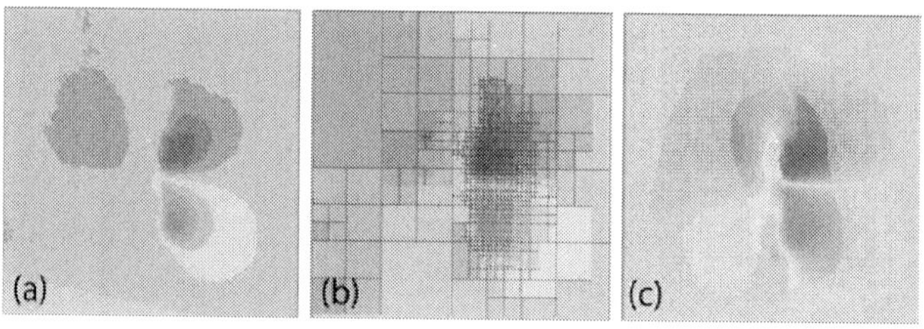

Figure 9. The 2003 Bam (Iran) earthquake displacement field (a), downsampled with the Quadtree algorithm (b) and a regular mesh with variable posting areas (c).

7. Uncertainty and trade-offs

InSAR measurements are always affected by different sources of uncertainty, as shown by the ample literature on this topic (see [23] and references therein). Here we discuss the strategies generally adopted to propagate the data uncertainty to the source parameters in the non-linear and linear inversions.

For convenience, linear inversion is firstly analyzed, where ordinary rules for the error propagation can be applied, as:

http://dx.doi.org/10.5772/57479

$$\sum_{m} = \mathbf{G}^{-g} \sum_{d} \mathbf{G}^{-gT} \quad (7)$$

where the full variance/covariance matrix $\mathbf{\Sigma}_d$ of the observed data is used as input to find the full variance/covariance matrix $\mathbf{\Sigma}_m$ of the model parameters. To build $\mathbf{\Sigma}_d$, the power spectra analysis of a displacement map containing only noise and artefacts is used. From this analysis, the covariance vs. distance scatter plot, fitted by means of some simple *ad hoc* function (Figure 10), as described in [23], is retrieved. One of the simplest way to express an InSAR covariogram is the exponential function of the type

$$C(r) = \begin{cases} \sigma^2 \text{for } r = 0 \\ C_0 e^{-\alpha r} \text{for } r > 0 \end{cases} \quad (8)$$

where C_0 is the covariance at zero distance, generally lower than the overall σ^2 variance, and α controls the distance at which data can be considered uncorrelated. More sophisticated functions can be found in literature, accounting also for a possible spatial anti-correlation [61].

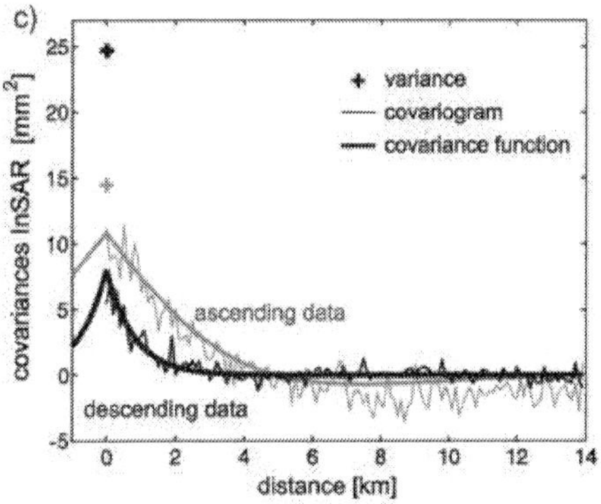

Figure 10. Empirical covariance function (from [61])

The full variance/covariance matrix $\mathbf{\Sigma}_d$ can be obtained by setting every off-diagonal position $\sigma_{i,j}$ to the covariance value obtained with (8), setting $r_{i,j}$, as distance between the *i*-th and the j-th point; diagonal values are set equal to σ^2. After that, equation (7) can be applied.

In the non-linear case, a formal expression of the uncertainty propagation is difficult to obtain and an empirical approach is commonly used: for tens or hundreds of times synthetic noise datasets $\mathbf{d}_{noise}$ are generated, added to the observed data $\mathbf{d}_{obs}$, then the inversion is performed and the results collected. To construct the $\mathbf{d}_{noise}$ dataset, a Cholesky decomposition of $\mathbf{\Sigma}_d$ must be preliminary carried out, such that

$$\sum_{d} = \mathbf{LL}^{T} \tag{9}$$

The **L** matrix is then multiplied by a vector $\mathbf{d}_{Gauss}$ of uncorrelated Gaussian noise

$$\mathbf{d}_{noise} = \mathbf{Ld}_{Gauss} \tag{10}$$

to get the synthetic noise dataset $\mathbf{d}_{noise}$ characterized by the same InSAR covariogram, as explained in detail in [62]. After the inversions have been carried out, the results can be efficiently visualized as a grid of scatter plots and histograms, describing trade-offs between coupled parameters and single parameter uncertainty (Figure 11).

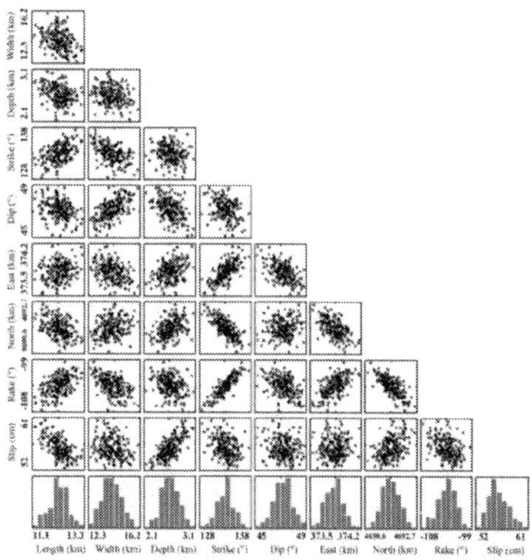

Figure 11. Uncertainty (red histograms) and trade-offs between parameters (from [59])

8. Interactions between sources

Models derived from InSAR data give important hints for the hazard assessment, as discussed later in this chapter, and in this respect an increasingly considered aspect is the way sources interact with each other. An earthquake occurs when the internal strength is exceeded by the surrounding stress, loaded during the interseismic phase. We generally do not know the absolute stress value for a given crust volume, primarily because the loading phase spans through centuries or millennia. On the contrary, we can quantitatively calculate the stress variation induced by a fault dislocation.

The stress released during a seismic event perturbs an area extending far from the source itself, where surrounding locked faults are likely to be present. The rearranged stress conditions following an earthquake may increase or decrease the current (unknown) shear stress level acting on a fault surface: we can therefore calculate if such a variation moved the receiver source closer to its failure.

The analytical solutions proposed in [63] allow to calculate the internal deformations induced by a dislocation over a fault plane; internal deformation can be then easily converted into stress variations, using the Coulomb Failure Function variation (ΔCFF), described in [64], and defined as

$$\Delta\mathrm{CFF} = \Delta\tau + \mu\cdot(\Delta\sigma_n + \Delta p) \tag{11}$$

where $\Delta\tau$ is the shear stress change over the fault, calculated for a given slip direction, μ is the friction coefficient, $\Delta\sigma_n$ is the normal stress variation and Δp the pore pressure change [65]. The latter is usually unknown and thus neglected in stress transfer calculation.

Despite its apparent simplicity, the application of the stress change analysis must be considered with care, because of the intrinsic unknown of the already existing background stress, the mislocation of possible receiver sources, the uncertainty of the triggering source parameters (derived from inversion) and the presence of spatial inhomogeneities. They all introduce a high level of uncertainty, as discussed in [66] and references therein.

The stress variation analysis has been successfully adopted also in a volcanic context, to study the interaction among sources of different nature: magma chambers, dykes, faults [67][68][50]. Though these analyses are always conducted *a posteriori*, after the event occurrence, they contribute to support the stress-transfer reliability.

However the warning issued by [64] is still holding: "much work remains before we can understand the complete story of how earthquakes work". The CFF analysis is a powerful tool to describe the interaction between sources, but it is still not adequate to deterministically state how close to the failure a receiver source has been pushed by any triggering event.

9. From science to operational risk management

The use of SAR data and techniques has greatly stimulated the progress of the Solid Earth science. Globally, over three hundreds deformation fields related to the earthquake cycle (including inter-, co-, and post-seismic deformation), and over a thousand volcanic edifices, have been studied using SAR data since the beginning of the "InSAR era" [30] [31] [32][70][71]. Important new knowledge has been acquired on processes such as fault dislocation, fault segmentation, magma and gas migration, volcanic spreading, stress transfer, strain accumulation and release, poro-elastic diffusion, and visco-elastic relaxation. Better descriptions of the seismic and volcanic cycles are today available thanks to these studies.

Soon after InSAR started to demonstrate its potential to sustain new scientific developments in crustal deformation studies, practitioners started to investigate the possible use of this new information in the risk management activities [72][73]. It was rapidly realized that the new geodetic "imaging" capabilities provided by satellite SAR sensors could strongly support two main components of the disaster risk management, namely the assessment/prevention and the response components.

However, the space and ground segments of the first satellite SAR systems (i.e. ERS, ENVISAT, JERS, ALOS) were targeted mainly to scientific use and, while they could provide delayed data to carry out pre- or post-disaster scientific analyses eventually supporting the hazard assessment component [74], they lacked the near real time capabilities needed for use in the response phase. Even after the launch of the first commercial SAR satellite in 1995 (Radarsat-1) the use of SAR data in disaster risk management did not flourish, due to the lack of a constant repeat pass, global coverage, and to the high costs required to maintain updated archives.

This situation will change radically in 2014, when the European Space Agency Sentinel-1 operational satellite (and its companion Sentinel-1 B in 2015) will start to provide a full InSAR coverage of nearly all land areas with pre-defined, constant repeat pass [75]. The Sentinel-1 data will be delivered in near real time to selected service providers to generate support products during natural disasters or emergencies. Sentinel-1 minimum revisit time will be 12 days, improving to 6 days after the launch of the second satellite, and the mission continuity will be guaranteed for many years [76].

While the "flat" data flow rate and the "full and open data access" policy of the Sentinels will certainly represent a breakthrough for the use of remote sensing data for operational risk management, other SAR satellite constellations have already started to demonstrate the potential for operational emergency response. The Italian Space Agency COSMO-SkyMed four-satellite constellation was in fact expressly developed to support the monitoring and assessment of natural and anthropogenic disasters [77], although it is a dual-use mission, also employed for defense purposes. This constellation presently allows much shorter revisit times than possible with Sentinel-1, down to 1 day depending on satellite.

In the following sections, with reference to Table 1, we will show examples and examine the operational capabilities of InSAR data, and of geophysical models they can constrain, to support activities in the two main components of seismic risk management: the risk assessment/prevention, and the response to a seismic crisis.

9.1. SAR-derived products to support seismic hazard assessment

Seismic Hazard Assessment (SHA) is the process of calculating, for a given area, the probability of the occurrence of a certain ground shaking level within a defined period of time. The typical SHA synthetic result for a region is a map showing the spatial variations of the horizontal Peak Ground Acceleration which have a probability of exceedance of 10% within a 50 or 75 year time frame. These maps (complemented by more detailed, site-specific seismic hazard curves) are generated using information on the existing seismic sources, their activity rates and maximum earthquake magnitude, and by estimating the relations between epicentral distance and ground motion for a given earthquake magnitude and fault type (attenuation relations).

InSAR data can provide important information regarding the earthquake source (Table 1). One field of analysis deals with the actual detection of faults by means of their geomorphological signature (e.g. linear scarps, triangular facets on mountain fronts, displaced terraces, drainage network offsets, etc.). This is a classical application of photo-interpretation techniques (structural and geomorphic), in which B&W SAR intensity images give a contribution comparable to optical images. The satellite image analysis can provide especially valuable support to the mapping of active faults and earthquake sources, in areas which are of difficult access and for which detailed geological data do not exist.

While intensity image analysis reveals the fault presence by investigating peculiar landforms created by surface deformation cumulated during geological times, multitemporal InSAR processing can provide a quantitative measurement of the ongoing deformation rates (and their spatial patterns) used to characterize the fault behavior. Many scientific results obtained in the last ten years [32] have contributed precious information for the parameterization of the seismic sources [79][80][81][82][83][84], but also for the definition of the present deformation rates in areas where multiple sources are present [37][85] [86][87][88][89], for the partitioning of strain among different faults [90][91], for the improvement of tectonic models in seismogenic areas [92][93][94].

Among the most important SAR-derived inter-seismic source parameters directly contributing to hazard estimates is the long-term slip rate. This is the yearly rate of slip which is modeled as occurring on the deep part of an active fault plane, below a given "locking depth". The latter is defined as the depth separating the upper brittle crust, where the fault plane is locked, from the lower visco-elastic crust where the fault is instead slowly creeping, as explained in Section 4. If the region tectonics is dominated by a single large fault, as often is the case in plate margin contexts, the inversion of InSAR ground velocity measurements can be used to estimate the inter-seismic slip rate at depth.

The fault creep below the locking depth is in turn related to the stress build up in the upper crust, stress which will be eventually released during the seismic dislocation. Considering an ideal earthquake cycle, and assuming that the friction along the fault plane does not vary much through time, each successive fault rupture should occur when the accumulated shear stress overcomes a similar level of fault strength. Thus, the knowledge of the inter-seismic slip rate and of the seismic history of the fault should allow to estimate a recurrence time for moderate to large earthquakes on a given fault. These recurrence interval estimates have in some cases large uncertainties (fault strength and slip rates are indeed not constant through time), nonetheless the inter-seismic slip rates obtained by geodetic data (mainly InSAR and GPS), integrated with paleoseismological and seismological data, are considered important parameters to support seismic hazard calculations [95][96][97][98]. However, while GNSS data have been more extensively used (see for instance the SHARE EC project, the Working Group on California Earthquake Probabilities, 2011; [97]; [99]) to

		User needs	SAR-derived supporting information	SAR data analysis techniques
RISK ASSESSMENT AND PREVENTION	Seismic hazard	Identification of active faults.	Structural maps.	Amplitude image analysis.
		Parameterization of activity rates.	Long-term ground displacement rates maps (inter-seismic ground velocity) at low spatial resolution. Fault models and long-term slip rates.	Time-series InSAR techniques as Persistent scatterers and Small Baseline. Non-linear inversion modeling.
		Definition of maximum magnitude earthquake.	Fault model geometry and kinematics. long-term slip rates.	Non-linear inversion modeling of deformation data.
	Induced hazard	Pre-event identification of gravitational slope deformations.	Geomorphological analysis. Long-term ground displacement rates at high spatial resolution.	Amplitude image interpretation and soil moisture analysis. Time series InSAR techniques as Persistent scatterers and Small Baseline. Geomechanical modeling
	Seismic vulnerability	Identification of structural weaknesses in man-made structures.	Long-term ground displacement rates at very high spatial resolution.	Time-series InSAR techniques as Persistent scatterers.
DISASTER RESPONSE	Loss estimation	Rapid (1-2 days) spatial assessment of damage to man-made structures	Maps of damage classes at the district scale. Maps of collapsed structures at the single building scale. Very high resolution co-seismic ground displacements.	Intensity-based change detection and classification. Coherence-based change detection.
	Environmental damage estimation	Rapid spatial assessment of environmental effects of earthquake: fault scarps, diffuse ground displacement, reactivated landslides, drainage reversal or interruption, soil liquefaction, sinkhole collapse, etc.	Co-seismic ground displacement maps at high spatial resolution. Geomorphological and structural analysis.	SAR Interferometry, pixel offset tracking, Multiple Aperture Interferometry, intensity image interpretation, coherence analysis.
	Event scenario	Rapid assessment of earthquake sources and possible evolution of the aftershock sequence.	Co-seismic ground displacement maps at medium-high spatial resolution. Geometrical and kinematic parameters of the earthquake sources. Stress increase on nearby faults. Post-seismic ground displacement for the next few months after the mainshock.	SAR Interferometry, pixel offset tracking, Multiple Aperture Interferometry. Time-series InSAR techniques. Non-linear and linear inversion modeling of deformation data. Coulomb stress analysis.

Table 1. Uses of SAR-derived information to support various components and actions of seismic risk management

http://dx.doi.org/10.5772/57479

constrain the occurrence models for operational hazard assessment, InSAR data have so far been employed only marginally.

The reason why the inversion modeling of InSAR ground velocities is still not operationally used for the estimation of inter-seismic slip rates and other important fault parameters (as maximum magnitude earthquake), has to do with deficiencies in the data and with gaps in the underlying science. The inadequacy of present SAR system to provide useful data to measure small ground velocities over large spatial wavelengths has been addressed above, and it is expected to be partially resolved by the Sentinel-1 satellites. The scientific issues concern instead the incomplete knowledge of the processes driving the stress accumulation and release in the crust, which imply that the results of the models used to estimate active fault parameters through the inversion of inter-seismic ground velocities are are dependent on scientific judgment. In SHA, the uncertainties arising from the incomplete knowledge of the earthquake processes are addressed by ensuring that the generation of seismic hazard maps is based on discussions within the scientific community, aiming at developing the widest possible consensus on input data, methods, and practices [100]. We expect that in less than a decade the continuous InSAR data flow from the Sentinel-1 operational mission will have promoted the development of new inter-seismic source models, and better procedures for their significance and uncertainty assessment, effectively spreading the InSAR data use in SHA.

Two further important activities in seismic risk assessment/prevention in which SAR data can give a valuable contribution are (Table 1): 1) the identification of gravitational slope deformations which can undergo reactivation or catastrophic collapse during seismic shaking, and 2) the evaluation of the vulnerability level of buildings or infrastructures.

In the first case SAR imagery is used for two different purposes. The classical geomorphic photo-interpretation carried out also on optical images to detect the landforms indicating gravitational slope deformations, can be enhanced by the capacity of SAR intensity images to provide estimates of soil moisture [101]. In general these techniques are used to provide a spatial mapping of the landslide, with little information on its activity [102]. Another use relies instead on high resolution multi-temporal InSAR data to investigate the ongoing rates of gravitational mass movements in high seismic risk areas, which may be used to evaluate more specific hazard levels due to seismically-triggered landslide collapse [103]. The additional hazards induced by landslides reactivated by seismic shaking is very important especially in areas with high topographic relief: in the Wenchuan, 2008 earthquake over 20,000 casualties were attributed to landslide collapse [104].

Finally, a new field of use for InSAR-derived, high resolution ground deformation maps in the practice of seismic risk prevention, is to support a detailed vulnerability analysis in areas affected also by other deformation phenomena with high spatial frequencies (i.e. subsidence, sinkholes). If these phenomena occur in high seismic risk areas, the classification of the severity of the static deformation of man-made structures can allow a more accurate assessment of their resilience to future dynamic actions caused by seismic shaking.

9.2. SAR-derived products to support earthquake emergency response

The response phase of seismic risk management concerns all activities needed to promptly respond to the effects of a damaging earthquake. It can be divided in two main, temporally linked sub-phases, the Immediate response and the Sustained response [105].

During the Immediate Response phase, which usually lasts from few to several days, depending on the dimensions of the disaster, the main priorities are search and rescue actions and the emplacement of immediate preliminary measures to save lives and protect the population

Chapter 10

Dikes Stability Monitoring Versus Sinkholes and Subsidence, Dead Sea Region, Jordan

Damien Closson and Najib Abou Karaki

Additional information is available at the end of the chapter

http://dx.doi.org/10.5772/57277

1. Introduction

This chapter features how radar remote sensing can improve an embankment dam security project, from the planning level, to the feasibility study, the building phase, and the operational stage. The Dead Sea region provides a meaningful test bed owing to dozens kilometers of earthen dikes built over soft and very soft sediments affected by strong subsidence and sinkholes. The first section describes the context in which the collapses and subsidence proliferation have taken place. The attention is drawn over the Lisan Peninsula, Jordan, where two salt evaporation ponds encompassed by embankment dams have been built over its western margin. Then, the geological framework and the chronology of the most important damages are presented. Representative results obtained from radar differential interferometry techniques applied to ERS, Envisat, ALOS and Cosmo-SkyMed images are shown and described. Interferograms and ground displacement maps complement the work of the security engineers in providing the spatial extension and the dynamics of the geo-hazards they are dealing with.

2. The Dead Sea, the sinkholes and subsidence

The Dead Sea is sitting in a pull-apart basin of the Jordan-Dead Sea Transform fault zone. It is the lowest emerged place on Earth. Early January 2014, the water level was-427.82 m (Hydrological Service of Israel). But fifty years ago, it was around-395 m. The difference of 32 m results in the over pumping of the main tributaries, such as the Jordan river, and in the siphoning of the brine itself by the Dead Sea Works, in Israel, and the Arab Potash Company, in Jordan. The decline is constantly accelerating. It exceeds one meter per year in 2014. In the 1960s, the

INTECH
open science | open minds

terminal lake was about 80 by 15 km. Since then, one third of its original surface disappeared, leading to major changes in the hydrogeological setting and in the landscape. Before the 1980s, the Dead Sea was made up of two sub-basins (depth:-730 m and-402 m respectively) connected by the Lynch Strait and separated by the Lisan Peninsula (Fig. 1). Ever since, the deepest basin remains.

The salinity of the Dead Sea is ten times higher than the ocean water one. This characteristic makes it an attractive place for tourism and industrial activities. Two profitable mineral companies developed their plants over the shallow southern sub-basin from the mid-1960s to the early 1980s. Their production and benefits depend on the evaporation ponds' area. Fig. 1 shows the landscape of the southern Dead Sea in February 2000, when the expansion of the solar evaporation systems was at the maximum. One month later, SP-0B (Fig. 1; 7) was destroyed and one year later SP-0A (Fig. 1; 8) was emptied for dike repair.

Salty minerals soak the environment. For millennia, arid climatic conditions allowed salt crystals to grow up in the open air. Holocene sediments of the coastal zone are thus prone to dissolution when in contact with unsaturated water with respect to salt. Concomitantly to the Dead Sea level decline, an hydraulic gradient appeared with the surrounding water tables. In consequence, an increasing amount of groundwater has been drained into the shrinking lake to compensate for the lowering [1]. Gradually, the interface configuration and the equilibrium state between the hyper-saline surface body and the adjacent fresh groundwater body receiving recharge modified. The areas underlying the coastal aquifers formerly occupied by the Dead Sea water became flushed and occupied by fresh water. The latter became salinized due to the residuals of Dead Sea water in the aquifer matrix. Dissolution of covered salty deposits (Lisan formation) caused subsidence and collapses along the shorelines in the form of sinkholes, tens of meters in diameter and depth [2]. Each year, this process is causing more damages.

In 2014, the cumulated number of sinkholes recorded since the 1980s ranges between 3000 and 4000. They are found from some meters below the lake level up to several kilometers landward. At least since the 1960s, sinkholes appeared around the former southern basin, maximum 32 km away from the present-day Dead Sea shoreline (Fig. 1, 10). They underline the great fragility of the past and present coastal zones. The precise moment and location of the very first ground collapses are unknown. Eli Raz (personal communication) mentioned that, since the late 1970's, sinkholes have been known from Ein Gedi southward (Fig. 1, 1). Between 1978 and 1981, the southern basin and the Lynch Strait emerged gradually. Itamar and Reizmann [3] identified sinkholes over aerial photographs dating back to 1982 along the eastern and western margins of the just-emerged Lynch Strait (Fig. 1, 6). Abelson et al. [4] computed the first graph of the cumulated number of sinkholes over time for the western coast. The curve shows a gradual increase from the 1980s, a first inflexion occurred between 1997 and 2000, and then a drastic increase from 2004.

Retrospectively, the oldest and most damaged places are located inside an area about 15 by 25 km, bounded by the resorts of Ein Gedi, Ein Boqeq, Al Mazra'a and the Wadi Shuqeiq delta [5] (Fig. 1, 1-4). At regional scale, from a tectonic point of view, this zone is characterized by a coalescence of faults [6], and is centered over the Lisan diapir, which is the largest salt dome

Figure 1. Dead Sea southern basin covered by salt evaporation ponds of the Arab Potash Company (East, Jordan) and Dead Sea Works (West, Israel). 1) Ein Gedi; 2) Wadi Shuqeiq delta; 3) Ein Boqeq; 4) Al Mazra'a; 5) Lisan Peninsula; 6) Lynch Strait (dried up); 7) $38 M SP-0B; 8) $32 M SP-0A; 9) SP-01; 10) southernmost sinkhole site; 11) Wadi Araba (braided river) flowing toward the northern Dead Sea basin; 12) Arab Potash Company brine intake station; 13) Truce line flood channel (borderline). Background: picture STS099-751-26 (Feb. 2000), Image Science and Analysis Laboratory, NASA-Johnson Space Center. "The Gateway to Astronaut Photography of Earth."

in the Dead Sea pull-apart basin [7]. Inside this polygon, two prominent sinkholes lineaments located at Ghor Al Haditha and the Lynch Strait, about 6 and 2 km long respectively, highlight the relationship between tectonics and ground collapses distribution. In [8], an agreement was found between their azimuths, and the one of the Jordan-Dead Sea fault system as illustrated by the focal mechanism of the April 23rd 1979 (mb=5.1) earthquake: N (20 ± 5)° E. Besides, both lineaments gather a significant part of the southern Dead Sea collapses and they are also the closest clusters to the April 23rd 1979 epicenter. The 6 km-long alignment is the longest in the whole Dead Sea area. On the western side, a similar geometrical agreement had been found between the sinkholes distribution and the main structural directions [9]. Besides, geophysical studies [e.g. 10] have shown that sinkholes appear when a particular layer of halite, deposited 10000 years ago [11], is present several decameters below the ground level. In 2013, Ezersky and Frumkin [12] synthesized most of the available sub-surface and surface observations. They have shown that a large number of sinkholes sites occurred where both the edge of this layer of halite and underground discontinuities (faults or fractures acting as preferential channel-ways) are simultaneously present.

The 135 km long coastline is variously affected by sinkholes clusters. Most of them are located along the western part. In terms of damages, the eastern part-especially the industrial zone built over the Lisan Peninsula-is the most concerned. The Arab Potash Company lost a $38 M production unit ("SP-0B") in 2000 [13] (Fig. 1, 7). Another saltpan "SP-0A" (Fig. 1, 8) is constantly threatened, necessitating lots of costly repairs. Hence, to illustrate how radar remote sensing can improve an embankment dam project, the western margin of the Lisan Peninsula where SP-0A had been built is investigated with radar/optical/thermal remote sensing techniques applied before, during, and after the construction.

3. Geological setting of the Lisan Peninsula

The Lisan Peninsula (Fig. 1, 5) is a massive stack of late Pleistocene uplifted salt and marls layers. The Lisan Formation is characterized by laminated, biogenic carbonates, and siliciclastic sediments. Investigations revealed several transgressive depositional cycles, all terminating with massive gypsum precipitation [14].

About 120 m of these deposits cap the Lisan salt diapir probably formed from an anticline of an en-echelon fold train [15]. The main structure is a 9 by 6 km dome, N-S elongated. Two faults, striking N-S and SW-NE, bound the Lisan Peninsula in the E and NW respectively. Two secondary domes exist in the southern part [16]. Numerous lineaments related to faults directions are observed. They are represented by straight or gently curved wadies. Their direction and length vary considerably [17]. For centuries, an original salt karst developed [18] owing to the meteoric water percolation and to the base level fluctuations resulting either from the rise of the salt diapir, or the strike-slip and vertical movements related to the pull-apart basin, or the variations of the Dead Sea water level.

3.1. Lisan Peninsula foreshore

From the early 1970s, a wave-cut platform surrounding the Lisan Peninsula appeared progressively. Nowadays, it extends over more than three kilometers and dips very gently with a typical slope of 1: 250. Prior to the building of SP-0A and SP-0B, the surface was covered with a thin salt crust forming rigid polygonal plates typically 0.3 m thick and 1-2 m across. Small pressure ridges were apparent between many of the plates [19, 20]. These pop-up structures resulted of the hydration of anhydrite ($CaSO_4$) to gypsum ($CaSO_4{:}2H_2O$) due to under-saturated groundwater circulation in the upper horizons.

Boreholes drilled close to the brine intake station (Fig. 1, 12) of the Arab Potash Company indicated that the platform consisted of disturbed clayey silts with bands of carbonate sand and some gravel at depth [21]. Down to 13 m, the strata were very soft. Then, they became soft, and firm from 16.5 m. In general, zones of unlaminated calcareous clayey silt were interbedded with zones of gray clayey silt (calcite) thinly laminated with white silt (aragonite), typical of the Lisan formation. Laminations were mostly disturbed and discontinuous with orientations varying from horizontal, sub-horizontal to sub-vertical in places, or presented as isolated pockets within the unlaminated clayey silt and sand strata. This disturbed nature of the soil might indicate previous liquefaction of the soil layers, possibly caused by earthquake loading.

Between 17.5 m and 18.0 m, a 4 mm vertical displacement in the horizontal laminations was observed, resulting from a shear discontinuity. Around various depths, zones of thin cemented aragonite sheets were encountered. It was presume that these zones might have higher horizontal permeabilities than the soil above and below.

Calcareous sand and gravel strata were detected at 6.5 m, 19.5 m and 21.5 m. Although very silty in places, they were suspected to serve as higher permeability drainage horizons for the formation aiding consolidation. A massive halite layer was found at 17-19 m and 25-50 m depth [19, 20]. This layer was supposed to be at the origin of the sinkholes that appeared in the early 1990s some height kilometers south of the intake station. Similar observations have been done all along the Dead Sea coast [12].

3.2. Hydrogeological conditions before SP-0A construction

Classification of Landsat 4 and 5 satellite images (1984-1992) collected before the setting up of SP-0A (1996-1997) allowed the delineation of local areas characterized by a specific reflectance (Fig. 2). Taking into account the context of the rapid emergence of the wave cut platform, the difference between classes is mainly related to the moisture variations in the superficial horizons. Two extreme classes (100% water=Dead Sea, river; 0% water=former Lisan Peninsula) allow a pertinent classification of the moisture content over the platform and in the Lynch Strait.

Fig. 2 describes the Lisan foreshore as a seepage (discharge) zone. The water that circulate below the platform comes from the lowering of the Gijben-Herzberg water lens located below the Lisan Peninsula [18], and from the percolation of the confined brine in saltpan SP-01, south of the future SP-0A (Fig. 2: see remarkable seepages). A thin lens of brackish water exists below the Peninsula (Elia Salameh, personal communication). It is fed from the Mazra'a graben (Fig 1, 4) with water coming from the Dhira basin, in connection with the Moab plateau, where annual precipitations range from 250 to 350 mm. For centuries, the thin lens was in hydrostatic equilibrium with the Dead Sea water body which extended below the Peninsula. This setting is similar to that of islands' freshwater lens in equilibrium with the ocean water [1, 2]. Two main differences exist. The first is the angle of the fresh/saline interface. Because of the Dead Sea salinity, the angle is about ten times less than the one existing with the ocean water. The second is the physical connection with a remote recharge area (less than 6 km). The temporal stability of the fresh/saline interface is attested by the development of sub-parallel caves (Fig. 3) whose floors are at the elevation of-390/-395 m. Caves are visible all along the Lisan cliff in contact with SP-0A. They extend over dozens of meters following lineaments ESE-WNW oriented.

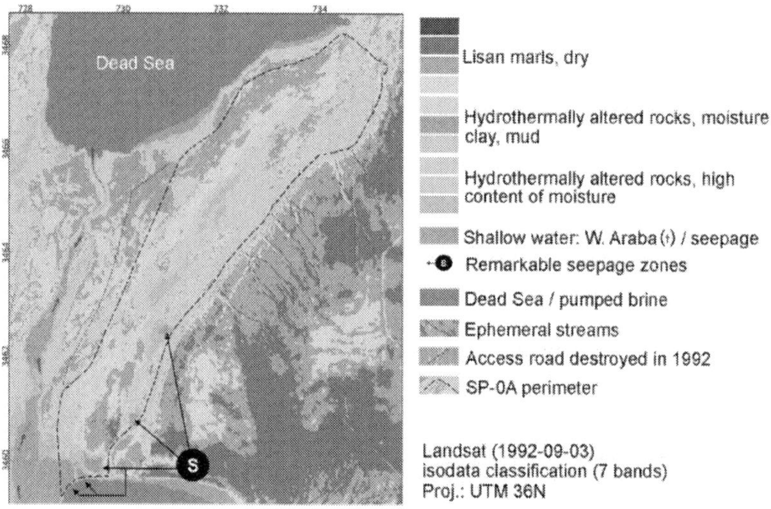

Figure 2. Isodata classification of a Landsat image acquired prior (1992-09-03) to the setting up of SP-0A (1997-1998).

From the 1960s, the thin water lens moved down to accommodate the lowering of the base level. As a consequence, the wave-cut platform turned progressively into a discharge area. When the Lynch Strait dried up, the elevation of the lens was controlled by the level of the Dead Sea in the northern Lisan, by the elevation of Wadi Araba crossing the former Strait in the southwestern part, and by the brine level occupying the former shallow sub-basin.

Figure 3. Picture taken from dike 18 (SP-0A) toward the east (Lisan Peninsula) and showing SP-0A filled (May 10th 2007) and cave entrances at the place where the level of the Dead Sea was in the 1960s (around-390 m). In the background are the sub-horizontal salty marls of the Lisan Formation uplifted by a massive salt diapir more than 120 m below the surface (Fig. 2; grey colors).

The easily erodible sediments of the Lynch Strait allowed a rapid entrenchment of Wadi Araba, and, consequently, a rapid drop down of the surrounding brackish water table. In 1982, the Arab Potash Company completed its solar evaporation pond system in the southern part of the Dead Sea (Fig 1; 9). The peripheral dike was about 29 km long with its crest at-395.0 m and an average height of 5 m. Reservoir SP-01-bounding the southern part of the Lisan Peninsula-was filled with Dead Sea brine in 1983. The operating brine level was-398.75 m, similar to the Dead Sea level in the early 1970s. Hence, the hydrological context was "reset" along the southern margin but not along the western and northern sides. The concentration of remarkable seepages zones (Fig. 2) in the southern part of SP-0A can be explained by the fact that the brine in SP-01 seep out through the dike and owing to numerous underground discontinuities caused by the rise of the Lisan diapir [17].

4. Method used: DInSAR with short perpendicular baseline and SBAS

4.1. DInSAR

In differential interferometry based on Synthetic Aperture Radar (SAR) imagery, the aim is to measure the differential fringe component to detect local displacements along the line of sight. After the introduction of the concept at the end of the 1980s [22] and the first demonstrations based on satellite data [23], the first dramatic examples [24] showed in the early 1990s the potential of this technique for large area measurement of small terrain deformations. Exploiting images acquired from about 700 km, with a resolution of about 20 meters, the coherent comparison of the radar backscattered signal phase allowed the measurements of displacements in the order of some fractions of the system wavelength (some cm in case of the satellite SAR systems of ALOS, Envisat and ERS).

With a precise and accurate knowledge of the orbital parameters-mainly the baseline components-orbital fringes can be adequately removed. Then, topographic and differential fringe components can only be separated if one of these components is known. Hence, a topographic phase reference is needed to be subtracted from the interferogram and to generate the differential one. This topographic phase reference may be obtained either from an external DEM or from another SAR pair known as free from any differential phase component.

For the selected pairs, differential interferometric processing was applied to derive unwrapped deformation phases using the following steps (Fig. 4 [25]): raw data processing and/or direct reading of Single Look Complex (SLC) data; co-registration of SLCs to common geometry; two-pass differential interferometry processing using an oversampled ASTER GDEM as height reference, slope adaptive common band filtering, and baseline refinement; finally, phase unwrapping and geocoded displacement map.

The evolution of the research in ground motions based on DInSAR and the larger availability of data and tools allowed in the 2000s to identify the strengths and the weaknesses of this approach, its potential accuracy and the path to follow to get into an operational exploitation phase. It was clear that, to improve the reliability and accuracy of the results, to distinguish between real displacements and effects due to other factors like atmospheric turbulences and temporal changes of the observed objects, and to add a temporal evolution component of the observed displacement, the approach should be extended to the analysis of several images acquired over a long time period over the same area (stacking techniques).

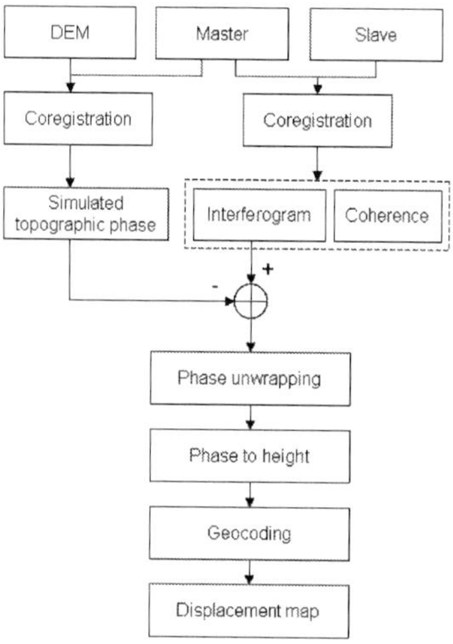

Figure 4. DInSAR flowchart [from 25]

4.2. SBAS

In the Dead Sea region, better results are obtained with large datasets of short baseline interferograms (see chapter "Mapping of Ground Deformations with Interferometric Stacking Techniques"). Hence, the basic idea for dikes stability monitoring is to derive deformation time series from a set of differential interferograms with "short" perpendicular baselines and time intervals keeping the level of phase gradients sufficiently low to be able to resolve the phase unwrapping stage. Using short spatial baselines optimizes the coherence and minimizes the topographic phase resulting from errors in the SRTM or ASTER GDEM height used as topographic references.

Stacking techniques have been used to get knowledge of the ground deformations from 1992 to 2010. ERS and Envisat images were stacked and processed to provide a single time series gathering all available information.

4.3. Radar image datasets

Several datasets have been processed to cover the period 1992-2012. The main one consisted of Envisat and ERS radar images (1992-2010): 40 Envisat ASAR images for the period 2003-2010, and 50 ERS AMI-SAR images acquired from 1992. Another set gathered 9 ALOS images

(2007-2010). The fourth set was made up of 11 Cosmo-SkyMed images acquired from December 2011 to June 2012. Table 1 summarizes the characteristics of the pairs used to show some relevant results of the investigations.

Fig. #	Sat.	Interferometric pairs		T.B.	P.B.	A.H.	DS level	Dike 18 of SP-0A living phases	
5	ERS	1992-06-11	1993-08-05	420	126	76	-407.4	Planned	... -1992
6	ERS	1995-07-29	1996-07-14	351	20	485	-409.6	Feasibility study	1993-1995
7	ERS	1995-12-17	1997-10-12	665	6	1485	-410.8	Building	1996-1997
		1997-12-21	1999-03-21	455	18	531	-412.0	Impounding and operation	1998-2002
8	Envisat	2004-06-27	2004-11-14	140	37	249	-417.2	Emptying and repairs	2003-2006
		2005-08-21	2006-03-19	210	16	571	-418.5		
9	ALOS	2008-04-01	2008-05-17	46	41	1551	-421.2	Impounding and operation	2007- ...
10	Cosmo-SkyMed	2011-12-14	2012-05-06	144	231	36	-426.0		

Table 1. Selection of interferometric pairs in the study of dike 18 living phases. T.B.=temporal baseline; P.B.=perpendicular baseline; A.H.=ambiguity height; DS level=level of the Dead Sea at the moment of the latest radar acquisition (e.g. 1993-08-05). During the period of observation, the lake level decreased by about 20 m, or one meter per year. The slope of the Lisan foreshore where SP-0A is located is 1:250.

Figures 5 to 10 are a selection of unwrapped and wrapped interferograms having in common the small perpendicular baseline leading to high ambiguity heights. It is noteworthy to mention that the area of interest is flat, without vegetation, and human activities confined to the top of the dike.

5. Results

To evaluate the usefulness of radar interferometry in dike projects, it is necessary to review and to synthesize the chronology of the hazardous events that have occurred since the 1990s until present over the Lisan foreshore. Data shown in Fig. 6 to 11 have been selected to highlight specific periods of time before the construction of dike 18, during the setting up, and during operation.

5.1. Chronology of the main events at the Arab Potash Company solar evaporation ponds system: SP-OA and SP-OB units

Figure 5 summarized the record of hazardous events for the different living stages of SP-0A and dike 18. Elements cited below come from direct observations, satellite images analysis, discussions with Arab Potash engineers, and bibliography (mainly Arab Potash Company unpublished reports, proceedings, and few articles).

1. In 1982, the Arab Potash Company completed its solar evaporation system in the dried-up southern Dead Sea basin. At the same period, the very first sinkholes appeared in three different places: on the eastern and western margins of the former Lynch Strait, and in Ghor Al Haditha [8].

2. During the 1980s, the yearly decline in the lake level discovered an abrasion platform around the Lisan Peninsula. Early 1990s, Arab Potash set up an expansion project consisting in the setting up of two major production units over the western and northern flanks of the Lisan: saltpans SP-0A and SP-0B, both encompassed by embankment dams over 10 meters high and numbered 18 for SP-0A, and 19-20 for SP-0B (Fig. 1; 7, 8).

3. On March 22nd 1991, a major flood occurred over a 24-hour period. It was caused by a short period of intense rainfall. The Dead Sea rose from-407.701 m on February 27 to-407.512 m, contrary to the long term trend of a steady decline. The dried-up Lynch Strait topography was markedly influenced by this flood. Some abandoned channels were suddenly reactivated either due to the rapid entrenchment of Wadi Araba or created during the emergence of the Lynch Strait. Subsurface water circulation was affected too.

4. In June 1991, the eruptions of Mount Pinatubo, Philippines, affected the climate of the whole planet for several years. Between 1992 and 1995, winters were particularly rainy, leading to a rise in the Dead Sea level. Winter 1991-92 was particularly remarkable; cold-air temperature in the Middle East was 3 to 4°C below average. The volume of rain provoked an addition of about 1.5 10^9 m^3 freshwater to the Dead Sea and an increase of two meters in the lake level [19, 20].

5. In October 1992, a wide sinkhole suddenly appeared in an access road along the elevation contour-404 m to the west of the Lisan Peninsula. This road was routed along the intended alignment of the future SP-0A dike 18 for the extension scheme, and was to be used as access for the site investigations. Further sinkholes were discovered close to the road, and inspection of aerial photographs revealed about 70 of similar holes following what appeared to be a 1.6 km channel reactivated in 1991. A second sinkholes cluster was detected two kilometers south, and 400 m north of SP-01 dike 1 (Fig 1; 9). Other collapses were also identified in the flood channel, along dike 1 (Fig 1; 13) [19, 20]. Investigations revealed that the cavities most probably developed from a massive halite layer some 15-20 m below the ground. At the same period, sinkholes sites spread over the western coast, between Ein Gedi and Ein Boqeq, and in cropped areas of Ghor Al Haditha. These sinkholes have modified the original shape of SP-0A. The new scheme avoided the collapsed area, thus reducing the volume of the basin.

6. Feasibility study started in 1993. At chainage 1+000, boreholes revealed artesian conditions.

7. SP-0A was built up from January 1996 to December 1997 [26]. $32 M dike 18 (13 km by 14 m) was designed to encompass a 95 M m^3 pond over a reactivated salt karst characterized

by soft to very soft silty clay and massive salt rock. Several incidents happened. They were related either to large vertical settlements (2-3m) of very soft clays, or artesian conditions where sand and salt layers were present, or to the development of sinkholes.

8. In 1997, when SP-0A was still empty, a leak in dike 18 was discovered [27]. Technical studies were conducted in collaboration with local and international experts, including Sir Alexander Gibb & Partners that designed the dike. They indicated that the problem could be either due to the formation of an artesian water basin whose high water level affected the dam or due to a sinkhole that formed back in 1996. From chainage 0+000 to 3+200, a 110 m wide berm was built inside SP-0A to protect the dike.

9. Early 1998, SP-0A impounding operation started. One Landsat image acquired in February 17th 1998 showed SP-0A nearly filled. At chainage 10+600, dike 18 showed an obvious curvilinear appendix corresponding to a landfill jetty 240 m long, 30 to 60 m wide, from dike 18 toward the basin center. It was maintained throughout 1998 following the appearance of new sinkholes.

10. All Landsat images acquired in 1998 revealed important seepage zones from chainage 0+000 to around 2+000.

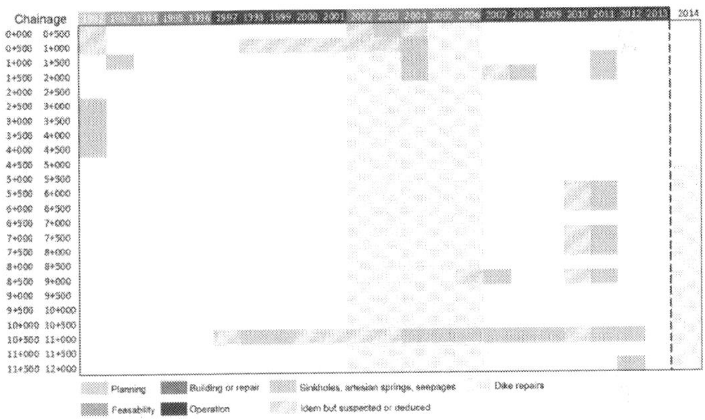

Figure 5. Hazards affecting SP-0A from planning to operation phase, distributed along dike 18 from chainage 0+000 to 12+000.

11. In February-March 1999, the jetty (chainage 10+600) was extended over 70 m to fill again a sinkhole site affecting the bottom of SP-OA.

Remarks:

- In 2004, closer inspections revealed that this structure had been made in haste. It suggested a problem caused by flow through a large channel in the vicinity of the dike. When such a problem happens, a whirlpool can appears at the surface of the pond. Once a whirlpool is observed, complete failure can follow rapidly (hence the need to act very quickly). It is the most serious condition that can be observed because it can enlarge until the dike is breached.

- Discussions with Arab Potash engineers confirmed that the jetty was an attempt to plug the entrance of a sinkhole site with riprap first, and then, when the plug attempt decreased the flow, with smaller materials. Since that time, this zone remained active [18]. It is the only place along the 12 km dike 18 where a dedicated road sign warns on the possible presence of sinkholes. Collapses are frequently observed and filled as soon as they appear.

- In May 2009, the remaining tracks of a former wide sinkhole were observed at the entrance of the jetty. No indication of seepage flow was apparent.

- Seepage is a normal phenomenon in dikes. However, it must be controlled in both velocity and quantity. If uncontrolled, it can progressively erode soil from the embankment dam or its foundation, resulting in rapid failure of the dike. Soil erosion begins at the downstream side, either in the dike proper or the foundation, and progressively works toward the reservoir. Eventually it develops a direct connection to the pond. This phenomenon of regressive erosion is known as "piping".

- Piping action can be recognized by an increased seepage flow rate, discharge of muddy or discolored water, and sinkholes on or near the dike. The above mentioned collapses are evidence of active piping in that zone from at least 1996, before the first filling of the reservoir.

12. These problems didn't stop the expansion of Arab Potash over the recently emerged wave-cut platform surrounding the Lisan Peninsula. Between March 16th 1998 and December 8th 1999, a $38 M SP-0B (11 km^2) production unit was established and came into service early 2000. This project required the building of two dikes (n°19 and 20; 11.6 km) with a height of 14 m. Sir Alexander Gibb & Partners designed the dikes and the construction contract of dike 19 was awarded to the Turkish firm ATA. Pumping brine began January 4th 2000. The pond had a capacity of 76 M m^3.

13. On March 22nd 2000, 4:30 PM, when the quantity of brine reached 56 M m^3, a breach occurred in dike 19 which caused about 2.3 km to collapse. The brine flowed back to the Dead Sea in 30 minutes [13].

Remarks:

- In 2003, Dar Al-Handasah Harza JV, was appointed to assess the damages sustained on Dike 19. They indicated that the repair costs of the dike exceeded its net book value of an amount of $24.4 M.

- Legal proceedings ensued between the Arab Potash Company and all parties involved in the construction. On May 12th 2010, the International Centre for Settlement of Investment Disputes ordered "that the ongoing Jordanian court proceedings in relation to the Dike 19 dispute be immediately and unconditionally terminated, with no possibility to engage further judicial proceedings in Jordan or elsewhere on the substance of the dispute" [28].

- Between 2004-2009, in parallel to those legal proceedings, the ruins of the abandoned dike 19 and 20 exhibited an increasing number of cracks, sinkholes and decameters to hectometers landslides. In a few places bushes of Tamarix were growing, underlying the fact that unsaturated water was flowing below the pond owing to major faults playing the role of conduits [18]. Some fractures became impassable by car without using a footbridge. These damages showed an increasing instability that was consistent with observations performed elsewhere around the Dead Sea, e.g. [11]. Despite its costly setback, in 2010-2011, Arab Potash requested a "Comparative Risk Analysis for Reconstruction of a Partially Failed Dike System" [29] in which the authors propose two alternatives for the reconstruction of the failed facility.

- Cracking, settlement, and slides are the more common signs of structural failure of embankments. Seepage and structural failure are often interrelated in a complex manner. For example, uncontrolled seepage may weaken the soil and lead to a structural failure. A structural failure may shorten the seepage path and lead to a piping failure.

14. In 2000, dike 18 was seriously damaged by sinkholes [30]. It was obvious that there were subversive holes in the dike floor, since it has been filled with seawater in the year 1998. In May 2001, an increase of the artesian pressure at the bottom of the dike was noted, a fact which was considered as an indicator of the safety factor decline. This necessitated the lowering of the water level in the pond by around two meters, to raise the safety coefficient and put the pond temporarily out of operation [30].

15. A 10 m resolution SPOT image acquired on April 11th 2003 indicated important seepage zones coming from SP-01 toward SP-0A and affecting the bottom from chainage 0+000 to around 1+000. Seepage extended outside SP-0A by passing below dike 18 and emerged between stations 1+000 and 2+000.

16. One high resolution satellite image acquired on October 20th 2004-when SP-0A was nearly emptied-showed that a berm had been built at chainage 10+600 (jetty) over a length of about 400 m and a distance of 150 m. Several sinkholes affected the bottom about 300 m basin-ward. The same image clearly showed several other circular depressions at chainage 0+700, about 800 m basin-ward. An artesian spring of very salty water was identified at chainage 1+200, about 200 m from the dike, basin-ward. One sinkhole perforated the berm inside SP-0A at chainage 1+500.

17. November 9th 2004 an agreement was signed with Sinohydro Corporation to rehabilitate Dike 18 for an amount approximating $19.5 M [31]. The most conspicuous work consisted in the setting up of a wide seepage berm outside dike 18. Its width varied from 90 m in the southern part to 150 m in the northern part, in front of the jetty area (10+600). The project was supervised by the Dutch consultant Royal Haskoning.

18. A satellite image acquired on November 30th 2004 confirmed the observations done 40 days before. An artesian salty spring was clearly identified at chainage 1+700. Outside dike 18, from around chainage 1+000 to 1+700, important seepages of very salty brine were visible.

19. Impounding of saltpan SP-0A started on March 14th 2006 and successfully completed its operation level on October 1st 2006. More than 87M m^3 of brine was pumped to reach the operation level of-393.6 m [31]. This elevation corresponded to the Dead Sea level in the mid-1930s. A Taking-Over-Certificate was issued by Royal Haskoning on December 6, 2006 and the project was considered substantially completed on November 23, 2006. The Defects Liability Period commenced on November 23rd 2006 and expired on November 22nd 2007 [32].

20. Field surveys carried out during and after repairs-in 2004, 2005, 2007, 2008, and 2009-have shown that if the rehabilitation works had actually increased the safety coefficient, so far, they failed to stop the cause of the problems. Many cracks and backfilled sinkholes were located [5, 18]. Indeed, dike 18 is constantly threatened by cracks and sinkholes. For example, end of July 2008, from chainage 1+600 to 2+000, a 400 m long dike segment was enlarged to increase its safety factor. Since 1997, its width tripled.

21. Early 2011, the company launched a bid to fill underground cavities between station 6+100 and 6+250 (i.e. 150 m dike segment) with cement and thus protect a fragile part of dike 18. The works included drilling of around 40 boreholes to an average depth of 40 m through the dike body and foundation soil. Thereafter, the subsurface cavities and boreholes were grouted.

22. During the 24th AFA International Technical Fertilizers Conference and Exhibition at Amman, Jordan (22-24 November 2011), Zaid Halasah, Senior Chemist at Arab Potash Company, presented two maps of sinkholes affecting SP-0A with an emphasis of the situation at station 10+600 (jetty dating back to 1998) [33]. Other sinkholes having perforated dike 18 were pointed out at chainage 5+800, 7+400, 8+000. In the vicinity of the dike, sinkholes sites were found from 5+500 to 8+800, and at 10+600.

23. In September 2012, during a visit taking place during the first EAGE workshop on sinkholes held in Amman, a collapse in development was located at chainage 11+500.

24. End of December 2012, a single elliptical structure having 250 by 300 m in diameter was identified within SP-0A from a Worldview-2 image acquired on April 2nd 2011 (chainage 1+000 to 1+600). The feature was unknown to the most aware security consultant (Royal Haskoning). Some other circular elements having a size compatible with the biggest sinkholes in the Dead Sea were also found. This "finding" raised lots of questions regarding the origin of the underlying cavity, regarding the capabilities of prediction of all models developed up to now in Jordan and in Israel about the Dead Sea sinkholes, as well as the strategies, approaches, and methods used by engineers/geophysicists the deal with such features.

25. In April 2013, a new tender announcement was launched to raise dike 18 between station 5+300 to 11+750 (i.e. 6450 m of dike segment) and risk control works.

5.2. Detection of ground displacements during planning stage

Line of sight (LOS) motions displayed in Fig. 6 derive from a pair of ERS images acquired in June 11th 1992 and August 5th 1993, i.e. more than two years before the building of dike 18

http://dx.doi.org/10.5772/57277

(1996-1997). The subsidence rate ranges from 1 to 12 cm/year. Wide areas are found at chainage 6+000 to 7+000; 7+500 to 8+500; 10+000; and especially 11+000 to 11+500. Sills appear at chainage 9+000 and 7+500. Linear subsidence outside dike 18 from chainage 3+000 to 4+000 correspond the 70 sinkholes that appeared in 1992, cutting an access road that should have turned into a dike segment.

The southern part of the future SP-0A is affected by seepages coming from SP-01 or to artesian pressure in relation with the water lens below the Lisan Peninsula in hydrostatic desequilibrium with the Dead Sea level. Two hectometre-long sags are found inside Sp-0A at around chainage 0+700, several hundreds of meters basin-ward.

In term of structural influences, the wide shallow subsidence area at chainage 9+500 to 11+500 (green colors) is a sag basin bounded by sub-parallel lineaments oriented N-S and SW-NE. These directions are also found in the Lisan Peninsula (dotted lines). The area located beyong chainage 11+000 is a very active zone of subsidence. During the 1990s its activity is attested in all interferograms computed with ERS-1/2 pairs [e.g. 34]. SSE-NNW directions are also found at chainage 5+000. They are evidenced by the limits between brown and yellow colors. In all the distribution of lineaments inside SP-0A is in agreement with the known strike-slip derived features evidenced by geophysical prospecting in that area [35].

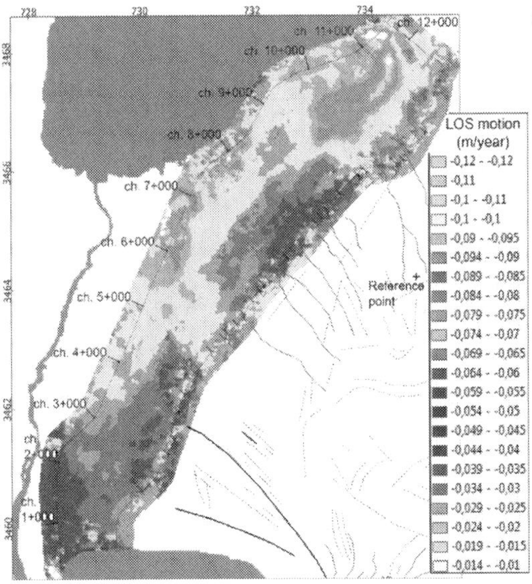

Figure 6. Line of sight motion expressed in meter/year. The period extends from June 11th 1992 to August 5th 1993 (420 days). The ERS-1 pair is characterized by a perpendicular baseline=126 m. Interferometric coherence (not shown) is well preserve exept along the cliff separating the Lisan Peninsula to the foreshore zone where SP-0A is located (Fig. 3). Lineaments and faults are represented as well as ephemeral streams and the cave entrances along SP-0A. The topographic phase was removed with data from ASTER GDEM. Projection UTM 36N, WGS84.

5.3. Detection of ground displacements during feasibility study

Comparison between Fig. 7 and Fig. 6 indicates that the dike segments around chainage 10+000 and 11+500 are still active. The same pattern than in 1992-1993 can be found. However, the most important subsidence extends from 6+000 to 8+5000. This area also existed in 1992-1993 but its perimeter enlarged basin-ward. The main difference is the wide uplifted areas all along the Lisan Peninsula. This zone does not affect dike 18.

The southern part of SP-0A is affected by uplift and subsidence in a way different than in the central and northern parts. This is most probably due to seepages coming from SP-01.

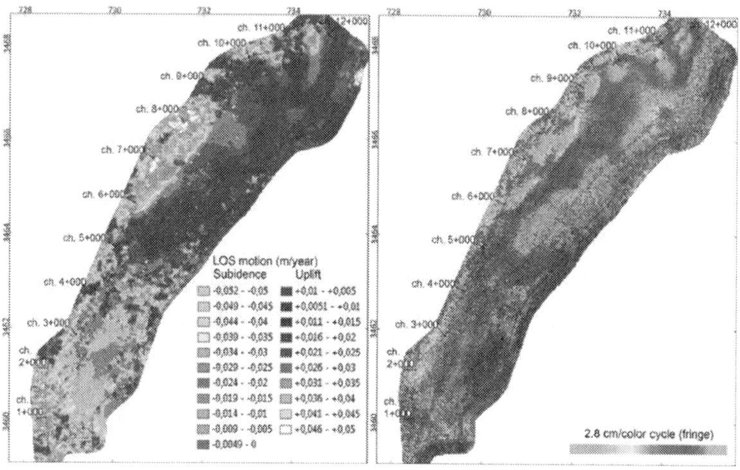

Figure 7. Motion map (left) and filtered interferogram (right) for the period 1995-07-29 to 1996-07-14 (351 days). Both images show the same information but through a classification, on the one hand, and through continuous fringes, on the other hand. Comparison between the two images allows the readers to appreciate the added value of the two sources of information. Like DEMs, motion data can be presented by mixing an hillshade representation that emphasises particular directions and slopes with a particular classification color palette. It allows an intuitive understanding of the uplift (dark brown-white) and subsidence (green) areas. On the right side, fringes reveal the continuity of the deformation fields and their interconnections. Both are usefull to catch the various dynamics.

5.4. Detection of ground displacements before and after impoundment (starting operation stage)

Fig. 8, left side, shows the ground motions from 1995-12-17 to 1997-10-12, i.e. during the building phase. The dike area is strongly affected by the lack of coherence leading to incomplete fringe pattern and the impossibility to generate a realistic displacement map like in Fig. 7, left. Inside SP-0A, coherence was well preserved and allowed the detection of uplift and subsidence areas. The patterns are consistent with the ones visible in Fig. 6 and 8. The most conspicuous extended from 11+000 to 12+000 and from 6+000 to 8+000.

Once in operation (Fig. 8, right side), no information can be retrieved from the bottom anymore (white patch). The deformation fields around saltpan are the only available information to deduce the phenomena occurring below the pond and the dike. Interferogram corresponding to the period 1997-12-21 to 1999-03-21 indicates a strong subsidence in the Lynch Strait. Three fringes representing 3 x 0.028 m of displacement along the line of sight can be easily delineated. This subsidence is caused by the continuous supply of sediments. Decorrelation occurs over the delta because the rate of subsidence is too important to be detected in C-band while it can be done in L-band (see Fig. 10). This external deformation field affect dike 18 from chainage 6+000 to 7+500.

Complex fringes are also visible between 9+500 and 11+500. They are localized in one of the most active zone detected in 1992-1993 (Fig. 6). They attest that underground water circulation is still active in that zone. This element is in agreement with the setting up of a jetty at 10+600 to seal sinkholes at the bottom of SP-0A.

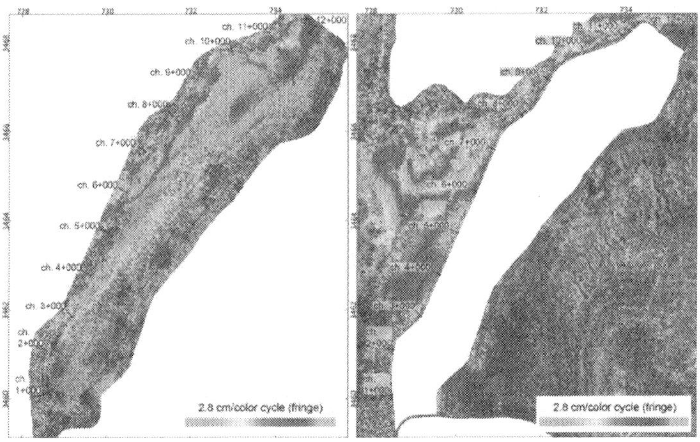

Figure 8. Left, filtered interferogram showing ground deformations recorded from 1995-12-17 to 1997-10-12 (665 days). The very short baseline of only 6 m allowed a clear delineation of the deformations where interferometric coherence was preserved. Obviously, during building stage, human activities modify the original distribution of the scatters on the ground. On the right, once filled, the deformations affecting the bottom become inacessible. The pair 1997-12-21 to 1999-03-21 (455 days) provided the deformation fields with a great precision and accuracy owing to the 18 m perpendicular baseline.

5.5. Detection of ground displacements during repairs

In 2001, sinkholes strongly affected the stability of dike 18. Saltpan Sp-0A was progressively emptied to raise the safety factor. As a result, a wide part of the bottom became again accessible to study the ground deformations with radar interferometry. Fig. 9 shows two informative inferograms. Comparison with Fig. 7-9 indicates many similarities. For example, both interferograms show an important subsidence between chainage 11+000 and 12+000.

In the southern part, from chainage 0+000 to 2+000, the deformations of the bottom in relation with seepage coming from SP-01 are always present. Subsidence and uplifted zones are in geographical agreement with the sinkholes and artesian springs observed over contemporary visible images.

In terms of displacement fringes, a clear contrast exists between the Lisan foreshore (now the bottom of SP-0A) and the Lynch Strait. This can be the result of water circulation below the platform caused by Wadi Araba supply. These complex deformation fields are affecting dike 18 over most of its length. Lynch Strait is a no-man's land between Israel and Jordan. There is no vegetation or activity and the area is flat. Therefore, the interferograms are showing with lots of details the ground movements. The recorded losses of coherence correspond to the presence of surface water (Wadi Araba, artesian water basins), or to areas where the speed of subsidence is too high to maintain coherence in C band. For example, this is the case of the Wadi Araba delta.

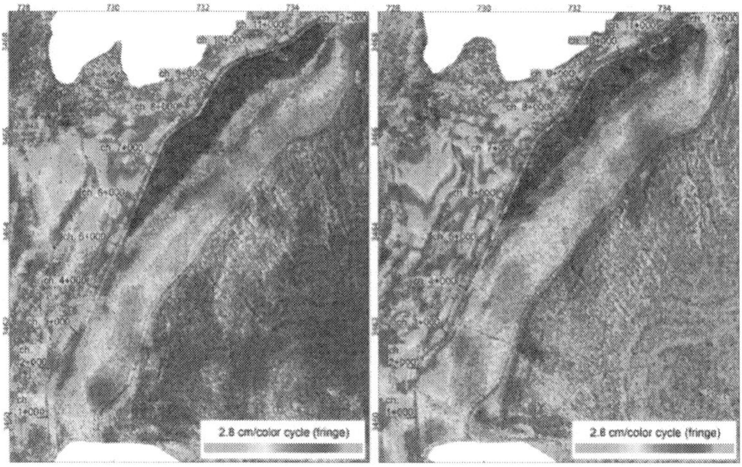

Figure 9. Left, filtered interferogram showing the deformation fields from 2004-06-27 to 2004-11-14 (140 days) with a perpendicular baseline of 37 m leading to an altitude of ambiguity of 249 m. On the right, ground movements recorded between 2005-08-21 and 2006-03-19 (210 days) with a perpendicular baseline of 16 m and an altitude of ambiguity of 571 m. Deformations inside SP-0A are quite similar and in agreement with the deformations recorded more than 10 years before.

5.6. Detection of ground displacements after repairs

Interferogram displayed in Fig. 10 seems essentially devoid of major displacements by comparison to the previous examples showing intricated fringe patterns. Two factors contribute to this appreciation. The very short period of observation (46 days) does not allow the necessary time to contrast uplift and subsidence. Secondly, L-band systems are four times less sensitive than C-band (2.8 cm versus 11.8 cm). This apparent weakness can be an advantage.

For example, L-band allows detection in place affected by movements too rapid to preserve coherence in C-band. Comparison with Fig. 9 shows how noisy areas in the Envisat data (e.g. delta) are correctly imaged with ALOS information.

At a closer inspection, all along dike 18, some tiny deformation fields are visible. Their geographical extensions fit with ground motions previously recorded with ERS and Envisat. From Fig. 10, dike 18 can be divided into different zones based on the relative displacement between each sections: 0+000 to 1+000 : uplift; 1+000 to 2+200 : important subsidence caused by the underground water flux coming from SP-01; 2+200 to 4+000 : uplift; 4+000 to 6+600 : subsidence; 6+600 to 8+500 : strong subsidence; 8+500 to 9500 : subsidence; 9+500 to 11+00 : important uplift; 11+000 to 11+700 : subsidence.

The color cycle represents 11.8 cm of displacement along the line of sight recorded in 46 days. Two places have recorded a full cycle: UTM coord. 730-3468 and 728-3460. The zone located at 728-3460, outside SP-0A, is in the extension of the sinkholes lineament that appeared in 1992. Its zigzagging shape displays directions similar the the ones of the sinkholes lineament too. One as to note that in these two places, the mean velocity is about 90 cm per year. About 15 years before, the rates ranged from 10 to 15 cm per year. The difference in intensity could be explained by the always increasing head difference between the Dead Sea level and the surrounding water tables leading to a growth in the underground water discharge.

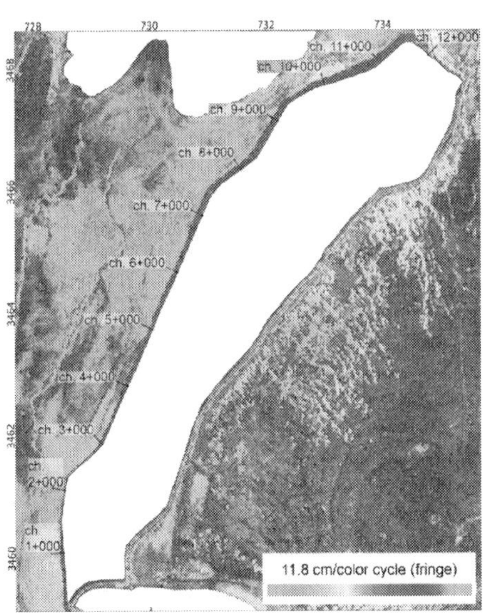

Figure 10. ALOS palsar filtered interferogram showing ground deformations recorded from 2008-04-01 to 2008-05-17 (46 days). The short baseline of 41 m leads to an altitude of ambiguity of 1551 m. The image image is therefore a "true" differential interferogram. Because of L-band, some deformation fields appear clearer than with C-band in the most active zones such as the delta of wadi Araba.

5.7. Detection of ground displacements during operation

Since 2007, the new high-resolution radar sensors allow a much more detailed analysis of surface displacements. Cosmo-SkyMed images, in standard mode, have a resolution between 2-3 meters, i.e. about 10 times better compared to ERS and Envisat images.

As an illustration, Fig. 11 shows the situation prevailling in the southern part of the SP-0A basin during the first half of 2012. On the right side, owing to its "coastal" (400-450 nm) spectral band, very high resolution Worldview-2 image enable the visualization of the Lisan foreshore through some meters of brine. Color changes correspond to subtle variations in the bathymetry."Topographic anomalies" such as dark circles can be detected. They are either sinkholes or artesian springs already detected at least five years earlier, when that part of SP-0A was dried up for dike repairs. In particular, a circle 300 m in diameter appears close to dike 18. Other circular structures occupy the bottom of an ESE-WNW oriented "valley".

Landsat images dating back to 1984 clearly show this longitudinal "corridor"as a dark polygon surrounded by bright sediments like those of the Lisan Peninsula. The context indicates that the difference in color results from a variation of moisture. Landsat data acquired from 1984 to 1992 also revealed that dike SP-01 was affected by seepage phenomena.

Based on these observations, this structure could be a former backfilled valley reactivated after the setting-up of evaporation ponds in the former southern shallow sub-basin. The number and size of the circular structures are indicators of the importance of subsurface water flow.

Beyond the dike, Worldview 2 image does not provide any relevant indications at the exeption of the sinkholes and of the most apparent cracks. These elements, although important in hazard mapping, provide a rough idea of the deformation fields threatening the integrity of dike18.

The unwrapped phase computed from a pair of Cosmo-SkyMed images reveals the staggering blow generated by the groundwater entering the Lynch Strait by passing under dike 18. The deformation field is made of ripples (anticlinal and synclinal) centered on the main impact zone (chainage 1+200). The affected dike section is more than one kilometer long. The threat is extremely diffuse so that the established security measures oriented towards stabilization of cracks and sinkholes ("point-like" events) can not be efficient in this case. Fig. 11 provides an overall picture of the situation and this one could not be described as accurately with traditional techniques even with a dam over-equipped with sensors.

5.8. Ground motions with SBAS technique

The previous examples illustrated the richness of the filtered interferograms and the usefulness of the unwrapped phase data in the quantitative evaluation of the deformations.

Interferogram Stacking Technique had been designed for monitoring scenes characterized by a distributed scattering at a low resolution scale. SBAS consists in an analysis of multilook interferograms. The multilooking is a spatial averaging which is carried out by exploiting the hypothesis that the scattering is distributed. It allows increasing the phase signal quality and thus reliability of measurements. The SBAS technique uses only interferograms generated by choosing thresholds on the spatial and temporal baselines, and on Doppler centroid difference.

http://dx.doi.org/10.5772/57277

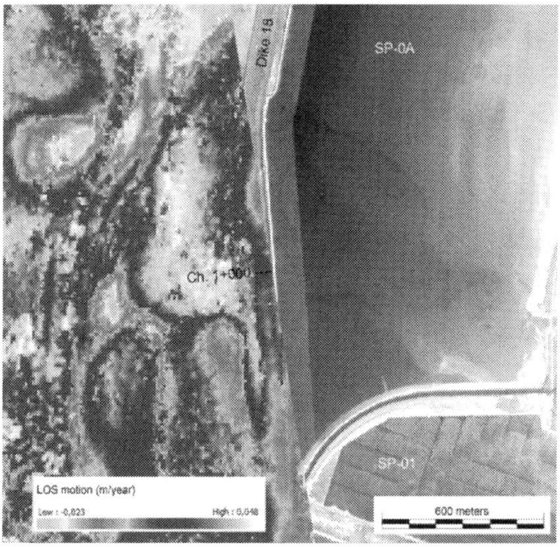

Figure 11. On the left, motion data computed from a pair of Cosmo-SkyMed images spanning from 2011-12-14 to 2012-05-06 (141 days). The deformations affecting dike 18 are related to the ongoing groundwater circulation below SP-0A. Sinkholes distribution affecting the bottom of SP-0A (right side) indicate the position of the underground stream. Worldview 2 image acquired the April 2nd 2011. Both data sets are in high resolution (about 2 m).

After the unwrapping, the interferograms are inverted to retrieve the phase signal over the stack of acquisitions.

Fig. 12 shows the total displacements expressed in cm computed from a stack of 31 Envisat images (2003-07-13 to 2010-06-06). The reference is 2003-07-13. During the period of observation, SP-0A was emptied for dike repairs (2003-2006). The bottom was almost accessible to measurements with radar interferometry. About 30% remained inaccessible, from chainage 4+000 to 12+000 (masked area).

Regarding ground displacements, SP-0A can be divided, from chainage 6+000, into a northern and southern compartment. The northern part is affected by strong subsidence from chainage 6+500 to 9+000. The maximum is reach at 8+500 (white pixels or no data). Another maximum is found at 10+600. The two maxima are identified in C-band from the absence of detection (white pixels) taking into account the context: no vegetation, no human activities, rapid subsidence detected in L-band (Fig. 10).

The southern compartment is characterized by three wide uplifted zones. A remarkable subsidence area is found from chainage 1+000 to 3+000. In this place, a wide berm (cross hatched) was build to reinforce the dike. Circular features observed in Figure 10 are also located in that zone. The results of SBAS are in agreement with the previous observations done with DInSAR approach.

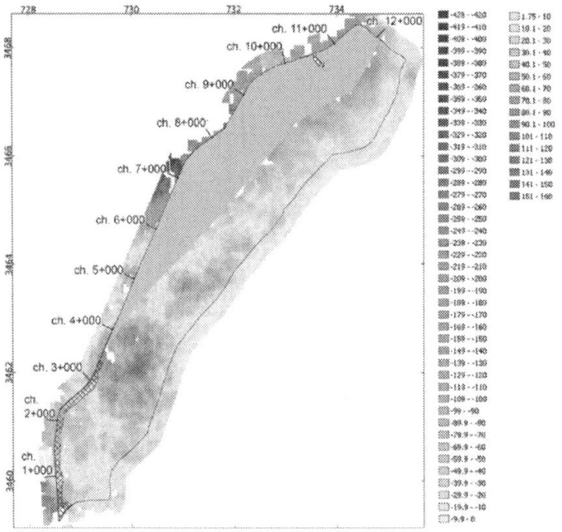

Figure 12. Total displacement in cm from 2003-01-19 to 2010-06-06 computed with SBAS technique. Blue zone corresponds to the part of the basin that remained covered by water during dike repair between 2003-2006. Cross hatched ribbon is the berm built in 1997 and designed to reinforce dike 18.

6. Discussion

Interferograms show the spatial extent of subsidence and uplift phenomena. At the field level or dam instruments, they seem localized, punctual, while in reality there is an important areal component, generally invisible to the eyes. It is this spatial extension that gives interest to the interferograms.

SBAS data operate a large number of interferograms. This technique allows a comparison between them and delivers a time series of ground motions that can be used to represent the dynamics of potential or actual hazards for dike 18. As an illustration, SBAS results allow a comparison with the hazards recorded during the two previous decades. Figure 13, right side column, summarizes the SBAS information in a qualitative way.

The circular features located between 1+000 and 2+000 are in relation with subsidence phenomena. They were identified in the early 2000s owing to SPOT images and VHR acquisitions. They probably result from water ingress along a former flood channel (Fig. 11).

Those recorded from 5+500 to 12+000 are also in connection with subsidence. The dike segments located on the limits between uplift and subsidence are exposed to fractures where sinkholes can eventually appear. Such fractures are common in the remaining dikes 19 and 20 of SP-0B.

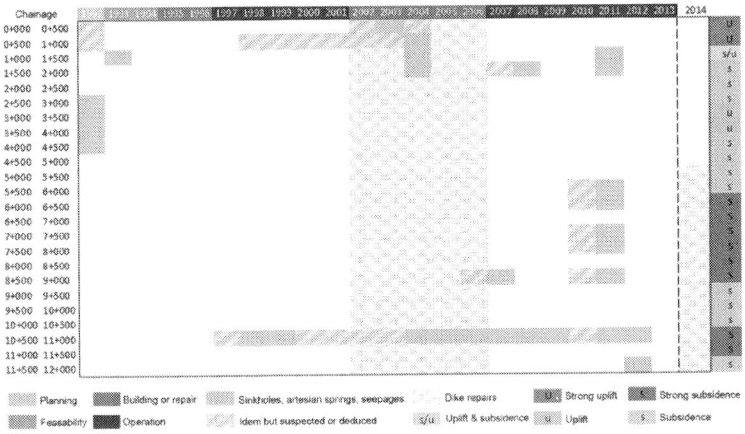

Figure 13. Comparison between sequences of hazard events along dike 18 and a qualitative assessment of the ground deformations computed from 1992 to 2010 with SBAS technique [36-38] ("SBAS" module of Sarscape software).

7. Conclusions

Sound interpretation of data from radar interferometry can provide the engineers in charge of dike safety with the areal component and the dynamic of hazardous phenomena, elements that are so essential to the setting up of a successful and optimally economic strategy to mitigate the damage.

The case study of dike 18 shows that an approach based on sensor densely located all along the dike provide satisfactory results when dealing with well delimited problems. The situation in the south of SP-0A, however, is different. When the hazardous events become too large, it is essential either to correlate the specific measures carried out on the dike, or use techniques such as DInSAR and SBAS which were designed for this kind of issue.

Acknowledgements

The work carried out is the fruit of a collaboration of three years with the company Sarmap SA. Special thanks go to Paolo Pasquali, Alessio Cantone and Paolo Riccardi. All results shown in this chapter have been generated with the SARscape® software package. The work of Najib Abou Karaki is supported by the Deanship of scientific research at the University of Jordan. Thanks are due to Arab Potash Company, KELLER, Sarmap, and ACES for supporting the first EAGE Workshop on Dead Sea Sinkholes (23-25.09.2012), Amman, Jordan (http://www.eage.org/events/index.php?evp=6716&ActiveMenu=2&Opendivs=s3).

Author details

Damien Closson[1] and Najib Abou Karaki[2]

*Address all correspondence to: Damien.closson@yahoo.fr

1 Royal Military Academy, CISS Department. Brussels, Belgium

2 The University of Jordan, Environmental and Applied Geology Department. Amman, Jordan

References

[1] Salameh E and El-Naser H. Changes in the Dead Sea Level and their Impacts on the Surrounding Groundwater Bodies. Acta Hydrochemica et Hydrobiologica 2000; 28 2-33.

[2] Salameh E and El-Naser H. The Interface Configuration of the Fresh / Dead Sea Water – Theory and Measurements. Acta Hydrochemica et Hydrobiologica 2000; 28(6) 323-328.

[3] Itanar A and Reizman Y. Air Photo Survey of Sinkholes in the Dead Sea Area. Geological Survey of Israel-Current Research 2000; 12 21-24.

[4] Abelson M, Yechieli Y, Crouvi O, Baer G, Wachs D, Bein A, Shtivelman V. Evolution of the Dead Sea sinkholes. Geol Soc Am Spec Pap 2006; 40 241–253.

[5] Closson D and Abou Karaki N. Human-induced geological hazards along the Dead Sea coast. Environmental Geology 2009; 58(2) 371–380.

[6] Abou Karaki N. Synthèse et carte sismotectonique des pays de la bordure de la Méditerranée: sismicité du système de faille du Jourdain-Mer Morte. PhD thesis. Université Louis Pasteur, Strasbourg, France; 1987.

[7] Al-Zoubi A and ten Brink U.S. Salt Diapir in the Dead Sea and their Relationship to Quaternary Extensional Tectonics. Marine and Petroleum Geology 2001; 18 779-797.

[8] Closson D and Abou Karaki N. Salt karst and tectonics: sinkholes development along tension cracks between parallel strike-slip faults, Dead Sea, Jordan. Earth Surface Processes and Landforms 2009; 34(10) 1408–1421.

[9] Abelson M, Calvo R, Gabay R, Yechieli Y. Potential levels for sinkholes collapse along the western coast of the Dead Sea-Updated maps (in Hebrew, English abstract). GSI/34/2009. Jerusalem: Geological Survey of Israel; 2009.

[10] Ezersky M G, Eppelbaum L V, Al-Zoubi A, Keydar S, Abueladas A, Akkawi E, Medvedev B. Geophysical prediction and following development sinkholes in two Dead Sea areas, Israel and Jordan. Environmental Earth Sciences 2013; 70 1463-1478.

[11] Yechieli Y, Abelson M, Bein A, Crouvi O. Sinkholes "swarns" along the Dead Sea coast: reflection of disturbance of lake and adjacent groundwater systems. GSA Bull 2006; 118(9/10) 1075–1087.

[12] Ezersky M and Frumkin A. Fault-dissolution front relation and the Dead Sea sinkhole problem. Geomorphology 2013; 201 35-44.

[13] Closson D, Abou Karaki N, Hansen H, Derauw D, Barbier C, Ozer A. Space-borne Radar Interferometric Mapping of Precursory Deformations of a Dyke Collapse-Dead Sea Area – Jordan. International Journal of Remote Sensing 2003; 24(4) 843-849.

[14] Landmann G, Abu Qudaira G.M, Shawabkeh K, Wrede V, Kempe S. Geochemistry of the Lisan and Damya Formations in Jordan, and implications for palaeoclimate. Quaternary International 2002; 89(1) 45-57.

[15] Smit J, Brun J-P, Fort X, Cloetingh S, Ben-Avraham Z. Salt tectonics in pull-apart basins with application to the Dead Sea Basin. Tectonophysics 2008; 449 1-16.

[16] Shimoni M, Hanssen R, Van der Meer F, Kampes B.M, Ben-Dor E. Salt diapir movements using SAR interferometry in the Lisan Peninsula, Dead Sea Rift. SPIE 4543, SAR Image Analysis, Modeling, and Techniques IV, 151 2002; doi:10.1117/12.453964.

[17] Sunna B.F. The geology of salt deposits in the Lisan Peninsula – Dead Sea: Seminar on salt in the Arab World, Ministry of Energy and Mineral Resources, Natural Resources Authorities, May 4-6, 1986, Amman, Jordan.

[18] Closson D, LaMoreaux P.E, Abou Karaki N, al-Fugha H. Karst system developed in salt layers of the Lisan Peninsula, Dead Sea, Jordan. Environmental Geology 2007; 52(1) 155-172.

[19] Knill J. Report on a Visit 8-11 January 1993 Extension West of Lisan Peninsula Sinkholes Along Access Road, The Arab Potash project, Amman, 1993. (Unpublished report)

[20] Knight D.J. Extension West of Lisan Peninsular Sinkholes along Access Road. The Arab Potash Company, DJK/A110/92235B, Amman, 1993. (Unpublished report)

[21] Al-Homoud A.S, Makinson C, Tal A.B, Penman J.G. In-situ geotechnical experimentation and stability evaluation of an instrumented 180 m x70 m x12.5 m fill dike constructed on laminated soft ground. Environmental Geology 1999; 39(2) 177-196.

[22] Gabriel A.K, Goldstein R.M, Zebker H.A. Mapping small elevation changes over large areas: differential radar interferometry. Journal of Geophysical Research 1989; 94 (B7) 9183-9191.

[23] Guarnieri M, Parizzi F, Pasquali P, Prati C, Rocca F. SAR Interferometry Experiments with ERS-1. In: Proceedings of IGARSS `93, August 1993, vol. 3, Tokyo, Japan.

[24] Massonnet D, Rossi M, Carmona C, Adragna, Peltzer G, Feigl K, Rabaute T. The displacement field of the Landers earthquake mapped by radar interferometry. Nature 1993; 364 138-142.

[25] Ge L, Chang H, Ng AH-M, Rizos C. Spaceborne radar interferometry for mine subsidence monitoring in Australia. In: Proceedings Future of Mining, presented at 1st International Future Mining Conference, 19-21 November 2008, UNSW, Sydney, Australia.

[26] Tabbal M, Mansour Z. Extensive geotechnical instrumentation program to control dike raising constructed on soft clay. Jurnal Ilmiah Semesta Teknica 2009; 12(2) 147-156.

[27] Jordan Equity Research. Arab Potash Co. Dead Sea Harvest; 2003.

[28] International Centre for Settlement of Investment Disputes. Washington, D.C., Case No. ARB/08/2. In the proceedings between ATA Construction, Industrial and Traiding Company (Claimant)-and-The Hashemite Kingdom of Jordan (Respondent), May 18, 2010. https://icsid.worldbank.org/ICSID/FrontServlet?requestType=CasesRH&actionVal=showDoc&docId=DC1491_En&caseId=C264 (accessed 15 July 2013).

[29] Djavid M, Mohammadi J. Comparative Risk Analysis for Reconstruction of a Partially Failed Dike System. American Society of Civil Engineers, Practice Periodical on Structural design and Construction 2011; 130-143.

[30] Arab Potash Company, Annual report 2001. www.arabpotash.com/_potash/App_Upload/PDF/2001.pdf (accessed 15 July 2013).

[31] Arab Potash Company, Annual report 2007, 80. www.arabpotash.com/_potash/App_Upload/pdf/2007_Annual_English.pdf (accessed 15 July 2013).

[32] Arab Potash Company, Annual report 2006. www.arabpotash.com/_potash/App_Upload/PDF/2006English.pdf (accessed 15 July 2013)

[33] Halasah Z. Utilization of Satellite Image to Improve Solar Ponds Production. 24th AFA International Technical Fertilizers Conference and Exhibition, 22-24 November 2011, Amman, Jordan.

[34] Baer G, Schattner U, Wachs D, Sandwell D, Wdowinski S, Frydman S. The Lowest Place on Earth is Subsiding-An InSAR (Interferometric Synthetic Aperture Radar) Perspective. Geological Society of America Bulletin 2002; 114(1) 12-23.

[35] Ben-Avraham Z. Geophysical Framework of the Dead Sea: Structure and Tectonics. In: Niemi T.M, Ben-Avraham Z, Gat J. (Eds) The Dead Sea, the Lake and its Setting. Oxford University Press; 1997. p22-35.

[36] Closson D, Pasquali P, Abou Karaki N, Milisavljevic N, Hallot F, Acheroy M, Holecz F. Dead Sea Karst System Dynamics Measured with Insar PS and SBAS Techniques.

Poster XY 383, EGU 2011. http://presentations.copernicus.org/EGU2011-1125_presentation.pdf (accessed 15 July 2013).

[37] Closson D, Abou Karaki N, Pasquali P, Riccardi P, Holecz F. Karst Dynamics Revealed by Small Baseline Subset Interferometric Technique. EAGE Workshop on Dead Sea Sinkholes – Causes, Effects & Solutions, 23 September 2012, Amman, Jordan.

[38] Pasquali P, Riccardi P, Abou Karaki N, Closson D, Holecz F. Very High Resolution ground displacement mapping and monitoring over the Lisan Peninsula: Preliminary Results. EAGE Workshop on Dead Sea Sinkholes – Causes, Effects & Solutions, 23 September 2012, Amman, Jordan.